GAS DYNAMICS

GAS DYNAMICS

Second Edition

JAMES E. A. JOHN
Dean, School of Engineering
University of Massachusetts

PRENTICE HALL, Upper Saddle River, NJ 07458

Printed in the United States of America
20 19 18 17 16 15 14 13

ISBN   0-205-08014-6

Prentice-Hall International (UK) Limited, London
Prentice-Hall of Australia Pty. Limited, Sydney
Prentice-Hall Canada Inc., Toronto
Prentice-Hall Hispanoamericana, S.A., Mexico
Prentice-Hall of India Private Limited, New Delhi
Prentice-Hall of Japan, Inc., Tokyo
Pearson Education Asia Pte. Ltd., Singapore
Editoria Prentice-Hall do Brasil, Ltda., Rio De Janeiro

TO MY PARENTS

CONTENTS

2

WAVE PROPAGATION IN COMPRESSIBLE MEDIA   26

3

ISENTROPIC FLOW OF A PERFECT GAS   39

4

NORMAL SHOCK WAVES   63

8

APPLICATIONS II   155

9

FLOW WITH FRICTION   177

13

EQUATIONS OF MOTION FOR MULTIDIMENSIONAL FLOW   266

14

LINEARIZED FLOWS   279

15

METHOD OF CHARACTERISTICS   312

16

RAREFIED GAS DYNAMICS 338

17

MEASUREMENTS IN COMPRESSIBLE FLOW 351

APPENDICES 371

PREFACE

This book is intended to provide the undergraduate and first-level graduate student in engineering with a knowledge and understanding of the basic fundamentals of compressible flow and gas dynamics. The material covered should serve to bridge the gap between sophomore or junior-level courses in thermodynamics and fluid mechanics and advanced courses in propulsion, turbomachinery, energy conversion, advanced fluid mechanics, and advanced aerodynamics.

Certain modifications and changes have been made to improve this new edition. These changes resulted from the suggestions of faculty using *Gas Dynamics* as a teaching text. Modifications made for the second edition include the following: A change from English units to all SI units, an increase in the number of problems at the ends of chapters to roughly twice those of the first edition, inclusion of conical shock charts, expanded tables for isentropic flow and normal shocks to cover gases with $\gamma = 1.3$ and $\gamma = 5/3$, as well as $\gamma = 1.4$, and more complete coverage of shock tube flows and one-dimensional unsteady flow. It was felt the use of all SI units would be far more satisfactory than the cumbersome dual-unit approach, especially for a senior or first-level graduate student. There has been an attempt, in the second edition, to present more of the algebraic equations used for problem solutions rather than to rely solely on using the numerical tables in rote fashion. At present, virtually all students have electronic calculators (many programmable) for use in treating complex algebraic equations. Several computer problems are provided in the text at the ends of chapters.

There is an emphasis in the text on the application of theory to real engineering systems. It is felt that an engineering student must be kept in contact with the real world and be shown the type of problems that he or she is capable of solving in a particular area. The success that this text has enjoyed has been dependent on an approach that emphasizes the clear, logical development of basic theory. This concept continues to be emphasized in the second edition.

Many worked-out examples are provided throughout the text to demonstrate the application of theory. At the conclusion of each chapter, problems are available for the student. In addition, two entire chapters, Chapters 5 and 8, are devoted to applications of the theory presented in the preceding chapters. Throughout the text, examples and problems have been chosen that involve up-to-date applications that the student might be reading about in current literature.

Organization of the text is such that it can serve a one- or two-term course in gas dynamics. The material usually covered in a first course, including isentropic flow, normal and oblique shock waves, Prandtl Meyer flow, and flows with friction and heat addition, is contained in Chapters 1 through 10. A possible addition might be Chapter 11, involving MHD flows. In Chapters 1 through 11, flow is restricted to one-dimensional flow of a perfect gas (oblique shocks and Prandtl Meyer expansions are treated, essentially, from a one-dimensional approach). Chapters 12 through 17 include specialized topics, such as real gas effects, experimental methods, and rarefied gas dynamics, as well as multidimensional flow. The latter includes linearized flows and the method of characteristics. These latter chapters would seem suitable for a second course in gas dynamics.

In conclusion, it is hoped that this text will serve the purpose of providing a good foundation in the field of gas dynamics, both for the engineering student going directly into industry upon graduation and for the student going on to graduate school.

ACKNOWLEDGMENTS

I would like to acknowledge the assistance given by my many students in the development of a teachable text. I would also like to thank my friends and faculty colleagues at the University of Maryland, University of Toledo, and Ohio State University, with whom I have had many fruitful discussions. Professor Frank Kreith of the University of Colorado, Professor John Boyd of Worcester Polytechnic Institute, Professor Colin Marks of the University of Maryland, Professor William Janna of the University of New Orleans, Professor Gerald Jakubowski of the University of Toledo, and Professors Michael Paolino and James Strozier of the United States Military Academy have provided many valuable suggestions. I would also like to thank Joan Scanlon at the Ohio State University and Oretta Taylor at the University of Massachusetts for their help with the preparation of the manuscript for this second edition.

In a more general sense, I am indebted to my wife Connie and to my children, Elizabeth, Jimmy, Thomas, and Constance, for their patience and understanding during the extensive time periods involved in preparation of both the original manuscript and that of the second edition. I thank my son, Thomas, for assisting in preparation of some of the tabular material in the Appendices. Finally, I wish to acknowledge with deep gratitude the many opportunities provided for me by my parents; it is these opportunities that have made the work possible.

GAS DYNAMICS

1

BASIC EQUATIONS
OF COMPRESSIBLE FLOW

1.1 Fluids, Liquids, and Gases

A fluid is a substance that cannot sustain a shear stress: it is, in other words, a substance that deforms continuously when subjected to shear. This is unlike a solid, which under the action of a shear stress will deform and then remain at equilibrium. This difference between solids and fluids shows up in the tendency of fluids to flow under the action of shear, whereas solids remain rigid.

From a molecular point of view, in solids there exist very strong intermolecular attractive forces; in fluids, these intermolecular forces are very much weaker. The definition of a fluid, as stated previously, includes both liquids and gases. From common experience, we know that the difference between these is that, whereas liquids tend to occupy a more or less well-defined volume, gases will fill the volume allotted to them. In other words, the intermolecular attractive forces between gas molecules are much less than the forces between liquid molecules.

Since water or other liquids tend to occupy a fixed or well-defined volume, it is generally assumed that the *density* of a liquid, defined as the mass per unit volume, remains constant under the action of an externally applied pressure. Actually, large pressures exerted on a liquid in a container can cause slight changes in the liquid volume (e.g., imposing a pressure of 200 times normal atmospheric pressure on liquid water raises the density by 1 percent). However, the rate of change of density with pressure is small enough so that, in a study of the flow of liquids, it is generally assumed that no changes in density occur as the result of the flow. This type of flow is called *incompressible flow*.

With a gas, however, large density changes can occur as a result of an externally applied pressure. We have only to look at an equation of state for a gas (e.g., $p = \rho RT$) to see how density is related to pressure. It follows that appreciable density changes can occur in the flow of a gas; such a flow is called *compressible*. The student has, by now, completed a study of the incompressible flow of liquids. This text will be devoted to a study of the dynamics of compressible gas flow. It should be understood that, under certain conditions, gas flows occur that do not bring about significant changes in gas density, and such flows can be handled with the usual equations of incompressible flow. Here we will be primarily interested in the more general case of flows that involve density changes occurring as the result of the flow.

1.2 Continuum

A gas or other substance consists of a large number of molecules. One way to analyze the behavior of a gas in motion is to consider the motions of the individual molecules, using a statistical approach. Such a microscopic approach is generally very cumbersome and time consuming, especially since what we are generally after are the gross macroscopic properties of a gas. In the continuum model, the gas is approximated as a continuous substance, with only the averaged effects of all the molecules in a finite region of the gas being considered. For example, using the continuum approach, let us define the density of a gas.

Let $\rho = \lim_{\Delta V \to \nu} \Delta M / \Delta V$, where ΔM is the total mass of all the molecules in a volume ΔV, and let ΔV approach a small volume ν. This definition of density, a gross fluid property, is valid as long as ν contains a sufficient number of molecules. If ν were to be too small, it could contain no molecules, and density would have no meaning. Thus it is impossible to speak of the existence of density at a point in a gas flow; rather, we must consider density as a property of at least a very small volume containing a sufficiently large number of molecules.

Similarly, if we define the *gross fluid property pressure* to be the normal force per unit area exerted on a surface,

$$p = \lim_{\Delta A \to a} \frac{\Delta F_{normal}}{\Delta A}, \qquad \text{with } a \text{ small} \tag{1.1}$$

then it must be understood that we cannot truly define pressure at a point, but must refer at least to a very small but finite area. To treat a gas as a continuum, then, we must keep in mind that the distances and volumes being considered must always be large in comparison to molecular sizes and intermolecular distances. A criterion for the adoption of the continuum approach is that the characteristic size of the bodies under investigation must be much greater than the mean free path between intermolecular collisions. Certainly, the continuum approach involves great simplifications in the analysis of gas flows. Unfortunately, in certain very low density flows this approach cannot be employed, and motions of individual molecules must be analyzed. This field of study, called rarefied gas dynamics, will be covered in Chapter 16. With this one exception, however, continuum flow will be assumed throughout this text.

1.3 Viscosity and the Boundary Layer

The coefficient of viscosity of a fluid relates the shearing stress applied to a fluid to the resultant rate of shearing strain. In a continuum, when a viscous fluid flows over a fixed surface, layers of fluid next to the surface are held back by the viscous forces and stick to the surface; in other words, the velocity of the fluid at the fixed wall is zero. As we move away from the wall, the velocity increases to its free stream value, and a velocity distribution is built up as shown in Figure 1.1. The effects of viscosity are dominant in the region near the surface. For most fluids, and certainly for gases, this viscosity is quite small, so viscous effects are confined to a very thin layer in the vicinity of the surface called the *boundary layer.* Outside the boundary layer, the fluid can be analyzed with inviscid theory.

In a boundary layer, the velocity components of a continuum flow at a fixed surface are zero, both normal and tangential to the wall surface. For inviscid flow (zero viscosity), the normal component at the wall is zero, yet the tangential component need not be zero. The boundary layer equations, unlike the inviscid flow equations, have terms containing the viscous forces, which makes them far more difficult to handle. Fortunately, the boundary layer is usually thin enough that it can be assumed that there is no pressure gradient in the direction normal to the wall surface. Thus the pressure distribution on a body, even in the presence of a boundary layer, can often be calculated using the simpler inviscid equations. Furthermore, in considering the flow of a gas through an internal passage or nozzle, usually the boundary layer thickness can be taken as small enough so as not to affect appreciably the area available to

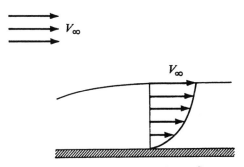

Figure 1.1 Boundary layer flow

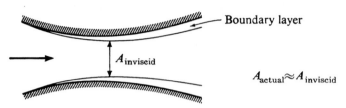

Figure 1.2   Nozzle flow

the inviscid flow outside the boundary layer, and thus not to require the tedious calculation of boundary layer thickness along the walls (see Figure 1.2).

For a wide variety of cases, then, viscous forces can be neglected in the analysis of gas flows. However, in other cases, such as the calculation of convective heat-transfer coefficients, skin friction drag, and flow separation, viscous forces are very important and play a large part in the resultant analysis.

1.4 Equations of Motion

In a study of basic fluid mechanics covering the flow of water and other liquids, incompressible flow can be assumed with density not a variable. For this type of flow, two equations are generally sufficient to solve the problems encountered: the continuity equation or conservation of mass, and a form of the Bernoulli equation, derivable from either momentum or energy considerations. Variables are generally pressure and velocity.

For a compressible flow, density becomes an additional variable; furthermore, significant variations in fluid temperature may occur as a result of density or pressure changes. There are four possible unknowns, and four equations are required for the solution of a problem in compressible gas dynamics: equations for the conservation of mass, momentum, and energy, and a thermodynamic equation of state for the substance involved. Since a study of compressible flow necessarily involves an interaction between thermodynamics and fluid mechanics, the remainder of this chapter will be devoted to a review of the basic principles and equations of these two sciences. To simplify the working of problems involving the fundamental equations, the next section will discuss very briefly the system of units to be used throughout the text.

1.5 Units of Measurement

In any engineering subject, considerable care must be exercised in the use of a consistent set of units. The primary dimensions for an engineer consist of mass, length, time, and temperature. The system of units to be used throughout this text will be the International System (SI). In SI units, the unit of mass is the kilogram (kg), the unit of length is the meter (m), the unit of time is the second (s), and the unit of temperature is the degree Kelvin (K). It is also common practice to express temperature in degrees Celsius (°C), where K = °C + 273.15.

The units of length, time, and temperature present little difficulty. However, the differentiation between units of force and mass is not as easily grasped and should be reviewed. Force and mass are related by Newton's law of motion, $F = ma$. In SI units, the unit of force is the newton (N), defined as the force required to accelerate a mass of 1 kilogram at the rate of 1 meter per second per second.

$$1 \text{ N} = 1 \text{ kg} \cdot \text{m/s}^2$$

Weight is the force with which a mass is attracted to the earth or some other body. From the law of conservation of mass, the mass of a body remains constant, independent of distance from the earth's surface. However, body weight decreases as it is moved away from earth.

The following multiplying prefixes will be used in conjunction with the various units:

Factor	Prefix	Symbol
10^6	Mega	M
10^3	Kilo	k
10^{-2}	Centi	c
10^{-3}	Milli	m
10^{-6}	Micro	μ

For example, 1 kN = 1000 N, 1 cm = 0.01 m, 1 μg = 0.000001 g.

Units of energy are joules (J) where 1 J = 1 N·m. Power, the rate of doing work, has units of watts (W), where 1 W = 1 J/s.

Pressure was defined in Section 1.2 as a normal force per unit area. Pressure in SI units is expressed in newtons per square meter (N/m²). One N/m² is called 1 pascal (Pa). For comparison, 1 standard atmosphere is equal to 101,325 Pa or 101.325 kPa or 0.101325 MPa. Pressure given relative to zero pressure is called *absolute pressure*; pressure given relative to the atmospheric pressure of the surroundings is called *gage pressure*. For example, a pressure gage connected to a compressed air tank registers 100 kPa; if the local atmospheric pressure is 101 kPa, the absolute pressure inside the tank is 201 kPa. Unless indicated otherwise, pressures given in pascals throughout this text will refer to absolute pressures.

1.6 Control Volume Approach

Two approaches are possible in writing the equations of motion of a fluid. The first follows a fixed mass of fluid particles as it moves throughout the flow field.

The other considers a fixed control volume in the flow field and relates the movements of mass, momentum, and energy across the control volume boundaries to changes taking place inside the control volume. A great majority of the problems that are encountered deal with steady flow, in which the flow properties at a point in the fluid do not change with time. Whereas a fluid particle can accelerate or decelerate as it moves from point to point in the flow field, steady flow requires that the velocity of all particles passing by a given point have the same value at that point. It is more convenient to use the control volume approach so that, for steady flow, time is not an independent variable. This approach will be used exclusively throughout the text.

To use the familiar form of Newton's laws of motion, in which a fixed mass of particles is considered, it is necessary to relate the fixed-mass system to the control volume as is shown in Figure 1.3. Let S_1 be the boundary of a system of particles at time t. After a time interval Δt, the mass of particles will have moved to a new location in the flow field, bounded by S_2. Let X_t be the total mass, momentum, or energy possessed by the system of particles at time t. Divide the region S_1 and S_2, as shown in Figure 1.3, so that $V_1 + V_2$ equals the volume bounded by S_1, and $V_2 + V_3$ equals the volume bounded by S_2. The control volume is selected as the volume bounded by S_1. Now

$$X_t = X_{V_1 t} + X_{V_2 t} \tag{1.2}$$

where, for example, $X_{V_1 t}$ represents the mass, momentum, or energy possessed by the fluid particles in V_1 at time t.

$$X_{(t+\Delta t)} = X_{V_2(t+\Delta t)} + X_{V_3(t+\Delta t)} \tag{1.3}$$

The change in X of the system of particles during Δt is

$$X_{(t+\Delta t)} - X_t = X_{V_2(t+\Delta t)} - X_{V_2 t} + X_{V_3(t+\Delta t)} - X_{V_1 t} \tag{1.4}$$

or

$$\frac{\Delta X}{\Delta t} = \frac{\Delta X_{V_2}}{\Delta t} + \frac{X_{V_3(t+\Delta t)} - X_{V_1 t}}{\Delta t} \tag{1.5}$$

Let Δt approach zero so that the volume V_2 approaches that of the control volume. The first term on the right-hand side of Eq. (1.5) becomes the rate of change of X within the control volume. The second term expresses the

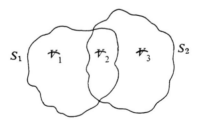

Figure 1.3   Control volume

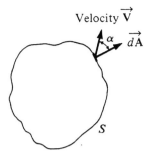

Figure 1.4 Efflux from control volume

difference between the rate at which X leaves the control volume and that at which it enters the control volume.

This net rate of efflux can be expressed in more familiar terms. Let x be equal to the value of X per unit mass. Consider an elemental area $d\mathbf{A}$ on the surface of the control volume in Figure 1.4. (*Note*: The vector $d\mathbf{A}$ is assumed positive with the normal to the differential area pointing outward.) The rate at which mass flows across this area is given by the dot product $\rho\mathbf{V}\cdot d\mathbf{A}$, so the rate of efflux of X across dA is given by $x\rho\mathbf{V}\cdot d\mathbf{A}$. Now Eq. (1.5) can be written as

$$\left(\frac{dX}{dt}\right)_{\substack{\text{for fixed mass} \\ \text{system of particles}}} = \left(\frac{dX}{dt}\right)_{\substack{\text{for fluid in} \\ \text{control volume}}} + \iint_{s} x(\rho\mathbf{V}\cdot d\mathbf{A}) \qquad (1.6)$$

integration around
surface S bounding
control volume

or

$$\frac{dX}{dt} = \frac{\partial}{\partial t}\iiint_{\text{c.v.}} x\rho\,d\mathcal{V} + \iint_{s} x(\rho\mathbf{V}\cdot d\mathbf{A}) \qquad (1.7)$$

Equation (1.7) relates the properties of a fixed mass system of fluid particles to the properties of the fluid inside of and crossing the boundaries of the control volume. The equations of conservation of mass, momentum, and energy for a fluid can now be related to the control volume.

1.7 Conservation of Mass

For the equation of conservation of mass, or the continuity equation, let X equal the total mass M of the system of fluid particles. For this fixed mass system, $dM/dt = 0$, since mass can be neither created nor destroyed. x refers to the total mass per unit mass, so x is equal to 1. Equation (1.7) reduces to

$$0 = \frac{\partial}{\partial t} \iiint\limits_{\text{c.v.}} \rho \, d\mathcal{V} + \iint\limits_{s} \rho \mathbf{V} \cdot d\mathbf{A} \tag{1.8}$$

The rate of increase of mass within the control volume must equal the rate of mass influx into the control volume. Note that the term

$$\iint\limits_{s} \rho \mathbf{V} \cdot d\mathbf{A}$$

is positive for efflux and negative for influx. For steady flow,

$$\frac{\partial}{\partial t} \iiint\limits_{\text{c.v.}} \rho \, d\mathcal{V} = 0$$

so

$$\iint\limits_{s} \rho \mathbf{V} \cdot d\mathbf{A} = 0 \tag{1.9}$$

If $\mathbf{V}$ does not vary in either magnitude or direction across a cross-sectional area A of the flow, then

$$\rho V A \cos a = \text{constant} \tag{1.10}$$

where a is the angle between the velocity vector $\mathbf{V}$ and the outer normal to the surface A.

Example 1.1

Ten kilograms per second of air enters a tank $100 \, \text{m}^3$ in volume while 2 kg/s is discharged from the tank (Figure 1.5). If the temperature of the air inside the tank remains constant at 300 K, and the air can be treated as a perfect gas ($p = \rho RT$), find the rate of pressure rise inside the tank. Take R for air as $0.287 \, \text{kJ/kg} \cdot \text{K}$.

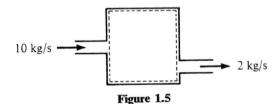

10 kg/s

2 kg/s

Figure 1.5

Solution

Select a control volume as shown in the sketch. For this case, $\iint_s \rho \mathbf{V} \cdot d\mathbf{A}$, the net rate of efflux of mass from the control volume, is equal to -8 kg/s. From Eq. (1.8),

$$100 \frac{\partial \rho}{\partial t} - 8 = 0$$

But

$$p = \rho RT$$

So

$$\frac{\partial p}{\partial t} = RT \frac{\partial \rho}{\partial t}$$

$$= (0.287 \text{ kJ/kg} \cdot \text{K})(300 \text{ K})\left(\frac{8 \text{ kg/s}}{100 \text{ m}^3}\right)$$

$$= 6.888 \frac{\text{kJ}}{\text{s} \cdot \text{m}^3}$$

$$= 6.888 \frac{\text{kN}}{\text{m}^2 \cdot \text{s}}$$

$$\underline{= 6.888 \frac{\text{kPa}}{\text{s}}}$$

Example 1.2

Two kilograms per second of liquid hydrogen and eight kg/s of liquid oxygen are injected into a rocket combustion chamber in steady flow (Figure 1.6). The gaseous products of combustion are expelled at high velocity through the exhaust nozzle. Assuming uniform flow in the rocket nozzle exhaust plane, determine the exit velocity. The nozzle exit diameter is 30 cm, and the density of the gases at the exit plane is 0.18 kg/m³.

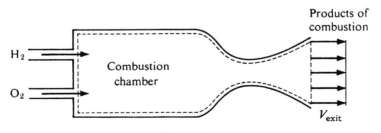

Figure 1.6

Solution

Select a control volume as shown in the sketch. For this case of steady flow, Eq. (1.9) is applicable.

$$\iint_s \rho \mathbf{V} \cdot d\mathbf{A} = 0$$

The rate of influx into the control volume is 10.0 kg/s. The rate of efflux is

$$\rho A V_{exit}$$

where

$$A = \frac{\pi}{4} D^2$$

$$= \frac{\pi}{4} (0.30)^2 \text{ m}^2$$

$$= 0.07069 \text{ m}^2$$

Therefore,

$$V_{exit} = \frac{10.0 \text{ kg/s}}{(0.18 \text{ kg/m}^3)(0.07069 \text{ m}^2)}$$

$$= \underline{785.9 \text{ m/s}}$$

1.8 Conservation of Momentum

Newton's second law of motion for an inertial reference, as applied to a fixed mass M, yields

$$\sum \mathbf{F} = M \frac{d\mathbf{V}}{dt}$$

or   **(1.11)**

$$\sum \mathbf{F} = \frac{d\mathbf{p}}{dt}$$

where $\mathbf{p}$ is the linear momentum of the system. This result can be applied to a control volume in a fluid by substituting $\mathbf{p}$ for X in Eq. (1.7), with $x = \mathbf{p}/M = \mathbf{V}$:

$$\frac{d\mathbf{p}}{dt} = \frac{\partial}{\partial t} \iiint_{c.v.} \rho \mathbf{V} \, d\mathcal{V} + \iint_s \mathbf{V}(\rho \mathbf{V} \cdot d\mathbf{A})$$

But   **(1.12)**

$$\sum \mathbf{F} = \frac{d\mathbf{p}}{dt}$$

So

$$\sum \mathbf{F} = \frac{\partial}{\partial t} \iiint_{c.v.} \rho \mathbf{V} \, d\Psi + \iint_{s} \mathbf{V}(\rho \mathbf{V} \cdot d\mathbf{A}) \tag{1.13}$$

The left-hand side represents a summation of all forces acting on the control volume. No restrictions have been placed on the nature of the forces, so $\sum \mathbf{F}$ may involve pressure forces, viscous forces, gravity, magnetic forces, electric forces, surface tension, and so on. The right-hand side represents the rate of increase of linear momentum within the control volume added to the net rate of efflux of linear momentum from the control volume. A restriction is imposed on Eq. (1.13) by the use of Eq. (1.11) in its derivation. The acceleration in Eq. (1.11) must be measured relative to an inertial reference. Since fluid velocities in Eq. (1.13) are taken relative to the control volume, this equation is only valid for fixed control volumes or control volumes translating at a constant velocity relative to an inertial reference.

For steady flow,

$$\frac{\partial}{\partial t} \left(\iiint_{c.v.} \rho \mathbf{V} \, d\Psi \right) = 0$$

so the momentum equation simplifies to

$$\sum \mathbf{F} = \iint_{s} \mathbf{V}(\rho \mathbf{V} \cdot d\mathbf{A}) \tag{1.14}$$

Example 1.3

An air stream at a velocity of 100 m/s and density of 1.2 kg/m³ strikes a stationary plate and is deflected by 90°. Determine the force on the plate. Assume standard atmospheric pressure surrounding the jet and an initial jet diameter of 2 cm.

Solution

Select a control volume as shown in Figure 1.7a. For this case of steady flow, Eq. (1.14) applies. From symmetry considerations, the force of the fluid on the plate is in the x direction, as shown in Fig. 1.7b. Writing the x component of Eq. (1.14),

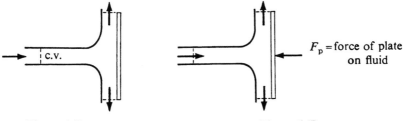

F_p = force of plate on fluid

Figure 1.7a   Figure 1.7b

we obtain

$$F_x = \iint_s V_x(\rho \mathbf{V} \cdot d\mathbf{A})$$

$$-F_p = (-100 \text{ m/s})\left[(1.2 \text{ kg/m}^3)(100 \text{ m/s})\frac{\pi}{4}(0.02^2 \text{ m}^2)\right]$$

$$F_p = 3.770 \text{ kg} \cdot \text{m/s}^2$$

$$= \underline{3.770 \text{ N}}$$

The force of the fluid on the plate is equal to F_p in magnitude, but opposite in direction.

Example 1.4

A rocket motor is fired in place on a test stand. The rocket exhausts 10 kg/s at an exit velocity of 800 m/s. Assume uniform steady conditions at the exit plane with an exit plane static pressure of 50 kPa. For an ambient pressure of 101 kPa, determine the rocket motor thrust transmitted to the test stand as shown in Figure 1.8a.

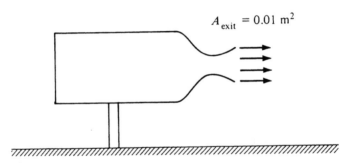

$A_{\text{exit}} = 0.01 \text{ m}^2$

Figure 1.8a

Solution

Select a control volume as shown in Figure 1.8b. The forces acting on this control volume are the thrust and the unbalanced pressure force acting on the exit plane. Applying Eq. (1.14), we obtain

$$\mathcal{J} - (p_e - p_a)A_e = \iint_s \mathbf{V}(\rho \mathbf{V} \cdot d\mathbf{A})$$

$$\mathcal{J} = (50 - 101)\frac{\text{kN}}{\text{m}^2}(0.01 \text{ m}^2) + (800 \text{ m/s})(10 \text{ kg/s})$$

$$= -0.51 \text{ kN} + 8.0 \text{ kN}$$

$$= \underline{7.49 \text{ kN}}$$

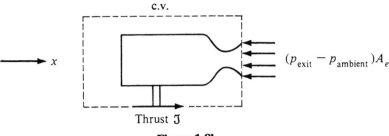

c.v.

x

$(p_{exit} - p_{ambient})A_e$

Thrust $\mathfrak{J}$

Figure 1.8b

1.9 Conservation of Energy—First Law of Thermodynamics

For a system consisting of a fixed mass of particles, the law of conservation of energy can be expressed as

$$\delta Q - \delta W' = dE \qquad (1.15)$$

where Q and W', heat and work, are forms of energy that are defined as they cross the system boundaries, and E is the total energy possessed by the system in a given state. W' is used in a broad sense to include not only simple mechanical work, but also, for example, electrical work and magnetic work done by the system. E includes the internal energy U associated with the random motions of the molecules possessed by the system, kinetic energy KE and potential energy PE due to the movement and position of the entire system mass, and other forms of storable energy that are characteristic of a fixed mass system in a given state, such as electrical energy that can be stored, for example, in a capacitor or chemical energy.

To apply these results to a control volume, use Eq. (1.7). Let e equal the total energy E of the fixed mass system per unit mass, so that

$$\frac{dE}{dt} = \frac{\partial}{\partial t} \iiint_{c.v.} e\rho \, d\forall + \iint_s e(\rho \mathbf{V} \cdot d\mathbf{A})$$

$$= \frac{dQ}{dt} - \frac{dW'}{dt} \qquad (1.16)$$

If the system can be assumed to possess only internal, kinetic, and potential energies, so that $E = U + KE + PE$, then $e = u + \frac{1}{2}V^2 + gz$, where u is equal to the internal energy per unit mass. Substituting into Eq. (1.16) yields

$$\frac{\partial}{\partial t} \iiint_{c.v.} e\rho \, d\forall + \iint_s \left(u + \frac{V^2}{2} + gz \right)(\rho \mathbf{V} \cdot d\mathbf{A}) = \frac{d}{dt}(Q - W') \qquad (1.17)$$

C.S.

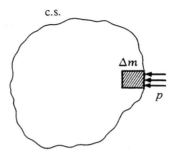

Figure 1.9 Flow work

For cases in which mass flows across the control volume boundaries, it is convenient to divide the work W' into two parts: (1) the work necessary to push the mass across the boundaries, and (2) all the other work W crossing the control surface, such as shaft work, electric and magnetic work, viscous shear work, and so on. A simple expression can be derived for the former type of work. Consider a mass Δm, shown in Figure 1.9, which is to flow across the control surface. It is desired to find the work done by the system in pushing the mass across the boundaries, that is, the work that acts against the external pressure at the boundaries. If the volume of the mass is $\Delta\mathcal{V}$, then the work required is $p\Delta\mathcal{V}$ or, since the mass density $\rho = \Delta m/\Delta\mathcal{V}$, the flow work done by the system per unit mass is p/ρ. For mass flowing into the control volume, the expression for flow work is negative, since this represents work done by the surroundings on the system. It is convenient to combine internal energy per unit mass u and flow work per unit mass p/ρ into the thermodynamic property enthalpy h:

$$h = u + \frac{p}{\rho} \tag{1.18}$$

Now the energy equation, Eq. (1.17), can be written as

$$\frac{\partial}{\partial t}\iiint_{\text{c.v.}} e\rho \, d\mathcal{V} + \iint_{s}\left(h + \frac{V^2}{2} + gz\right)(\rho\mathbf{V}\cdot d\mathbf{A}) = \frac{d}{dt}(Q - W) \tag{1.19}$$

where it is understood that W includes all work except for flow work.

Example 1.5

A rigid, well-insulated vessel is initially evacuated. A valve is opened in a pipeline connected to the vessel, which allows air at 3 MPa and 300 K to flow into the vessel. The valve is closed when the pressure in the vessel reaches 3 MPa. Determine the final equilibrium temperature of the air in the vessel. Over the temperature range of interest, assume that $u = c_v T$, with $c_v = 0.716$ kJ/kg · K, and $h = c_p T$, with $c_p = 1.005$ kJ/kg · K.

Solution

Select a control volume as shown in Figure 1.10. With no heat transfer, no work, and negligible ΔKE and ΔPE, Eq. (1.19) simplifies to

$$\frac{\partial}{\partial t} \iiint_{c.v.} u\rho \, dV + \iint_{s} h(\rho \mathbf{V} \cdot d\mathbf{A}) = 0$$

or

$$\frac{dU}{dt} - h\frac{dM}{dt} = 0$$

where M represents the mass inside the control volume and h is the enthalpy of the air crossing the control surface. (*Note:* Mass crossing the control surface from

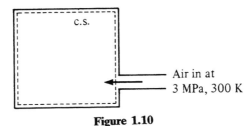

Air in at
3 MPa, 300 K

Figure 1.10

the pipe is an influx, so, according to the sign convention, it is a negative quantity.)

Integrating the preceding equation yields

$$\Delta U = h \, \Delta M$$

Initially, the tank is evacuated, so $\Delta M = M$ and $\Delta U = U_{\text{final}}$.

$$Mc_v T_{\text{final}} = Mc_p T_{\text{pipe}}$$

$$T_{\text{final}} = \frac{1.005 \text{ kJ/kg} \cdot \text{K}}{0.716 \text{ kJ/kg} \cdot \text{K}} (300 \text{ K})$$

$$= 1.4036 \, (300 \text{ K})$$

$$= 421.1 \text{ K}$$

Example 1.6

Steam enters a turbine with a velocity of 30 m/s, with enthalpy of 3000 kJ/kg. At the outlet, the velocity is 100 m/s and enthalpy is 2600 kJ/kg. Assume a heat loss from the turbine of 0.60 kJ/s, with negligible changes of potential energy. Determine the power output of the turbine for a steady flow of 0.1 kg/s.

Solution

Select a control volume as shown in Figure 1.11. For a steady flow with uniform conditions at intake and outlet, Eq. (1.19) yields

$$\underbrace{\left[\left(h_2 + \frac{V_2^2}{2}\right) - \left(h_1 + \frac{V_1^2}{2}\right)\right]\dot{m}}_{} = \frac{dQ}{dt} - \frac{dW}{dt}$$

$$\frac{V_2^2 - V_1^2}{2} = \frac{100^2 - 30^2}{2}\, m^2/s^2$$

$$= 4550\ \text{kg} \cdot m^2/\text{kg} \cdot s^2$$

$$= 4550\ \text{N} \cdot \text{m/kg}$$

$$= 4.550\ \text{kJ/kg}$$

$$\frac{dQ}{dt} - \frac{dW}{dt} = (2600\ \text{kJ/kg} - 3000\ \text{kJ/kg} + 4.550\ \text{kJ/kg})(0.1\ \text{kg/s})$$

$$= (-395.45\ \text{kJ/kg})(0.1\ \text{kg/s})$$

$$\frac{dW}{dt} = -0.6\ \text{kJ/s} + 39.545\ \text{kJ/s}$$

$$= 38.945\ \text{kW}$$

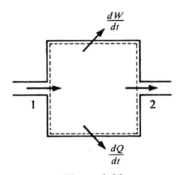

Figure 1.11

1.10 Second Law of Thermodynamics

Two concepts that are important to a study of compressible fluid flow are derivable from the second law of thermodynamics: the *reversible process* and the *property entropy*. For a thermodynamic system, *a reversible process is one after which the system can be restored to its initial state and leave no change in either system or surroundings.* As a consequence of this definition, it can be shown that a reversible process is quasi-static; changes occur infinitely slowly, with no energy being dissipated. Since thermodynamics is a study of equilibrium states, definite thermodynamic equations for changes taking place during processes can be derived only for reversible processes; irreversible processes can only be described thermodynamically with the use of inequalities. Irreversible processes involve, for example, the following: friction, heat transfer through a finite temperature difference, sudden expansion, magnetization with hysteresis, electrical resistance heating, and mixing of different gases. In general, any natural process is irreversible, so the assumption of reversibility, while it may simplify the thermodynamic equations, necessarily yields an approximation. For many cases, the assumption of reversibility leads to very accurate results; yet it is well to keep in mind that the reversible process is always an idealization.

The thermodynamic property derivable from the second law is entropy, which is defined for a system undergoing a reversible process by

$$dS = \frac{\delta Q}{T} \tag{1.20}$$

where S denotes entropy. For an irreversible process,

$$dS > \frac{\delta Q}{T} \tag{1.21}$$

According to Eq. (1.21), entropy changes can arise from heat transfer or from irreversibilities. During an *adiabatic process*, entropy either increases or remains constant. A process that is adiabatic and reversible involves no change in entropy and is called *isentropic.*

A useful thermodynamic equation for a pure substance, derivable from the first and second laws, is

$$T\,ds = dh - \frac{dp}{\rho} \tag{1.22}$$

where s represents entropy per unit mass. This equation contains only thermodynamic properties; it is independent of the path of a process. For example, Eq. (1.22) can be integrated between given end states to determine the entropy change regardless of whether the thermodynamic process involved is reversible or irreversible and of whether the process takes place in a closed container or in steady flow.

1.11 Equation of State

An equation of state for a pure substance is a relation between pressure, density, and temperature for that substance. Depending on the phase of the substance and on the range of conditions to which it is subjected, one of a number of different equations of state is applicable. However, for liquids or solids, these equations become so cumbersome and have such a limited range of application that it is generally more convenient to use tables of thermodynamic properties. For gases, an equation exists that does have a reasonably wide range of application, the *perfect gas law*; in its usual form, it is expressed as

$$p = \rho RT \tag{1.23}$$

where R is a gas constant, having its value dependent on the gas in question. For any perfect gas,

$$R = \frac{\bar{R}}{\bar{M}}$$

where

$$\bar{R} = \text{universal gas constant}$$
$$= 8314.3 \text{ J/kg-mole} \cdot \text{K}$$
$$\bar{M} = \text{molecular mass of gas}$$

For example, for hydrogen with $\bar{M} = 2.016$ kg/kg-mole,

$$R = \frac{8314.3 \text{ J/kg-mole} \cdot \text{K}}{2.016 \text{ kg/kg-mole}}$$
$$= 4.124 \text{ kJ/kg} \cdot \text{K}$$

In the derivation of the perfect gas law from kinetic theory, the volume of the gas molecules and the forces between the molecules are neglected. These assumptions are satisfied by a real gas only at very low pressures. However, even at reasonably high pressures, a real gas approximates a perfect gas as long as the gas temperature is great enough.

For example, for steam at 6 MPa and 500°C, the deviation of a perfect gas from a real gas is only about 5 percent. As the steam is cooled at this pressure so that its state approaches the saturation line, the deviation becomes more marked. More exact equations of state have been derived that account for molecular volumes (the *Clausius equation of state*) and intermolecular forces (the *van der Waals equation of state*). They will be discussed in more detail in a later chapter; however, these equations are more complex than the perfect gas law and lend added complications to the solutions of the flow equations. The perfect gas law, simple as it is, yields uncomplicated expressions for the various thermodynamic properties and can be applied over a wide range of pressures and temperatures with a high degree of accuracy.

J. P. Joule (1818–1889) demonstrated that the specific internal energy of a perfect gas is a function of temperature only:

$$u = u(T)$$

According to the definition of specific heat,

$$c_v = \left(\frac{\partial u}{\partial T}\right)_v$$

so that, for a perfect gas,

$$du = c_v \, dT \tag{1.24}$$

Enthalpy is defined as

$$h = u + \frac{p}{\rho}$$

or

$$dh = du + d\left(\frac{p}{\rho}\right) \tag{1.25}$$

and, for a perfect gas,

$$dh = du + R \, dT$$

Therefore, the enthalpy of a perfect gas is also a function of temperature only. From the definition of specific heat,

$$c_p = \left(\frac{\partial h}{\partial T}\right)_p$$

so that, for a perfect gas,

$$dh = c_p \, dT \tag{1.26}$$

$$c_p - c_v = R \tag{1.27}$$

c_p and c_v are not constants but vary with temperature.

If a perfect gas undergoes a thermodynamic process between two equilibrium states, then

$$h_2 - h_1 = \int_1^2 c_p \, dT \quad \text{and} \quad u_2 - u_1 = \int_1^2 c_v \, dT \tag{1.28}$$

If a constant specific heat is assumed (or else a mean specific heat over the temperature range of interest), then

$$h_2 - h_1 = c_p(T_2 - T_1) \quad \text{and} \quad u_2 - u_1 = c_v(T_2 - T_1) \tag{1.29}$$

An expression for the entropy change of a perfect gas can be derived from Eq. (1.22):

$$s_2 - s_1 = \int_1^2 ds = \int_1^2 c_p \frac{dT}{T} - R \ln \frac{p_2}{p_1} \tag{1.30}$$

For an isentropic process,

$$\int_1^2 c_p \frac{dT}{T} = R \ln \frac{p_2}{p_1}$$

If, in addition, the perfect gas is assumed to possess constant specific heats, then

$$\frac{c_p}{R} \ln \frac{T_2}{T_1} = \ln \frac{p_2}{p_1}$$

Let

$$\gamma = \frac{c_p}{c_v}$$

so that

$$\frac{c_p}{R} = \frac{\gamma}{\gamma - 1}$$

For a perfect gas with constant specific heats undergoing an isentropic process,

$$\frac{T_2}{T_1} = \left(\frac{p_2}{p_1}\right)^{(\gamma-1)/\gamma} = \left(\frac{\rho_2}{\rho_1}\right)^{\gamma-1}$$

or   (1.31)

$$\frac{p_2}{p_1} = \left(\frac{\rho_2}{\rho_1}\right)^{\gamma}$$

Example 1.7

Hydrogen is expanded isentropically in a nozzle from an initial pressure of 500 kPa, with negligible velocity, to a final pressure of 100 kPa. The initial gas temperature is 500 K. Assume steady flow with the hydrogen behaving as a perfect gas with constant specific heats, $c_p = 14.5$ kJ/kg · K. Determine the final gas velocity and the mass flow through the nozzle for an exit area of 500 cm².

Solution

With no heat or work terms, the energy equation, Eq. (1.19), for steady flow reduces to

$$\iint_s \left(h + \frac{V^2}{2}\right)(\rho \mathbf{V} \cdot d\mathbf{A}) = 0$$

Select a control volume as shown in Figure 1.12. With $V_1 = 0$, the preceding

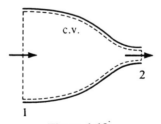

Figure 1.12

reduces to

$$h_1 = h_2 + \frac{V_2^2}{2}$$

$$V_2 = \sqrt{2c_p(T_1 - T_2)}$$

To find T_2, use Eq. (1.31):

$$\frac{T_2}{T_1} = \left(\frac{p_2}{p_1}\right)^{(\gamma-1)/\gamma}$$

For hydrogen, $R = 4.124 \text{ kJ/kg} \cdot \text{K}$, so

$$\gamma = c_p/c_v = c_p(c_p - R) = 1.397$$

Now

$$\frac{T_2}{T_1} = \left(\frac{100 \text{ kPa}}{500 \text{ kPa}}\right)^{(\gamma-1)/\gamma} = (0.2)^{0.2842} = 0.6329$$

or

$$T_2 = (500 \text{ K})(0.6329) = 316.45 \text{ K}$$

$$V_2 = \sqrt{2(14.5 \text{ kJ/kg} \cdot \text{K})(500 \text{ K} - 316.45 \text{ K})}$$

$$= \sqrt{5322.95 \text{ kJ/kg}}$$

$$= \sqrt{5.32295 \times 10^6 \text{ N} \cdot \text{m/kg}}$$

$$= \sqrt{5.32295 \times 10^6 \text{ kg} \cdot \text{m}^2/\text{s}^2 \cdot \text{kg}}$$

$$= 2307 \text{ m/s}$$

The mass flow rate at the nozzle exit is given by

$$\dot{m}_2 = (\rho A V)_2$$

$$= \frac{p_2}{RT_2} A V_2$$

$$= \frac{(100 \text{ kN/m}^2)(500 \times 10^{-4} \text{ m}^2)2307 \text{ m/s}}{(4.124 \text{ kN} \cdot \text{m/kg K})(316.45 \text{ K})}$$

$$= 8.8388 \text{ kg/s}$$

1.12 One-Dimensional Flow

A complete solution of a problem in compressible fluid mechanics requires a three-dimensional analysis. However, even for incompressible flow a complete solution in three dimensions is possible only for cases in which simple geometries are involved. Fortunately, a great many compressible flow problems can be solved, to a good engineering approximation, with the use of a one-dimensional analysis. One-dimensional flow implies that the flow variables are functions of only one space coordinate.

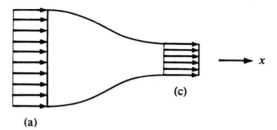

Figure 1.13   One-dimensional flow

Consider the flow in a varying area channel. One-dimensional flow requires that the velocity profiles be as shown in Figure 1.13. Actually, due to viscosity, the flow velocity at the fixed wall must be zero. The velocity profiles in a real fluid are as shown in Figure 1.14.

One-dimensional flow, by definition, prohibits velocity components in the y or z directions, as at (b) in Figure 1.14. In true one-dimensional flow, area changes are not allowed. However, the more gradual the area change with x, the more exact becomes the one-dimensional approximation. The real case, illustrated by Figure 1.14, can be reduced to the one-dimensional case by assuming a mean velocity at each cross section. Note that the one-dimensional approach can yield information only on variations in the x direction; variations normal to the flow direction are assumed to be very small. The value of the one-dimensional approximation lies in its simplifying the equations of fluid flow. Satisfactory engineering answers can be obtained for many complex problems that would otherwise be insoluble.

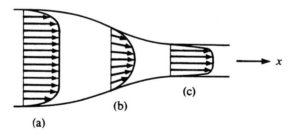

Figure 1.14   Real flow in varying area channel

1.13 Summary

Compressible fluid mechanics is a study of flow in which significant density variations occur throughout the fluid. With density and temperature as additional variables, the equations of incompressible fluid mechanics must be supplemented with those of thermodynamics. A solution of a problem in compressible flow requires utilization of the continuity and momentum equations, as well as the first and second laws of thermodynamics and an equation of state for the substance involved. The additional complexities introduced by compressible flow require that approximations be made in order to simplify the problems so that satisfactory engineering answers can be obtained. The one-dimensional flow and perfect gas approximations, affording considerable simplicity to the equations involved, yield results that, for many engineering problems, are within a tolerable degree of accuracy.

REFERENCES

General References

1. JOHN, J.E.A., AND HABERMAN, W.L., *Introduction to Fluid Mechanics*, 2nd ed., Englewood Cliffs, N.J., Prentice-Hall, Inc., 1980.
2. FOX, R.W., AND MCDONALD, A.T., *Introduction to Fluid Mechanics*, 2nd ed., New York, John Wiley & Sons, Inc., 1978.
3. STREETER, V.L., AND WYLIE, E.B., *Fluid Mechanics*, 6th ed., New York, McGraw-Hill Book Company, 1975.
4. HABERMAN, W.L., AND JOHN, J.E.A., *Engineering Thermodynamics*, Boston, Allyn and Bacon, Inc., 1980.
5. VAN WYLEN, G.J., AND SONNTAG, R.E., *Fundamentals of Classical Thermodynamics*, 2nd ed., New York, John Wiley & Sons, Inc., 1976.
6. REYNOLDS, W.C., AND PERKINS, H.C., *Engineering Thermodynamics*, 2nd ed, New York, McGraw-Hill Book Company, 1977.

PROBLEMS

1. Determine the weight force on an object of mass 2.0 kg located at the earth's surface at a point where the acceleration due to gravity, g, is 9.81 m/s^2. Determine the mass of the object on the surface of the moon at which the gravitational attraction is only one-sixth that on the earth's surface. Determine the weight force on the moon's surface.

2. Using SI units, show that $\frac{1}{2}\rho V^2$ has units of pressure.

3. Air at a pressure of 105 kPa, a velocity of 0.1 m/s, and a temperature 300 K flows steadily in a 10-cm-diameter duct. After a transition, the duct is exhausted uniformly through a rectangular slot 3 cm × 6 cm in cross section. Assume the air behaves as a perfect gas; $R = 0.287$ kJ/kg · K. Determine the exit velocity through the slot, assuming incompressible flow.

4. Air flows steadily in the circular pipe of Figure P1.4 with a velocity of 20 m/s. Surrounding the pipe, in an annulus, is a second flow of air, with a velocity of

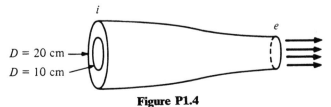

i

e

$D = 20$ cm

$D = 10$ cm

Figure P1.4

40 m/s. Both flows are exhausted into a pipe of 15-cm diameter. If the flow at e is uniform, determine the flow velocity at e. Assume the air density is constant.

5. Oil flows steadily in a 1.0-cm-diameter circular tube with a uniform velocity of 1.0 cm/s. At a cross section farther down the 1.0-cm tube, the velocity distribution is given by $V = U_0(1-r^2)$, with r in centimeters. Find U_0, assuming the oil density to be constant.

6. For the rocket of Figure 1.6, determine the rocket thrust. Assume that exit plane pressure is equal to ambient pressure.

7. Water enters a horizontal U-tube at i with a velocity of 1.6 m/s; the water exhausts through a nozzle to atmospheric pressure. Neglecting friction, determine the exit water velocity and the force exerted by the water on the nozzle and U-tube (see Figure P1.7). Exit nozzle diameter is 1.5 cm, U-tube diameter is 3 cm, and the density of water is 1000 kg/m³. Pressure at cross-section i is 150 kPa and ambient pressure is 101 kPa.

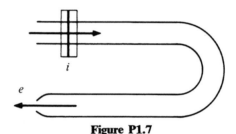

Figure P1.7

8. Determine the force F required to push the flat plate of Figure P1.8 against the round air jet with a velocity of 10 cm/s. The air jet velocity is 100 cm/s, with a jet diameter of 5.0 cm. Air density is 1.2 kg/m³.

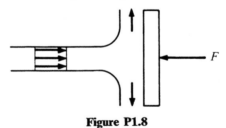

Figure P1.8

9. A jet engine (Figure P1.9) is traveling through the air with a forward velocity of 300 m/s. The exhaust gases leave the nozzle with an exit velocity of 800 m/s with respect to the nozzle. If the mass flow rate through the engine is 10 kg/s, determine the jet engine thrust. Exit plane static pressure is 80 kPa, inlet plane static pressure

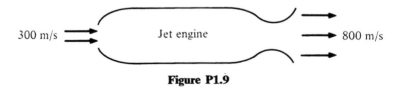

Figure P1.9

is 20 kPa, ambient pressure surrounding the engine is 20 kPa, and the exit plane area is 4.0 m².

10. Air is expanded isentropically in a horizontal nozzle from an initial pressure of 1.0 MPa, of a temperature of 800 K, to an exhaust pressure of 101 kPa. If the air enters the nozzle with a velocity of 100 m/s, determine the air exhaust velocity. Assume the air behaves as a perfect gas, with $R = 0.287$ kJ/kg · K and $\gamma = 1.4$. Repeat for a vertical nozzle with exhaust plane 2.0 m above the intake plane.

11. Nitrogen is expanded isentropically in a nozzle from a pressure of 2000 kPa, at a temperature of 1000 K, to a pressure of 101 kPa. If the velocity of the nitrogen entering the nozzle is negligible, determine the exit nozzle area required for a nitrogen flow of 0.5 kg/s. Assume the nitrogen to behave as a perfect gas with constant specific heats, mean molecular mass of 28.0, and $\gamma = 1.4$.

12. Air enters a compressor at a pressure of 100 kPa at a temperature 20°C, the flow rate is 0.25 m³/s. Compressed air is discharged from the compressor at 800 kPa and 50°C. Inlet and exit pipe diameters are 4.0 cm. Determine the exit velocity of the air at the compressor outlet and the compressor power required. Assume air to behave as a perfect gas with constant specific heat; $c_p = 1.005$ kJ/kg·K and $R = 0.287$ kJ/kg·K.

13. Hot gases enter a jet engine turbine with a velocity of 50 m/s, a temperature of 1200 K, and a pressure of 600 kPa. The gases exit the turbine at a pressure of 250 kPa and a velocity of 75 m/s. Assume isentropic steady flow and that the hot gases behave as a perfect gas with constant specific heats (mean molecular mass 25, $\gamma = 1.37$). Find the turbine power output.

14. Hydrogen is stored in a tank at 1000 kPa and 30°C. A valve is opened, which vents the hydrogen and allows the pressure in the tank to fall to 200 kPa. Assuming that the hydrogen that remains in the tank has undergone an isentropic process, determine the amount of hydrogen left in the tank. Take hydrogen as a perfect gas with constant specific heats; $\gamma = 1.4$, and $R = 4.124$ kJ/kg · K. Tank volume = 2.0 m³.

15. Methane enters a constant-diameter, 3-cm duct at a pressure of 200 kPa, a temperature of 250 K, and a velocity of 20 m/s. At the duct exit, the velocity reaches 25 m/s. For isothermal steady flow in the duct, determine the exit pressure, mass flow rate, and rate at which heat is added to the methane. Assume methane to behave as a perfect gas; $\gamma = 1.32$ (constant) and the mean molecular mass is 16.0.

2

WAVE PROPAGATION
IN COMPRESSIBLE MEDIA

2.1 Introduction

The method by which a flow adjusts to the presence of a body can be shown
visually by a plot of the flow streamlines about the body. Figures 2.1 and 2.2
depict the streamline patterns obtained for uniform, steady, incompressible
flow over an airfoil and over a circular cylinder, respectively. Note that the
fluid particles are able to sense the presence of the body before actually
reaching it. At points 1 and 2, for example, the fluid particles have been
displaced vertically, yet 1 and 2 are points in the flow field well ahead of the
body. This result, true in the general case of any body inserted in an incompressible flow, suggests that a signaling mechanism exists whereby a fluid
particle can be forewarned of a disturbance in the flow ahead of it. The velocity
of signal waves sent from the body, relative to the moving fluid, apparently is
greater than the absolute fluid velocity, since the flow is able to start to adjust
to the presence of a body before reaching it. Thus, when a body is inserted into
incompressible flow, smooth, continuous streamlines result, which indicate
gradual changes in fluid properties as the flow passes over the body. If the fluid
particles were to move faster than the signal waves, the fluid would not be able
to sense the body before actually reaching it, and very abrupt changes in
velocity vectors and other properties would ensue.

In this chapter, the mechanism by which the signal waves are propagated
through incompressible and compressible flows will be studied. An expression
for the velocity of propagation of the waves will be derived. From this result,

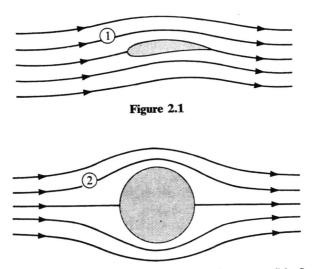

Figure 2.1

Figure 2.2 Streamline patterns for steady incompressible flow

significant conclusions can be drawn concerning the basic differences between incompressible and compressible flow.

2.2 Wave Propagation in Elastic Media

Let us examine what happens when a solid elastic object such as a steel bar is subjected to a sudden, uniformly distributed compressive stress applied at one end. In the first instant of time, a thin layer next to the point of application is compressed, while the remainder of the bar is unaffected. This compression is then transmitted to the next layer, and so on down the bar. Thus a disturbance created at the left side of the bar is eventually sensed at the opposite end. The compression wave initiated at the left side of the bar takes a finite time to travel to the right side, the wave velocity being dependent on the elasticity and density of the medium (see Figure 2.3).

Gases and liquids also are elastic substances, and longitudinal waves can be propagated through these media in the same way that waves are propagated through solids. Let a gas be confined in a long tube with a piston at the

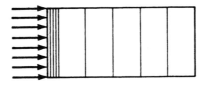

Figure 2.3 Compression wave in solid bar

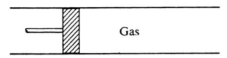

Figure 2.4   Piston motion into gas

left-hand side. The piston is given a sudden push to the right. In the first instant, a layer of gas piles up next to the piston and is compressed; the remainder of the gas is unaffected. The compression wave created by the piston then moves through the gas until eventually all the gas is able to sense the movement of the piston. If the impulse given to the gas is infinitesimally small, the wave is called a *sound wave*, and the resultant compression wave moves through the gas at a velocity equal to the velocity of sound.

For an incompressible medium, no changes in density are allowed. If the piston in Figure 2.4 were moved to the right in an incompressible medium, no piling up of fluid, or density changes, would occur at any point in the fluid. All the fluid would have to move instantaneously with the piston. Thus, the velocity of wave propagation in an incompressible fluid is infinite. A disturbance created at any point in an incompressible fluid is sensed instantaneously at all other points in the fluid. However, no medium is truly incompressible, so the velocity of sound has a finite value in solid, liquid, or gas. The more compressible the substance through which the wave propagates, the smaller will be the velocity of sound in that substance. The velocity of sound in water, for example, is much greater than it is in air.

2.3 Velocity of Sound

If an infinitesimal disturbance is created by the piston in Figure 2.4, the wave propagates through the medium at the velocity of sound relative to the gas into which the wave is moving. Let the piston be given a steady velocity to the right, of magnitude dV, with the resultant sound wave moving at a velocity a. As a result of the compression created by the piston, the pressure and density next to the piston are infinitesimally greater than the pressure and density of the gas at rest ahead of the wave. The gas between piston and wave must move with the piston velocity dV (see Figure 2.5). It is required to derive an expression for the velocity of sound, utilizing the flow equations reviewed in Sections 1.7 through 1.9.

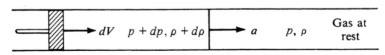

Figure 2.5   Sound wave

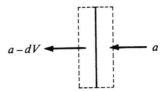

Figure 2.6 Control volume for sound wave

Choose a control volume containing the wave; all velocities must be expressed relative to this control volume (see Figure 2.6). This is now a steady-flow problem, which results in simplifications in the equations of motion. Equations (1.9) and (1.14) are applicable:

$$\iint_s \rho \mathbf{V} \cdot d\mathbf{A} = 0 \qquad (1.9)$$

$$\sum \mathbf{F} = \iint_s \mathbf{V}(\rho \mathbf{V} \cdot d\mathbf{A}) \qquad (1.14)$$

Since the sound wave initiated by the piston is a plane wave, changes in flow properties occur only in the flow direction. For this one-dimensional flow, Eq. (1.9) becomes

$$(\rho + d\rho)(a - dV) - \rho a = 0$$

Simplifying and dropping second-order terms yields

$$a \, d\rho - \rho \, dV = 0 \qquad (2.1)$$

The only forces acting on the control surface are pressure forces (Figure 2.7). Equation (1.14) yields

$$pA - (p + dp)A = [(a - dV) - a]\rho A a$$

Simplifying, we obtain

$$dp = \rho a \, dV \qquad (2.2)$$

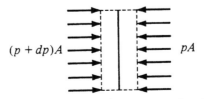

Figure 2.7 Forces acting on control surface

Combining with Eq. (2.1) yields

$$dp = a^2 \, d\rho$$

or

$$a^2 = \frac{dp}{d\rho} \qquad \qquad \textbf{(2.3)}$$

In Eq. (2.3), $dp/d\rho$ must be written as a partial derivative, since the manner in which pressure varies with density is dependent on the process occurring in the sound wave. For example,

$$\left(\frac{\partial p}{\partial \rho}\right)_T \neq \left(\frac{\partial p}{\partial \rho}\right)_s$$

The sound wave is a weak compression wave, across which occur only infinitesimal changes in fluid properties. Therefore, the process occurring in the wave satisfies the definition of reversibility from Section 1.10. Furthermore, the wave itself is extremely thin, and changes in properties occur very rapidly. The rapidity of the process precludes the possibility of any heat transfer between the system of fluid particles and its surroundings. The sound wave process is reversible and adiabatic, or isentropic.

$$a = \sqrt{\left(\frac{\partial p}{\partial \rho}\right)_s} \qquad \qquad \textbf{(2.4)}$$

Equation (2.4) has been derived for the case of a weak compression wave. It is well known that an audible sound wave consists of rarefactions or expansions, as well as compressions. The velocity of a weak expansion wave can be calculated using the same equations as were used with the compression wave. Allow the piston of Figure 2.5 to be drawn to the left with an infinitesimal velocity dV. This creates in the first instant a decrease in density at the piston face. The resulting weak expansion wave then moves to the right through the gas, traveling at the velocity of sound (see Figure 2.8).

To calculate the wave velocity, select a control volume that includes the wave, and express velocities relative to the control volume (see Figure 2.9.) Using the continuity and momentum equations, the same expression is obtained for the velocity of the expansion sound wave as was obtained for the compression sound wave:

$$a = \sqrt{\left(\frac{\partial p}{\partial \rho}\right)_s}$$

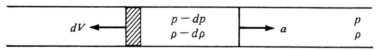

Figure 2.8   Expansion wave

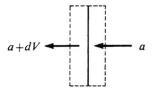

Figure 2.9 Control volume for expansion wave

The velocity of sound for a perfect gas can be evaluated by using Eq. (1.31), which expresses the variation of pressure with density for a perfect gas undergoing an isentropic process:

$$\frac{p_2}{p_1} = \left(\frac{\rho_2}{\rho_1}\right)^{\gamma} \tag{1.31}$$

or

$$\frac{p}{\rho^{\gamma}} = \text{constant}$$

Therefore, for a perfect gas,

$$\left(\frac{\partial p}{\partial \rho}\right)_s = \frac{\gamma p}{\rho}$$

and

$$a = \sqrt{\frac{\gamma p}{\rho}}$$
$$= \sqrt{\gamma R T} \tag{2.5}$$

Example 2.1

Calculate the velocity of sound in air at 0°C, with R for air $=0.287$ kJ/kg · K and $\gamma = 1.4$.

Solution

From Eq. (2.5),

$$a = \sqrt{(1.4)(0.287 \text{ kN} \cdot \text{m/kg} \cdot \text{K})(273 \text{ K})}$$
$$= \sqrt{(1.4)(287 \text{ kg} \cdot \text{m}^2/\text{kg} \cdot \text{K} \cdot \text{s}^2)(273 \text{ K})}$$
$$= 331.2 \text{ m/s}$$

For a substance that is not a perfect gas, it is desirable to express the velocity of sound in terms of a physical property of the substance. Compressibility is defined by[1]

$$k_s = \frac{1}{\rho}\left(\frac{\partial \rho}{\partial p}\right)_s$$

$$k_T = \frac{1}{\rho}\left(\frac{\partial \rho}{\partial p}\right)_T$$

where k_s and k_T are respectively, the isentropic and isothermal compressibilities, related by[1]

$$\frac{k_T}{k_s} = \gamma$$

Therefore,

$$a = \sqrt{\frac{1}{\rho k_s}} \qquad (2.6)$$

Example 2.2

At 0°C, one atmosphere, liquid water has an isothermal compressibility of $0.51 \times 10^{-6}\,(kPa)^{-1}$ and a density of $1000\,kg/m^3$. Determine the velocity of sound in water under these conditions.

Solution

For water, $c_p \approx c_v$, so that $\gamma \approx 1.0$.

$$a = \sqrt{\frac{1}{\rho k_s}}$$

$$= \sqrt{\frac{1}{(1000\,kg/m^3)0.51 \times 10^{-6}(kPa)^{-1}}}$$

$$= \sqrt{\frac{(1000\,Pa)10^6}{510\,kg/m^3}}$$

$$= \sqrt{(1.96 \times 10^6)\,kg/m \cdot s^2/kg/m^3}$$

$$= \underline{1400\,m/s}$$

For solids, the change in density with pressure is usually expressed in terms of β_s, the *bulk modulus*, defined by

$$\beta_s = \rho\left(\frac{\partial p}{\partial \rho}\right)_s$$

so that

$$a = \sqrt{\frac{\beta_s}{\rho}} \qquad (2.7)$$

Example 2.3

Calculate the velocity of sound in copper at 300 K. Take $\rho = 8950\,kg/m^3$ and $\beta_s = 1.32 \times 10^8\,kPa$ for copper at 300 K.

Solution

From Eq. (2.7),

$$a = \sqrt{\frac{\beta_s}{\rho}}$$

$$= \sqrt{\frac{1.32 \times 10^{11}\,\text{Pa}}{8950\,\text{kg/m}^3}}$$

$$= \sqrt{\frac{1.32 \times 10^{11}\,\text{kg/m}\cdot\text{s}^2}{0.8950 \times 10^4\,\text{kg/m}^3}}$$

$$= \underline{3840\,\text{m/s}}$$

2.4 Subsonic and Supersonic Flow

It has been established that the presence of a small disturbance is signaled throughout a fluid by means of a wave traveling at the local velocity of sound relative to the fluid into which the wave is propagating. If a body travels through a fluid at a velocity greater than the velocity of sound, the fluid ahead of the body is unable to sense the presence of the body, and abrupt changes in flow properties occur as the flow passes over the body. If a body travels through a fluid at a velocity less than the velocity of sound, the presence of the body is signaled to the fluid ahead of it, the fluid has a chance to adjust to the movement of the body, and gradual changes in flow properties result, with smooth, continuous streamlines.

Consider the following simplified representation of this discussion. A point projectile, acting as an infinitesimal disturbance, is injected with velocity V into a stationary fluid. Just as the piston of Figure 2.6 produced a compression wave ahead of it traveling at the velocity of sound, so the disturbance here creates a pressure wave propagating throughout the fluid in all directions at the velocity of sound. For a subsonic projectile velocity, the wave moves farther ahead of the projectile as time passes. This is exemplified in Figure 2.10. For this case, the point projectile has been made to travel to the right with a velocity V equal to one-half the velocity of sound. The location of the projectile and a wave emitted by the projectile are depicted at one, two, and three equal time intervals after the projectile has emitted the wave. The spherical wave precedes the projectile into the fluid, with the distance between projectile and wave increasing with time. Eventually, all the fluid ahead of the projectile is able to sense its motion and thereby adjust to it.

If the projectile travels at a velocity greater than the velocity of sound, it continually moves into a fluid that has not sensed its presence. At each instant, the disturbance must push aside undisturbed fluid. This case is illustrated in Figure 2.11.

The projectile is assumed to travel to the right with a velocity equal to

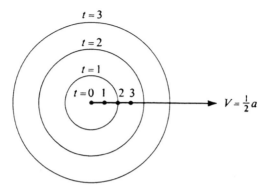

Figure 2.10   Subsonic velocity

twice the velocity of sound. Points 0, 1, 2, and 3 denote the locations of the projectile after 0, 1, 2, and 3 equal time intervals. The projectile is continually issuing waves during its motion; the waves move radially outward from the point of emission at the velocity of sound. The locations of the waves emitted by the projectile while at 0, 1, 2, and 3 are depicted after three time intervals. For example, while the projectile moves from 0 to 3, the wave emitted at 0 travels a distance $3a\,\Delta t$. Lines have been drawn tangent to the spherical waves, with the projectile at the apex of the cone thus formed. This cone defines the flow region in which the fluid has sensed the projectile motion. Everywhere outside the cone, the fluid is unaware of the presence of the moving projectile. For this reason, von Karman[2] termed the region inside the cone the *zone of action* and the region outside the cone the *zone of silence*. The nearness of the waves at any time signifies the strength of the resultant pressure disturbance at a point in the fluid. From Figure 2.11, it can be seen that the pressure

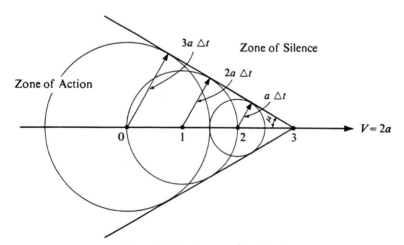

Figure 2.11   Supersonic velocity

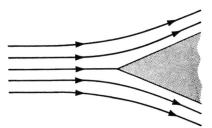

Figure 2.12   Subsonic wedge flow

disturbance is greatest in the vicinity of the cone. The lines at which the pressure disturbance is concentrated and which generate the cone are called *Mach lines* or waves. The angle between the Mach line and the projectile direction is the *Mach angle* μ:

$$\mu = \sin^{-1} \frac{a}{V} \qquad (2.8)$$

For subsonic flow, no such zone of silence exists. The entire fluid is able to sense the projectile moving through it, since the signal waves move faster than the projectile. No concentration of pressure disturbances can occur for subsonic flow; Mach lines cannot be defined.

Let us now compare steady, uniform, subsonic and supersonic flow over a finite wedge-shaped body. If the fluid velocity is less than the velocity of sound, flow ahead of the body is able to sense its presence. As a result, gradual changes in flow properties take place, with smooth, continuous streamlines (see Figure 2.12).

If the fluid velocity is greater than the velocity of sound, the approach flow, being in the zone of silence, is unable to sense the presence of the body. Unlike the point projectile discussed previously, the body now presents a finite disturbance to the flow. The wave pattern obtained is a result of the addition of

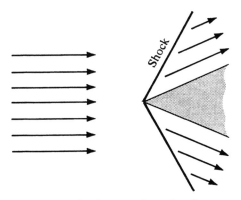

Figure 2.13   Supersonic wedge flow

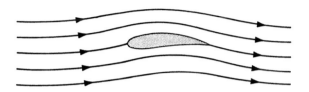

Figure 2.14 Subsonic flow over airfoil

the individual Mach waves emitted from each point on the wedge. This nonlinear addition yields a compression shock wave across which occur finite changes in velocity, pressure, and other flow properties. A typical flow pattern obtained for supersonic flow over the wedge is shown in Figure 2.13.

In this case, the adjustment of the flow to the body is not gradual but takes place entirely in the shock wave itself, which is of infinitesimal thickness. As a result, discontinuous changes in flow direction, pressure, temperature, density, and so on, occur across the wave.*

The design of an airfoil or other body to operate in subsonic flow is inherently different than the design for supersonic flow. For subsonic flow, a smooth shape is generally selected (Figure 2.14). However, for supersonic flow, the existence of shocks must be allowed for in design. A thin body with a pointed nose is preferable in order to minimize the strength of the shock in the vicinity of the front of the body. A diamond-shaped airfoil has met with considerable success in supersonic flow (Figure 2.15).

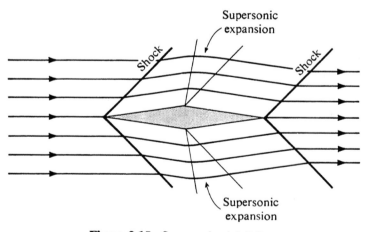

Figure 2.15 Supersonic airfoil flow

---

* In a real fluid, discontinuities cannot exist, since real fluid effects (viscosity and the like) tend to smooth them out. However, the measured thickness of a shock is on the order of the mean molecular free path, a small fraction of a millimeter.[3]

2.5 Mach Number

From the discussion in Section 2.4, the significance of the difference between subsonic and supersonic flow can be appreciated. The criterion for the type of flow is *Mach number*, defined by

$$M = \frac{V}{a} \qquad (2.9)$$

Mach number is an extremely important parameter in the study of compressible fluid flow. In the development of the equations of motion of a compressible fluid, much of the analysis will appear in terms of Mach number. The significance of the point at which Mach number is equal to 1 will be demonstrated again and again in future chapters.

2.6 Summary

As a body moves through a stationary fluid, waves are emitted from each point on the body and travel outward at the velocity of sound. In an incompressible fluid, the velocity of sound is infinite, so an entire body of fluid is able to sense instantaneously the motion of an object passing through it. In a compressible fluid, the velocity of sound has a finite value. If a body travels through a compressible fluid at a velocity *less* than that of sound, waves emitted by the body are able to move ahead of the body and signal the fluid to adjust to the oncoming disturbance. If a body travels at a velocity *greater* than that of sound, the waves are not able to signal the fluid ahead of the body. For the subsonic case, the fluid is able to adjust gradually to a moving object, and smooth, continuous streamline patterns result. For the supersonic case, the fluid must adjust rapidly to a moving object. Shock waves result with discontinuous changes in fluid properties. The importance of Mach number for compressible flow is established.

REFERENCES

Specific References

1. ZEMANSKY, M.W., ABBOTT, M.M., AND VAN NESS, H.C., *Basic Engineering Thermodynamics*, 2nd ed., New York, McGraw-Hill Book Company, 1975, pp. 32 and 254.
2. VON KARMAN, T., "Supersonic Aerodynamics—Principles and Applications," *Journal of Aeronautical Sciences*, Vol. 14, No. 7 (1947), p. 373.
3. SHAPIRO, A., *The Dynamics and Thermodynamics of Compressible Fluid Flow*, Parts I and II from Volume 1, New York, Ronald Press, 1958, pp. 131–134.

General References

4. Mironer, A., *Engineering Fluid Mechanics*, New York, McGraw-Hill Book Company, 1979.
5. Shapiro, A., *The Dynamics and Thermodynamics of Compressible Fluid Flow*, Vol. 1, New York, Ronald Press, 1953.

PROBLEMS

1. Using the expansion wave depicted in Figures 2.8 and 2.9, derive Eq. (2.4).
2. Derive an expression for k_s for a perfect gas, substitute your result into Eq. (2.6), and thereby demonstrate Eq. (2.5).
3. Derive an expression for β_s for a perfect gas, substitute your result into Eq. (2.7), and thereby demonstrate Eq. (2.5).
4. Using the data provided in Appendix G, Table G.1, determine the velocity of sound in helium, hydrogen, oxygen, and argon at 0°C and one atmosphere pressure.
5. Find the velocity of sound in air at 25°C and 101 kPa. Repeat for air at 25°C, 50 kPa, and 25°C, 150 kPa.
6. A jet plane is traveling at Mach 1.8 at an altitude of 10 km. Determine the speed of the plane in meters per second (see Appendix H.)

3

ISENTROPIC FLOW
OF A PERFECT GAS

3.1 Introduction

In this and subsequent chapters, the effect on compressible flow of area change, friction, heat transfer, and electromagnetic fields will be studied. In most physical situations, more than one of these effects occur simultaneously; for example, flow in a rocket nozzle involves area change, friction, and heat transfer. However, one of the effects is usually predominant; in the rocket nozzle, area change is the factor having greatest influence on the flow. The frequent predominance of one factor provides a justification for separating the effects, including them one at a time in the equations of motion, and studying the resultant property variations.

Whereas a certain loss of generality is incurred by treating each effect individually, this procedure does simplify the equations of motion so that the result of each effect can be easily appreciated. Furthermore, this simplification enables approximate solutions to be derived for a wide range of problems in compressible flow; such solutions are sufficiently accurate for many engineering applications. Attempts to include all the effects simultaneously in the equations of motion lead to mathematical complexities that mask the physical situation. In many cases, exact solutions to these generalized equations of motion are impossible.

Chapter 3 is concerned with compressible, isentropic flow through varying area channels, such as nozzles, diffusers, and turbine-blade passages. Friction and heat transfer are negligible for this isentropic flow; variations in properties are brought about by area change. One-dimensional, steady flow of a perfect gas is assumed in order to reduce the equations to a workable form. For gas flows, changes in potential energy and gravitational forces are neglected. To understand fully the limitations imposed by the various assumptions, reference should be made to Chapter 1. The applicability and accuracy of the equations derived in this and subsequent chapters depend to a large extent on the manner in which the assumptions fit the physical situation involved in a given engineering problem.

3.2 Equations of Motion

For one-dimensional steady flow through a varying area channel, select a control volume as shown in Figure 3.1. The continuity equation for steady flow yields

$$\iint_s \rho \mathbf{V} \cdot d\mathbf{A} = 0. \tag{1.9}$$

For this case of one-dimensional flow,

$$(\rho + d\rho)(A + dA)(V + dV) - \rho A V = 0$$

or

$$\frac{d\rho}{\rho} + \frac{dA}{A} + \frac{dV}{V} = 0 \tag{3.1}$$

For steady flow, the momentum equation is

$$\sum \mathbf{F} = \iint_s \mathbf{V}(\rho \mathbf{V} \cdot d\mathbf{A}) \tag{1.14}$$

In the absence of electromagnetic forces, and with friction negligible, the only forces acting on the control surface are pressure forces. Assume that a

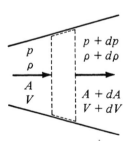

Figure 3.1   Control volume for varying area flow

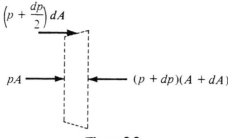

Figure 3.2

pressure $p + dp/2$ acts on the side surfaces of the control volume (Figure 3.2). Then Eq. (1.14) yields

$$pA + \left(p + \frac{dp}{2}\right) dA - (p + dp)(A + dA) = (\rho A V) \, dV$$

Simplifying yields

$$dp + \rho V \, dV = 0 \tag{3.2}$$

The energy equation, Eq. (1.19), with no external heat transfer and no work, becomes, for steady one-dimensional flow,

$$\iint_s \left(h + \frac{V^2}{2}\right)(\rho \mathbf{V} \cdot d\mathbf{A}) = 0$$

or

$$dh + d\frac{V^2}{2} = 0 \tag{3.3}$$

An expression for the second law of thermodynamics is given by Eq. (1.22):

$$T \, ds = dh - \frac{dp}{\rho} \tag{1.22}$$

For isentropic flow,

$$dh = \frac{dp}{\rho}$$

Combining with Eq. (3.3), we obtain

$$\frac{dp}{\rho} = -d\frac{V^2}{2}$$

or

$$dp + \rho V \, dV = 0$$

the same as Eq. (3.2).

3.3 Subsonic and Supersonic Isentropic Flow Through a Varying Area Channel

Combining the continuity and momentum equations for isentropic flow derived in Section 3.2, one obtains

$$dp + \rho V^2 \left[-\frac{d\rho}{\rho} - \frac{dA}{A} \right] = 0 \tag{3.4}$$

But

$$\left(\frac{\partial p}{\partial \rho} \right)_s = a^2$$

Therefore, for isentropic flow,

$$dp + \rho V^2 \left(-\frac{dp}{\rho a^2} - \frac{dA}{A} \right) = 0$$

$$dp(1 - M^2) = \rho V^2 \frac{dA}{A} \tag{3.5}$$

Equation (3.5) demonstrates the influence of Mach number on this flow. For $M < 1$, subsonic flow, the term $1 - M^2$ is positive. Therefore, an increase in area results in an increase in pressure and, from Eq. (3.2), a decrease in velocity. Likewise, a decrease in area results in a decrease in pressure and an increase in velocity (Figure 3.3). For supersonic flow, the term $1 - M^2$ in Eq. (3.5) is negative, and opposite variations occur (Figure 3.4).

The results illustrated in Figures 3.3 and 3.4 have many ramifications. Subsonic flow cannot be accelerated to a velocity greater than the velocity of sound in a converging nozzle. This is true irrespective of the pressure difference imposed on the flow through the nozzle. If it is desired to accelerate a stream from negligible velocity to supersonic velocity, a convergent–divergent channel must be used (Figure 3.5).

Note that the equation of state for a perfect gas has not yet been used.

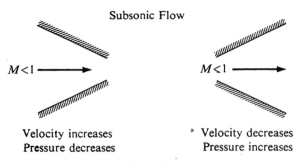

Subsonic Flow

$M < 1$ ⟶ $M < 1$ ⟶

Velocity increases   Velocity decreases
Pressure decreases   Pressure increases

Figure 3.3

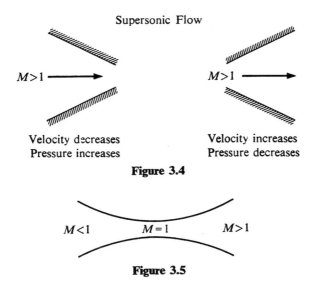

Figure 3.4

Figure 3.5

Equation (3.5) and the results provided by Figures 3.3 and 3.4 are valid, for example, for liquids as well as gases.

3.4 Stagnation Properties and the Use of Tables

Stagnation properties are useful in that they define a reference state for compressible flow. *Stagnation enthalpy*, or *total enthalpy*, at a point in a flow is defined as the enthalpy attained by bringing the flow adiabatically to rest at that point.

From Eq. (3.3),

$$h_t = h + \frac{V^2}{2}$$

where h_t is the stagnation or total enthalpy per unit mass. Likewise, *stagnation temperature* or *total temperature* T_t can be defined as the temperature measured by bringing a flow adiabatically to rest at a point. For a perfect gas with constant specific heats,

$$h_t - h = c_p(T_t - T)$$

or

$$T_t = \frac{V^2}{2c_p} + T$$
$$= T\left(1 + \frac{V^2}{2c_p T}\right)$$

But

$$c_p = \frac{\gamma R}{\gamma - 1}$$

so that

$$T_t = T\left[1 + \frac{(\gamma - 1)V^2}{2\gamma RT}\right]$$

where

$$\gamma RT = a^2$$

Combining terms, we obtain

$$T_t = T\left(1 + \frac{\gamma - 1}{2}M^2\right) \tag{3.6}$$

for a perfect gas with constant specific heats.

Equation (3.6) is tabulated for $\gamma = 1.4$, $\gamma = 1.3$, and $\gamma = \frac{5}{3}$ in Appendix A, as T/T_t versus M. For example, if a perfect gas with $\gamma = 1.4$ is traveling at Mach 3 with a static temperature of 250 K, the stagnation temperature is

$$T_t = T\left(\frac{T_t}{T}\right)$$

$$= 250\frac{1}{0.3571}$$

$$= 700.1 \ K$$

Stagnation pressure p_t at a point in a flow is defined as the pressure attained if the flow at that point is brought isentropically to rest. From Section 1.8, for a perfect gas with constant specific heats undergoing an isentropic process,

$$\frac{p_2}{p_1} = \left(\frac{T_2}{T_1}\right)^{\gamma/(\gamma - 1)} \tag{1.30}$$

Let state 2 be the stagnation state; then

$$\frac{p_t}{p} = \left(\frac{T_t}{T}\right)^{\gamma/(\gamma - 1)}$$

Using Eq. (3.6), we obtain

$$\frac{p_t}{p} = \left(1 + \frac{\gamma - 1}{2}M^2\right)^{\gamma/(\gamma - 1)} \tag{3.7}$$

This relationship is tabulated in Appendix A as p/p_t versus M, for $\gamma = 1.4$, $\gamma = 1.3$, and $\gamma = \frac{5}{3}$.

For example, if an airstream is moving at Mach 3 with a static pressure of 101 kPa, the stagnation or total pressure is

$$p_t = p\frac{p_t}{p}$$

$$= 101\frac{1}{0.02722}$$

$$= 3710.5 \text{ kPa}$$

(assuming the air to behave as a perfect gas with constant $\gamma = 1.4$)

To better understand these stagnation conditions, consider steady flow through a varying area channel, emanating from a large reservoir in which the velocity is negligible (Figure 3.6). Since the flow is at rest in the reservoir, the actual pressure and temperature of the gas in the reservoir are p_t and T_t. As the flow is accelerated through the channel, the static pressure and temperature decrease. However, if the flow is adiabatic,

$$h + \frac{V^2}{2} = \text{constant}$$

Furthermore, for a perfect gas with constant specific heats,

$$h + \frac{V^2}{2} = c_p T_t$$

or the stagnation temperature at any cross section of the flow remains equal to the reservoir temperature T_t. If the flow is reversible as well as adiabatic, both the stagnation pressure and stagnation temperature are constant and equal to the reservoir values.

Frequently, there is difficulty in understanding the difference between static and stagnation properties. Static pressure and static temperature of a moving stream are properties experienced by an observer moving with the same velocity as the stream. Stagnation pressure and stagnation temperature are properties experienced by a fixed observer, the fluid having been brought to rest at the observer. The difference between static and stagnation properties is due to the velocity or kinetic energy of the flow.

Mass flow rate at a flow cross-sectional area A can be expressed in terms

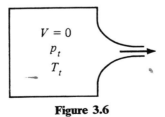

Figure 3.6

of stagnation pressure and temperature:

$$\dot{m} = \rho A V$$

For a perfect gas with constant specific heats,

$$\dot{m} = \frac{p}{RT} A M \sqrt{\gamma R T}$$

where

$$p = \frac{p_t}{\left(1 + \dfrac{\gamma - 1}{2} M^2\right)^{\gamma/(\gamma-1)}}$$

$$T = \frac{T_t}{1 + \dfrac{\gamma - 1}{2} M^2}$$

Thus

$$\dot{m} = \frac{p_t}{\sqrt{RT_t}} A \sqrt{\gamma} M \left(1 + \frac{\gamma - 1}{2} M^2\right)^{(\gamma+1)/(2-2\gamma)} \tag{3.8}$$

or, more simply,

$$\dot{m} = \frac{p_t A}{\sqrt{RT_t}} f(\gamma, M) \tag{3.9}$$

where $f(\gamma, M)$ is given by

$$f(\gamma, M) = M \sqrt{\gamma} \left(1 + \frac{\gamma - 1}{2} M^2\right)^{(\gamma+1)/(2-2\gamma)} \tag{3.10}$$

For isentropic flow, in which p_t and T_t are constant, cross-sectional area A can be related directly to Mach number. Select the area at which $M = 1$ as a reference area, A^*. For steady flow, the mass flow rate at area A, $\dot{m}_A$, is equal to the mass flow rate at A^*, $\dot{m}_{A^*}$; so

$$\frac{p_t}{\sqrt{RT_t}} A f(\gamma, M) = \frac{p_t}{\sqrt{RT_t}} A^* f(\gamma)$$

or

$$\frac{A}{A^*} = g(M, \gamma) \tag{3.11}$$

where

$$g(M, \gamma) = \frac{\sqrt{\gamma}\left(\dfrac{\gamma + 1}{2}\right)^{(\gamma+1)/(2-2\gamma)}}{M \sqrt{\gamma}\left(1 + \dfrac{\gamma - 1}{2} M^2\right)^{(\gamma+1)/(2-2\gamma)}} \tag{3.12}$$

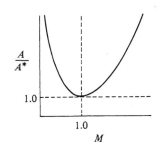

Figure 3.7

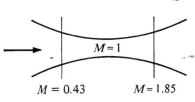

Figure 3.8

The result of Eq. (3.11) is plotted in Figure 3.7 for $\gamma = 1.4$. Numerical values of A/A^* versus M for $\gamma = 1.4$, $\gamma = 1.3$, and $\gamma = \frac{5}{3}$ are presented in Appendix A.

For each value of A/A^*, there are two possible isentropic solutions, one subsonic and the other supersonic. The minimum area, or throat area, occurs at $M = 1$. This concurs with the results of Eq. (3.5), illustrated in Figures 3.3 and 3.4; a converging–diverging nozzle is required to accelerate a slowly moving stream to supersonic velocities. For example, from Appendix A, with $\gamma = 1.4$, $A/A^* = 1.5$, $M = 0.43$, and also $M = 1.85$ (Figure 3.8).

Example 3.1

An airstream flows in a converging duct (Figure 3.9) from a cross-sectional area A_1 of $50\ \text{cm}^2$ to a cross-sectional area A_2 of $40\ \text{cm}^2$. If $T_1 = 300\ \text{K}$, $p_1 = 100\ \text{kPa}$, and $V_1 = 100\ \text{m/s}$, find M_2, p_2, and T_2. Assume steady, one-dimensional isentropic flow.

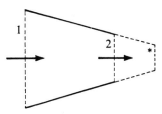

Figure 3.9

Solution

Over the temperature range of interest, air behaves as a perfect gas with $\gamma = 1.4$ (see Appendix G, Table G.3).

$$M_1 = \frac{V_1}{a_1} = \frac{100 \text{ m/s}}{\sqrt{(1.4)\left(0.287 \dfrac{\text{kN} \cdot \text{m}}{\text{kg} \cdot \text{K}}\right) 300 \text{ K}}}$$

$$= 0.288$$

If isentropic flow in the converging duct were to be maintained (with the duct extended, as shown), the flow would be accelerated to $M = 1$, with the area at this point equal to the reference area A^*.

At $M = 0.288$, from Appendix A, with $\gamma = 1.4$,

$$\frac{A_1}{A^*} = 2.111$$

But

$$\frac{A_2}{A_1} = \frac{40}{50} = 0.80$$

so that

$$\frac{A_2}{A^*} = 1.689$$

From Appendix A, $M_2 = 0.372$.

For isentropic flow, p_t and T_t are constant. At $M = 0.288$,

$$\frac{p_1}{p_{t_1}} = 0.944$$

$$\frac{T_1}{T_{t_1}} = 0.984$$

Therefore,

$$p_{t_1} = \frac{100}{0.944} = 105.9 \text{ kPa} = p_{t_2}$$

$$T_{t_1} = \frac{300}{0.984} = 304.9 \text{ K} = T_{t_2}$$

At $M = 0.372$,

$$\frac{p_2}{p_{t_2}} = 0.909$$

$$\frac{T_2}{T_{t_2}} = 0.973$$

or

$$p_2 = 0.909(105.9 \text{ kPa}) = \underline{96.3 \text{ kPa}}$$
$$T_2 = 0.973(304.9 \text{ K}) = \underline{296.7 \text{ K}}$$

3.5 Isentropic Flow in a Converging Nozzle

A fluid stored in a large reservoir is to be discharged through a converging nozzle. For a constant reservoir pressure p_r, it is desired to determine the rate of mass flow through the nozzle as a function of the back pressure p_b imposed on the nozzle.

Figure 3.10 depicts the pressure distribution obtained in the nozzle for six different values of back pressure. Figure 3.11 shows the mass flow rate through the nozzle for each of the values of back pressure. For $p_b = p_r$ (curve 1) there is no flow in the nozzle, and pressure is invariant with x. As p_b is reduced below p_r, more and more flow is induced through the nozzle (curves 2 and 3), and the static pressure decreases with x for this subsonic flow. The velocity at the nozzle exit plane increases as p_b is reduced, until eventually the velocity of sound is reached at the nozzle exit plane (curve 4).

The results of Section 3.3, illustrated by Figures 3.3 and 3.4, indicate that the flow in a converging nozzle cannot be accelerated to a velocity greater than the velocity of sound. To understand the phenomena that occur after sonic velocity is reached at the nozzle exit plane, let us discuss physically how the flow is able to adjust to changes in back pressure. The presence of a disturbance, such as a change in back pressure, is signaled through a compressible fluid by means of a wave traveling at the velocity of sound relative to the moving fluid. For subsonic flow in the nozzle, the signal wave propagates at a greater velocity than the flow. Changes in back pressure can be communicated back to the fluid in the reservoir, in the same way that the motion of the point

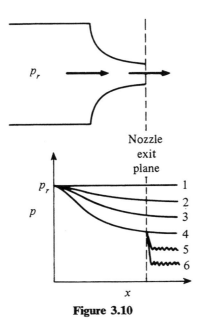

Figure 3.10

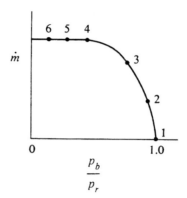

Figure 3.11

projectile in Chapter 2 is felt in the fluid ahead of it. For example, when a decrease in back pressure occurs, the change is telegraphed back through the fluid to the reservoir, and the reservoir sends out more flow. The subsonic flow in the nozzle is able to adjust gradually to the back pressure; so for all values of p_b/p_r greater than that corresponding to curve 4, the back pressure is equal to the exit-plane pressure.

For p_b/p_r equal to that of curve 4, sonic velocity occurs at the nozzle exit plane. The velocity of the signal waves is equal to the velocity of sound relative to the fluid into which the wave is propagating. If the fluid at a cross section is moving at the velocity of sound, the absolute velocity of the signal wave at this cross section is zero. In other words, the signal wave cannot travel past this cross section (Figure 3.12).

After sonic flow is attained at the nozzle exit plane, signal waves are unable to propagate from the back pressure region to the reservoir. Therefore, as the back pressure is decreased below that of curve 4, with the reservoir fluid not able to sense the decrease, flow through the nozzle remains the same as that of curve 4. Since the entire flow in the nozzle upstream of the exit plane is unable to sense changes in back pressure, the pressure distribution in the nozzle, p versus x, likewise remains the same as curve 4. Under these conditions, flow inside the nozzle cannot adjust to the changes in back pressure. Therefore, for back pressures less than that of curve 4, the exit plane

Velocity of wave relative to fluid $= a$
Absolute velocity of wave $= a - V$

Figure 3.12

pressure is not equal to the back pressure; instead, the flow must adjust to the back pressure by means of an expansion occurring outside the nozzle.

Reductions in back pressure below that of curve 4 cannot cause any more flow to be induced through the nozzle. Under these conditions, the nozzle is said to be *choked*. The pressure ratio, p_b/p_r, below which the nozzle is choked can be calculated for isentropic flow through the nozzle. For a perfect gas with constant specific heats,

$$p_t = p\left(1 + \frac{\gamma - 1}{2}M^2\right)^{\gamma/(\gamma - 1)} \tag{3.7}$$

For isentropic flow in the nozzle, p_t is constant and equal to p_r. Therefore, to just choke the nozzle, let the Mach number at the exit plane equal 1:

$$\frac{p_r}{p_b} = \left(1 + \frac{\gamma - 1}{2}\right)^{\gamma/(\gamma - 1)} \tag{3.8}$$

As examples, for $\gamma = 1.4$, $p_b/p_r = 0.5283$; for $\gamma = 1.3$, $p_b/p_r = 0.5457$; and for $\gamma = \frac{5}{3}$, $p_b/p_r = 0.4871$.

This ratio below which the nozzle is choked is termed the *critical pressure ratio*. For a perfect gas, the critical pressure ratio is only dependent on the ratio of specific heats.

Example 3.2

Air is allowed to flow from a large reservoir through a converging nozzle with an exit area of 50 cm². The reservoir is large enough so that negligible changes in reservoir pressure and temperature occur as fluid is exhausted through the nozzle. Assume isentropic, steady flow in the nozzle, with $p_r = 500$ kPa and $T_r = 500$ K; assume also that air behaves as a perfect gas with constant specific heats, $\gamma = 1.4$. Determine the mass flow through the nozzle for back pressures of 0, 125, 250, and 375 kPa.

Solution

For $\gamma = 1.4$, the critical pressure ratio is 0.5283; therefore, for all back pressures below 264.15 kPa, the nozzle is choked. Under these conditions, the Mach number at the exit plane is 1, the pressure at the exit plane is 264.15 kPa (not equal to the back pressure), and the temperature at the exit plane is $T = (T/T_t) \cdot T_t = 0.8333(500) = 416.7$ K, where T/T_t is obtained from Appendix A.1 for $\gamma = 1.4$. The mass flow rate for back pressures of 0, 125, and 250 kPa is

$$\dot{m} = \rho A V$$

$$= \frac{p_e}{RT_e} A_e M_e \sqrt{\gamma R T_e}$$

$$= \frac{(264.15 \text{ kN/m}^2)(50 \times 10^{-4} \text{ m}^2)(1.0)}{(0.287 \text{ kN} \cdot \text{m/kg} \cdot \text{K})(416.7 \text{ K})} \sqrt{1.4(0.287 \text{ kN} \cdot \text{m/kg} \cdot \text{K})(416.7 \text{ K})}$$

$$= 4.519 \text{ kg/s},$$

for back pressures of 0, 125, and 250 kPa (see Figure 3.13).

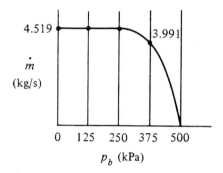

Figure 3.13

For a back pressure of 375 kPa, the nozzle is not choked, and the exit plane pressure is equal to the back pressure. For $p/p_t = 0.75$, $M_e = 0.654$ (Appendix A.1, $\gamma = 1.4$), and $T_e/T_t = 0.921$.

$$\dot{m} = \frac{(375)(50 \times 10^{-4})(0.654)}{(0.287)(0.921 \times 500)} \sqrt{1.4(287)(0.921 \times 500)}$$

$$= 3.991 \text{ kg/s}$$

for a back pressure of 375 kPa (see Figure 3.13).

Example 3.3

Nitrogen is stored in a tank 2 m^3 in volume at a pressure of 3 MPa and a temperature of 300 K (Figure 3.14). The gas is discharged through a converging nozzle with an exit area of 12 cm^2. For a back pressure of 101 kPa, find the time required for the tank pressure to drop to 300 kPa. Assume isentropic nozzle flow, with nitrogen behaving as a perfect gas, with $\gamma = 1.4$ (constant) and $R = 0.2968 \text{ kJ/kg} \cdot \text{K}$. Assume quasi-steady flow through the nozzle, with the steady-flow equations applicable at each instant of time; assume also that T_r is constant.

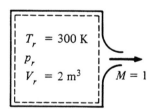

Figure 3.14

Solution

As the reservoir pressure drops from 3 MPa to 300 kPA, the ratio p_b/p_r remains below the critical pressure ratio, and the nozzle exit Mach number is 1.

$$\dot{m} = \rho A V$$

$$= \frac{p}{RT} A V$$

$$= \frac{(0.5283p_r)(12 \times 10^{-4} \text{ m}^2)\sqrt{(1.4)(296.8 \text{ J/kg} \cdot \text{K})(0.8333 \times 300 \text{ K})}}{(0.2968 \text{ kJ/kg} \cdot \text{K})(0.8333 \times 300 \text{ K})}$$

$$= (0.002754 p_r) \text{ kg/s with } p_r \text{ in kN/m}^2$$

From Eq. (1.8),

$$\frac{\partial}{\partial t} \iiint_{\text{c.v.}} \rho \, d\mathcal{V} + \iint_{s} \rho \mathbf{V} \cdot d\mathbf{A} = 0$$

$$\iiint_{\text{c.v.}} \rho \, d\mathcal{V} = m$$

where m is the mass inside the tank at any time

$$= \frac{p_r \mathcal{V}_r}{RT_r}$$

By substituting in Eq. (1.8), we obtain

$$\frac{\mathcal{V}_r}{RT_r} \frac{dp_r}{dt} + 0.002754 p_r = 0$$

$$\frac{2 \text{ m}^3}{(0.2968 \text{ kJ/kg} \cdot \text{K})(300 \text{ K})} \frac{dp_r}{dt} + 0.002754 p_r = 0$$

$$\frac{dp_r}{p_r} = -0.1226 \, dt$$

$$t = -\int_{3000}^{300} 8.1554 \frac{dp_r}{p_r}$$

$$= 8.1554 \ln 10$$

$$= \underline{18.78 \text{ s}}$$

3.6 Isentropic Flow in a Converging–Diverging Nozzle

Fluid stored in a large reservoir is to be discharged through a converging–diverging nozzle. It is desired to determine the mass flow and pressure distribution in the nozzle over a range of values of p_b/p_r. The reservoir pressure p_r is maintained constant, with one-dimensional isentropic flow in the nozzle.

Figure 3.15 depicts the pressure distribution obtained in the nozzle for five different values of back pressure. Figure 3.16 shows the mass flow rate for each value of p_b/p_r. For p_b equal to p_r (curve 1), there is no flow in the nozzle, and

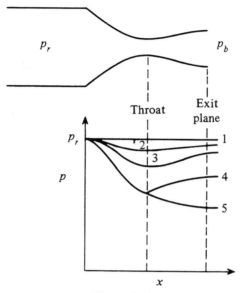

Figure 3.15

pressure is invariant with x. For p_b slightly less than p_r (curve 2), flow is induced through the nozzle, with subsonic velocities in both converging and diverging sections of the nozzle. Equation (3.5) tells us that, for subsonic flow, pressure decreases in the converging section and increases in the diverging section. As the back pressure is decreased, more and more flow is induced through the nozzle (curve 3) until eventually sonic flow occurs at the throat (curve 4). Further decreases in back pressure cannot be sensed upstream of the throat; so for all back pressures below that of curve 4, the reservoir continues to send out the same flow rate as curve 4, and the pressure distribution in the

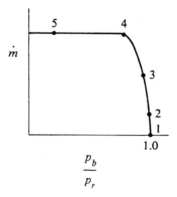

Figure 3.16

nozzle up to the throat remains the same. For all back pressures below that of curve 4, the converging–diverging nozzle is choked. Note that for the same reservoir pressure, a converging–diverging nozzle is choked at a greater back pressure than a converging nozzle.

There are two possible isentropic solutions for a given area ratio A/A^*, one subsonic and the other supersonic. For a throat Mach number of 1, isentropic flow can either decelerate to a subsonic exit velocity or continue to accelerate to a supersonic exit velocity. Curve 4 corresponds to the case of subsonic flow at the nozzle exit plane; curve 5 corresponds to supersonic flow at the exit plane. Thus, if the back pressure is lowered to that of curve 5, pressure decreases in both converging and diverging portions of the nozzle, with supersonic flow at the exit plane.

For back pressures between those of curves 4 and 5, one-dimensional isentropic solutions to the equations of motion are not possible. These flows involve shock waves, which are irreversible processes. Nonisentropic solutions for flows through a converging–diverging nozzle will be discussed in some detail after the shock process has been introduced and the equations of motion for a shock wave have been derived.

Example 3.4

A converging–diverging nozzle is designed to operate isentropically with an exit Mach number of 1.5. The nozzle is supplied from an air reservoir in which the pressure is 500 kPa; the temperature is 500 K. The nozzle throat area is 5 cm^2. Assume air to behave as a perfect gas, with $\gamma = 1.4$ and $R = 0.2870$ kJ/kg·K.

(a) Determine the ratio of exit area to throat area.
(b) Find the range of back pressure over which the nozzle is choked.
(c) Determine the mass flow rate for a back pressure of 450 kPa.
(d) Determine the mass flow rate for a back pressure of 0 kPa.

Solution

(a) To produce a supersonic Mach number of 1.5 at the nozzle exit, the Mach number at the throat must be 1. Therefore, the throat area is equal to A^*. From Appendix A for $M = 1.5$, $A/A^* = 1.176$, so the ratio of exit area to throat area to produce Mach 1.5 is <u>1.176</u>, or $A_e = 5.88$ cm^2.

(b) For all back pressures below that corresponding to curve 4 of Figure 3.15, the nozzle is choked. For curve 4, sonic flow is attained at the throat, followed by subsonic deceleration. The subsonic solution for $A/A^* = 1.176$ is found from Appendix A: $M = 0.61$. At this Mach number, $p/p_t = 0.778$; therefore, the greatest back pressure at which the nozzle is choked is 0.778(500 kPa) = 389 kPa. In other words, over the range $0 \le p_b \le 389$ kPa, the nozzle is choked.

(c) For a back pressure of 450 kPa, the nozzle is not choked; subsonic flow occurs throughout the nozzle. For this condition, the exit-plane pressure is equal to the back pressure. From Appendix A, for $p/p_t = 0.90$, $M = 0.39$ and $T/T_t = 0.971$. Exit-plane pressure p_e and temperature T_e are, respectively, 450 kPa and

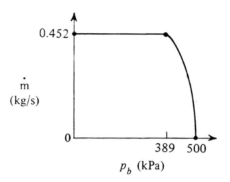

Figure 3.17

485.5 K.

$$\dot{m} = \rho_e A_e V_e$$

$$= \frac{p_e}{RT_e} A_e M_e \sqrt{\gamma RT_e}$$

$$= \left[\frac{450 \text{ kN/m}^2}{(0.287 \text{ kNm/kg} \cdot \text{K})(485.5 \text{ K})} \right] [5.88 \times 10^{-4} \text{ m}^2]$$

$$[0.39\sqrt{1.4(287 \text{ Nm/kg} \cdot \text{K})(485.5 \text{ K})}]$$

$$= (3.230 \text{ kg/m}^3)(5.88 \times 10^{-4} \text{ m}^2)(0.39 \times 441.7 \text{ m/s})$$

$$= \underline{0.327 \text{ kg/s}}$$

(d) For a back pressure of 0 kPa, the nozzle is choked, with the exit-plane pressure not equal to the back pressure. For this condition, the Mach number at the throat is 1, with the throat pressure and temperature equal, respectively, to 264.2 kPa and 416.7 K.

$$\dot{m} = \rho_{th} A_{th} V_{th}$$

$$= \frac{264.2 \text{ kN/m}^2(5 \times 10^{-4} \text{ m}^2)\sqrt{1.4(287 \text{ Nm/kg} \cdot \text{K})(416.7 \text{ K})}}{(0.287 \text{ kNm/kg} \cdot \text{K})(416.7 \text{ K})}$$

$$= 0.452 \text{ kg/s}$$

The results of this example are plotted in Figure 3.17.

Example 3.5

A nozzle is to be designed for a supersonic helium wind tunnel. Test section specifications are as follows:

Diameter, 10 cm
Mach number, 3.0
Static pressure, 12.1 kPa at 15-km altitude
Static temperature, 216.7 K at 15-km altitude

Determine the mass flow that must be provided, the nozzle throat area, and the reservoir temperature and pressure. Assume isentropic flow in the nozzle at the

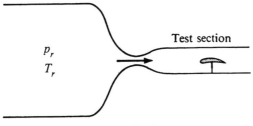

Figure 3.18

design condition, and neglect boundary layer effects (Figure 3.18). Assume that helium behaves as a perfect gas, with $\gamma = \frac{5}{3}$ (constant) and $R = 2.077$ kJ/kg·K.

Solution

Test section mass flow is given by

$\dot{m} = \rho A V$

$= \left(\dfrac{p}{RT}\right)\left(\dfrac{\pi}{4}D^2\right)(M\sqrt{\gamma RT})$

$= \left[\dfrac{(12.1 \text{ kN/m}^2)}{(2.077 \text{ kNm/kg} \cdot \text{K})(216.7 \text{ K})}\right]\left(\dfrac{\pi}{4}0.01 \text{ m}^2\right)3.0\sqrt{\dfrac{5}{3}(2.077 \text{ kNm/kg} \cdot \text{K})(216.7 \text{ K})}$

$= (0.0269 \text{ kg/m}^3)(0.00785 \text{ m}^2)(2598 \text{ m/s})$

$= \underline{0.5487 \text{ kg/s}}$

From Appendix A, Table A.3, at $M = 3.0$,

$$\frac{A}{A^*} = 3.00$$

$$\frac{p}{p_t} = 0.3125$$

$$\frac{T}{T_t} = 0.2500$$

Therefore,

$$\text{throat area} = \frac{0.00785 \text{ m}}{3.0} = \underline{0.002617 \text{ m}^2}$$

$$p_r = \frac{12.1 \text{ kPa}}{0.03125} = \underline{387.2 \text{ kPa}}$$

$$T_r = \frac{216.7}{0.2500} = \underline{866.8 \text{ K}}$$

3.7 Summary

In this chapter the equations of continuity, momentum, and energy are presented for one-dimensional isentropic flow through a varying area channel.

The effect of Mach number on the flow is demonstrated; the variations of pressure, velocity, and temperature with area for supersonic flow are entirely different than for subsonic flow. For supersonic flow, the increase of velocity and decrease of pressure as area is increased seems to violate our intuition; however, our intuition is based mainly on liquid water flow, in which density does not vary. For flow in a diverging channel at Mach numbers greater than 1, density decreases are great enough to outweigh the increase in area so that, with $\rho A V$ = constant, the result is an increase in flow velocity.

The phenomenon of choking also seems contrary to our intuition. As the back pressure on a nozzle is lowered for a constant reservoir pressure, more and more flow is induced through the nozzle until a maximum flow rate is reached. Further decreases in back pressure then have no effect on the mass flow rate; the nozzle is choked. Choking occurs when changes in back pressure can no longer be sensed in the reservoir. For subsonic flow, the reservoir is able to sense back pressure changes by means of sonic signal waves sent back through the fluid. Once the velocity of sound is reached at a point in the flow, the signal wave is not able to travel back to the reservoir; hence, the reservoir is unable to sense further decreases in back pressure.

It should be clear that certain intuitive notions gained from a knowledge of water flow cannot be applied to a study of compressible fluid mechanics. One should now begin to appreciate the significance of Mach number and the velocity of sound for compressible flow.

REFERENCES

1. JOHN, J.E.A., and HABERMAN, W.L., *Introduction to Fluid Mechanics*, 2nd ed., Englewood Cliffs, N.J., Prentice-Hall, Inc., 1980.
2. FOX, R.W., and McDONALD, A.T., *Introduction to Fluid Mechanics*, 2nd ed., New York, John Wiley & Sons, Inc., 1978.
3. ZUCROW, M.J., and HOFFMAN, J.D., *Gas Dynamics*, Vol. 1, New York, John Wiley & Sons, Inc., 1976.
4. MIRONER, A., *Engineering Fluid Mechanics*, New York, McGraw-Hill Book Company, 1979.
5. STREETER, V.L., and WYLIE, E.B., *Fluid Mechanics*, 6th ed., New York, McGraw-Hill Book Company, 1975.

PROBLEMS

1. Air flows at Mach 0.25 through a circular duct with a diameter of 60 cm. The stagnation pressure of the flow is 500 kPa; the stagnation temperature is 175°C. Calculate the mass flow rate through the channel, assuming $\gamma = 1.4$ and that the air behaves as a perfect gas with constant specific heats.

2. Helium flows at Mach 0.50 in a channel with cross-sectional area of 0.16 m². The stagnation pressure of the flow is 1 MPa, and stagnation temperature is 1000 K. Calculate the mass flow rate through the channel, with $\gamma = \frac{5}{3}$.

3. In Problem 2, the cross-sectional area is reduced to $0.12\ m^2$. Calculate the Mach number and flow velocity at the reduced area. What percent of further reduction in area would be required to reach Mach 1 in the channel?

4. An airflow at Mach 0.6 passes through a channel with a cross-sectional area of $50\ cm^2$. The static pressure in the airstream is 50 kPa; static temperature is 298 K.
   (a) Calculate the mass flow rate through the channel.
   (b) What percent of reduction in area would be necessary to increase the flow Mach number to 0.8? to 1.0?
   (c) What would happen if the area were reduced more than necessary to reach Mach 1?

5. A converging nozzle with an exit area of $1.0\ cm^2$ is supplied from an oxygen reservoir in which the pressure is 500 kPa and the temperature is 1200 K. Calculate the mass flow rate of oxygen for back pressures of 0, 100, 200, 300, and 400 kPa. Assume that $\gamma = 1.3$.

6. Compressed air is discharged through the converging nozzle as shown in Figure P3.6. The tank pressure is 500 kPa, and local atmospheric pressure is 101 kPa. The inlet area of the nozzle is $100\ cm^2$; the exit area is $34\ cm^2$. Find the force of the air on the nozzle, assuming the air to behave as a perfect gas with constant $\gamma = 1.4$. Take the temperature in the tank to be 300 K.

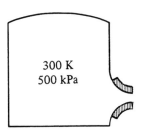

300 K
500 kPa

Figure P3.6

7. A converging nozzle has an exit area of $56\ cm^2$. Nitrogen stored in a reservoir is to be discharged through the nozzle to an ambient pressure of 100 kPa. Determine the flow rate through the nozzle for reservoir pressures of 120 kPa, 140 kPa, 200 kPa, and 1 MPa. Assume isentropic nozzle flow. In each case, determine the increase in mass flow to be gained by reducing the back pressure from 100 to 0 kPa. Reservoir temperature is 298 K.

8. Liquid water is allowed to flow from a large reservoir through a converging nozzle. Assuming isentropic nozzle flow with a back pressure of 101 kPa, calculate the reservoir pressure necessary to choke the nozzle. Assume the isothermal compressibility of water to be $5 \times 10^{-7}\ (kPa)^{-1}$.

9. Calculate the stagnation temperature in an airstream traveling at Mach 5 with a static temperature of 273 K (see Figure P3.9). An insulated flat plate is inserted into this flow, aligned parallel with the flow direction, with a boundary layer building up along the plate. Since the absolute velocity at the plate surface is zero, would you expect the plate temperature to reach the free stream stagnation temperature? Explain.

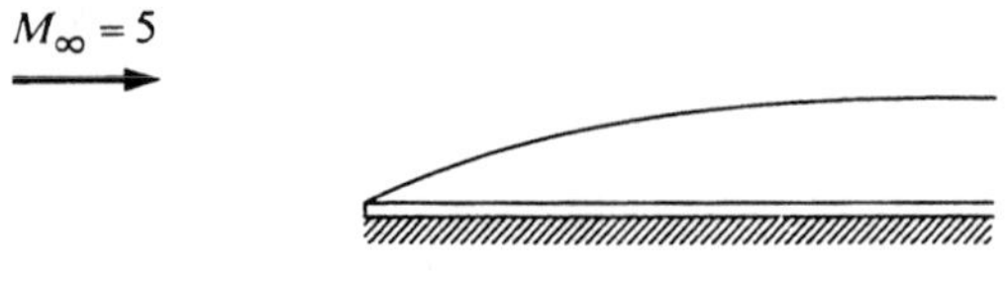

$M_\infty = 5$

Figure P3.9

10. A gas stored in a large reservoir is discharged through a converging nozzle. For a constant back pressure, sketch a plot of mass flow rate versus reservoir pressure. Repeat for a converging–diverging nozzle.

11. A converging–diverging nozzle is designed to operate isentropically with air at an exit Mach number of 1.75. For a constant chamber pressure and temperature of 5 MPa and 200°C, respectively, calculate the following:
    (a) Maximum back pressure to choke nozzle
    (b) Flow rate in kilograms per second for a back pressure of 101 kPa
    (c) Flow rate for a back pressure of 1 MPa
    Nozzle exit area is 0.12 m².

12. A supersonic flow is allowed to expand indefinitely in a diverging channel. Does the flow velocity approach a finite limit, or does it continue to increase indefinitely? Assume a perfect gas with constant specific heats.

13. A converging–diverging frictionless nozzle is used to accelerate an airstream emanating from a large chamber. The nozzle has an exit area of 30 cm² and a throat area of 15 cm². If the ambient pressure surrounding the nozzle is 101 kPa and the chamber temperature is 500 K, calculate the following:
    (a) Minimum chamber pressure to choke the nozzle
    (b) Mass flow rate for a chamber pressure of 400 kPa
    (c) Mass flow rate for a chamber pressure of 200 kPa

14. Sketch p versus x for the case shown in Figure P3.14.

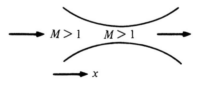

Figure P3.14

15. Steam is to be expanded to Mach 2.0 in a converging–diverging nozzle from an inlet velocity of 100 m/s. The inlet area is 50 cm²; inlet static temperature is 500 K. Assuming isentropic flow, determine the throat and exit areas required. Assume the steam to behave as a perfect gas with constant $\gamma = 1.3$.

16. Write a computer program that will yield values of T/T_t, p/p_t, and A/A^* for isentropic flow of a perfect gas with constant $\gamma = 1.35$. Use Mach number increments of 0.05 over the range $M = 0$ to $M = 5.0$.

17. A gas is known to have a molecular mass of 18, with $c_p = 2.0$ kJ/kg·K. The gas is expanded from negligible initial velocity through a converging–diverging nozzle with an area ratio of 5.0. Assuming an isentropic expansion in the nozzle with initial

stagnation pressure and temperature 1 MPa and 1000 K, respectively, determine the exit nozzle velocity.

18. A jet plane is flying at 10 km with a cabin pressure of 101 kPa and a cabin temperature of 20°C. Suddenly a bullet is fired inside the cabin and pierces the fuselage; the resultant hole is 2 cm in diameter. Assuming the flow through the hole to behave as that through a converging nozzle with an exit diameter of 2.0 cm, calculate the time for the cabin pressure to decrease to one-half the initial value. Assume the cabin volume to be 100 m³. At 10 km, $p = 26.5$ kPa and $T = 223.3$ K (see Appendix H).

19. A rocket nozzle is designed to operate isentropically at 20 km with a chamber pressure of 2.0 MPa and chamber temperature of 3000 K. If the products of combustion are assumed to behave as a perfect gas with constant specific heats ($\gamma = 1.3$ and $\bar{M} = 20$), determine the design thrust for a nozzle throat area of 0.25 m².

20. A converging nozzle has a rectangular cross section of a constant width of 10 cm. For ease of manufacture, the side walls of the nozzle are straight, making an angle of 10° with the horizontal, as shown in Figure P3.20. Determine and plot the variation of M, T, and p with x, taking $M_1 = 0.4$, $p_{t_1} = 200$ kPa, and $T_{t_1} = 350$ K. Assume the working fluid to be air, which behaves as a perfect gas with constant specific heats $(\gamma = 1.4)$, and that the flow is isentropic.

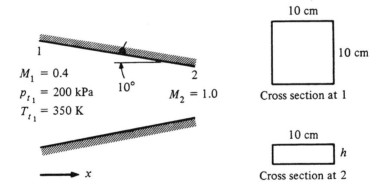

Figure P3.20

21. A spherical tank contains compressed air at 500 kPa; the volume of the tank is 20 m³. A 5-cm burst diaphragm in the side of the tank ruptures, causing air to escape from the tank. Find the time required for the tank pressure to drop to 200 kPa. Assume the temperature of the air in the tank is 280 K, and that the air flow through the opening can be treated as isentropic flow through a converging nozzle with a 5-cm exit diameter.

22. A converging–diverging nozzle has an area ratio of 3.3 to 1. The nozzle is supplied from a tank containing helium at 100 kPa and 270 K (see Figure P3.22). Find the maximum mass flow possible through the nozzle and the range of back pressures over which the mass flow can be attained. Repeat with hydrogen as the working fluid (use data from Table G.1 in Appendix G).

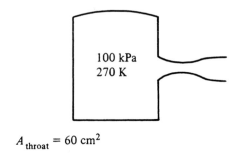

$$A_{\text{throat}} = 60 \text{ cm}^2$$

Figure P3.22

23. Superheated steam is stored in a large tank at 6 MPa and 800°C. The steam is exhausted isentropically through a converging–diverging nozzle. Determine the velocity of the steam flow when the steam starts to condense, assuming the steam to behave as a perfect gas with $\gamma = 1.3$.

4

NORMAL SHOCK WAVES

4.1 Introduction

The shock process represents an abrupt change in fluid properties, in which finite variations in pressure, temperature, and density occur over a shock thickness comparable to the mean free path of the gas molecules involved. It has been established that supersonic flow adjusts to the presence of a body by means of such shock waves, whereas subsonic flow can adjust by gradual changes in flow properties. Shocks may also occur in the flow of a compressible medium through nozzles or ducts and thus may have a decisive effect on these flows. An understanding of the shock process and its ramifications is essential to a study of compressible flow.

Chapter 4 is devoted to a consideration of the normal shock wave, a plane shock normal to the flow direction. This case represents the simplest example of a shock, in that changes in flow properties occur only in the direction of flow; thus, it can be treated with the equations of one-dimensional gas dynamics. Chapter 6 will cover the oblique shock wave, positioned at an angle to the flow direction.

It was pointed out in Section 2.4 that a series of weak compression waves can coalesce to form a finite compression shock wave. In this chapter, the mechanism by which this process occurs will be discussed in detail. The thermodynamics of the shock process will be reviewed, and the one-dimensional equations of continuity, momentum, and energy applied to the normal shock. Solutions of these equations will be presented to enable the working of practical engineering problems.

4.2 Formation of a Normal Shock Wave

It was shown in Section 2.2 that, when a piston in a tube is given a steady velocity to the right of magnitude dV (Figure 2.5), a sound wave travels ahead of the piston through the medium in the tube. Suppose the piston is now given a second increment of velocity dV, causing a second wave to move into the compressed gas behind the first wave. The location of the waves and the pressure distribution in the tube, after a time t_1, are shown in Figure 4.1. Each wave travels at the velocity of sound with respect to the gas into which it is moving. Since the second wave is moving into a gas that is already moving to the right with velocity dV, and since it is moving into a compressed gas having a slightly elevated temperature, the second wave travels with a greater absolute velocity than the first wave and gradually overtakes it. After a time t_2 (t_2 greater than t_1), the situation in the tube is as shown in Figure 4.2.

Now suppose the piston is accelerated from rest to a finite velocity increment of magnitude ΔV to the right. This finite velocity increment can be thought to consist of a large number of infinitesimal increments, each of magnitude dV. Figure 4.3 shows the velocity of the piston versus time, with the incremental velocities dV superimposed. As was demonstrated in Figure 4.2, the waves next to the piston tend to overtake those farther down the tube (see Figure 4.4).

As time passes, the compression wave steepens. The tendency of the higher density parts of the wave to overtake the lower density parts is finally counteracted by heat conduction and viscous effects taking place internal to the wave. The resultant constant-shape compression shock wave produced by the addition of the weak compression waves then moves through the undisturbed gas ahead of the piston. The slopes of temperature and pressure versus distance in the wave itself are very large (yet not truly infinite), and so the shock can be approximated by a discontinuity (Figure 4.5).

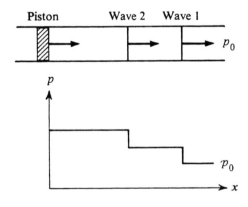

Figure 4.1

After time t_2:

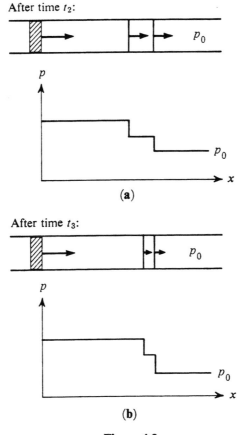

(a)

After time t_3:

(b)

Figure 4.2

If the piston in Figure 4.5 is suddenly given an incremental velocity dV to the left, a weak expansion wave propagates to the right at the velocity of sound. When the piston is given a second increment of velocity, a second expansion wave moves into the expanded gas behind the first wave, as shown in Figure 4.6. Again, each wave travels at the velocity of sound with respect to

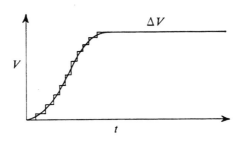

Figure 4.3

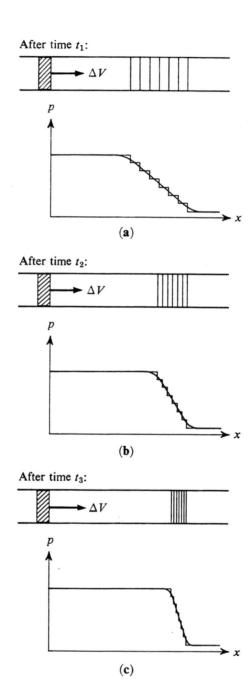

Figure 4.4

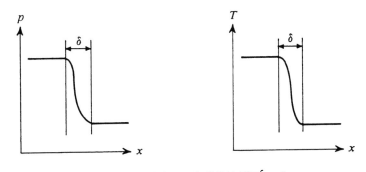

δ = shock thickness ($\approx 2.5 \times 10^{-5}$ cm)

Figure 4.5

the gas into which it is moving. In this case, the waves and gas are moving in opposite directions. Furthermore, the second wave is traveling into a gas that has already been expanded and cooled, which lowers the sound velocity. Both effects reduce the absolute wave velocity, and cause the second wave to fall farther and farther behind the first. In this manner, expansion waves spread out; they are not able to reinforce one another (see Figure 4.7). The creation of a finite expansion shock wave is impossible.

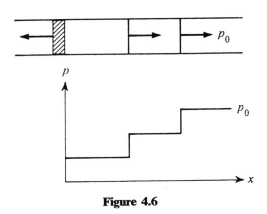

Figure 4.6

4.3 Equations of Motion for a Normal Shock Wave

A shock involves finite changes in pressure and temperature, and these changes occur very rapidly. The processes taking place inside the wave itself are extremely complex and cannot be studied on the basis of equilibrium thermodynamics. Temperature and velocity gradients internal to the shock provide heat conduction and viscous dissipation that render the shock process internally irreversible. In a practical sense, however, primary interest is not generally

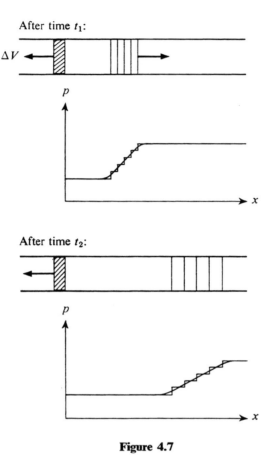

Figure 4.7

focused on the interior details of the shock wave, but on the net changes in fluid properties taking place across the entire wave. If one chooses a control volume encompassing the shock wave, the flow equations can be written without regard to the complexities of the internal processes. For this purpose, it is sufficient to note that the shock process is thermodynamically irreversible. Furthermore, with the shock temperature gradient inside the control volume, there is no external heat transfer across the control volume boundaries, so the shock process is adiabatic. Suppose a fixed plane shock occurs in one-dimensional, steady flow, as shown in Figure 4.8. The shock is assumed thin enough so that, even though it may occur in a varying area channel, as in Figure 4.9, there is no area change across the wave. If one refers to the control volume indicated in Figure 4.8, the continuity equation, Eq. (1.9), yields

$$\rho_1 V_1 = \rho_2 V_2 \tag{4.1}$$

With pressure forces the only external forces acting on the control volume, the

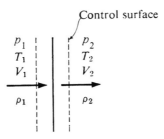

Figure 4.8

momentum equation, Eq. (1.14), becomes

$$p_1 A_1 - p_2 A_2 = \iint_s V_x (\rho \mathbf{V} \cdot d\mathbf{A})$$

$$= \rho_2 A_2 V_2^2 - \rho_1 A_1 V_1^2$$

or

$$p_1 + \rho_1 V_1^2 = p_2 + \rho_2 V_2^2 \tag{4.2}$$

For an adiabatic process, the energy equation, Eq. (1.19), for steady flow simplifies to

$$h_1 + \frac{V_1^2}{2} = h_2 + \frac{V_2^2}{2} \tag{4.3}$$

To obtain a solution for this system of equations, it is necessary to have a knowledge of the equation of state for the medium in which the shock occurs. Because of its reasonably wide range of application and its inherent simplicity, the perfect gas equation of state with constant specific heats will be assumed for the medium:

$$p = \rho R T$$

$$dh = c_p dT$$

and, from Eq. (2.5),

$$a^2 = \gamma R T$$

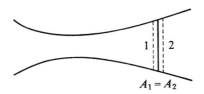

Figure 4.9

For a perfect gas with constant specific heats undergoing an adiabatic process, it was shown in Section 3.4 that the stagnation temperature remains constant. In other words, Eq. (4.3) simplifies to

$$T_1\left(1 + \frac{\gamma - 1}{2}M_1^2\right) = T_2\left(1 + \frac{\gamma - 1}{2}M_2^2\right) \tag{4.4}$$

For a perfect gas, Eq. (4.2) becomes

$$p_1 + \rho_1 V_1^2 = p_1(1 + \gamma M_1^2)$$

or

$$p_1(1 + \gamma M_1^2) = p_2(1 + \gamma M_2^2) \tag{4.5}$$

Combining Eqs. (4.5) and (4.4) with Eq. (4.1) yields

$$\rho_1 V_1 = \rho_2 V_2$$

$$\frac{p_1}{RT_1}M_1\sqrt{\gamma RT_1} = \frac{p_2}{RT_2}M_2\sqrt{\gamma RT_2}$$

$$\frac{M_1}{1 + \gamma M_1^2}\sqrt{1 + \frac{\gamma - 1}{2}M_1^2} = \frac{M_2}{1 + \gamma M_2^2}\sqrt{1 + \frac{\gamma - 1}{2}M_2^2} \tag{4.6}$$

By inspection, it is evident that one solution to Eq. (4.6) is the trivial one, $M_2 = M_1$. This solution, involving no change in properties in a constant area flow, corresponds to isentropic flow and is not of interest for the irreversible normal shock. Equation (4.6) can be solved to yield M_2 in terms of M_1. Squaring both sides gives

$$\frac{M_1^2\left(1 + \frac{\gamma - 1}{2}M_1^2\right)}{(1 + \gamma M_1^2)^2} = \frac{M_2^2\left(1 + \frac{\gamma - 1}{2}M_2^2\right)}{(1 + \gamma M_2^2)^2} \tag{4.7}$$

Expressed in terms of a quadratic in M_2^2, the result is

$$M_2^4\left(\frac{\gamma - 1}{2} - \gamma^2 L\right) + M_2^2(1 - 2\gamma L) - L = 0$$

where L is equal to the left-hand side of Eq. (4.7). Solving the quadratic equation for M_2^2, we obtain

$$M_2^2 = \frac{M_1^2 + \dfrac{2}{\gamma - 1}}{\dfrac{2\gamma}{\gamma - 1}M_1^2 - 1} \tag{4.8}$$

The result of Eq. (4.8) is plotted in Figure 4.10 for $\gamma = 1.4$.

For $M_1 > 1$, M_2 is less than 1, and vice versa. From Eq. (4.5), $M_1 > 1$ is the case of a compression shock; $M_1 < 1$ is the case of an expansion shock.

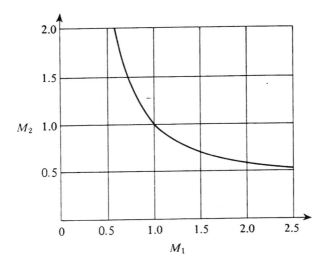

Figure 4.10

Equations (4.4) and (4.8) can be combined to yield T_2/T_1 as a function of M_1 and γ; in the same way, Eqs. (4.5) and (4.8) can be combined to yield p_2/p_1 as a function of M_1 and γ. The entropy change for a perfect gas with constant specific heats is given by Eq. (1.30):

$$s_2 - s_1 = c_p \ln \frac{T_2}{T_1} - R \ln \frac{p_2}{p_1} \qquad (4.9)$$

With p_2/p_1 and T_2/T_1 functions of M_1, Δs can be expressed in terms of M_1 alone. The result is plotted in Figure 4.11. The shock process has been determined to be adiabatic and irreversible.

From the second law of thermodynamics, for an irreversible process,

$$dS > \frac{\delta Q}{T} \qquad (1.21)$$

For the shock process, with $\delta Q = 0$,

$$ds > 0$$

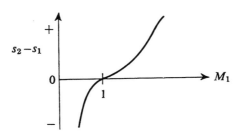

Figure 4.11

From Figure 4.11, $s_2 - s_1$ is greater than 0 for M_1 greater than 1. Thus the only solutions to the normal shock equations that do not violate the second law are those for which M_1 is greater than 1; the approach flow must be supersonic with respect to the wave. An expansion shock wave violates the second law of thermodynamics and is therefore impossible.

From the normal shock equations for a perfect gas with constant specific heats, as derived in Eqs. (4.8), (4.5), and (4.4), M_2, p_2/p_1, T_2/T_1, and ρ_2/ρ_1 can be expressed as a function of M_1 alone, for a given γ.

For example,

$$\frac{T_2}{T_1} = \frac{1 + \dfrac{\gamma - 1}{2} M_1^2}{1 + \dfrac{\gamma - 1}{2} M_2^2} \tag{4.4}$$

Combining with Eq. (4.8) yields

$$\frac{T_2}{T_1} = \frac{\left(1 + \dfrac{\gamma - 1}{2} M_1^2\right)\left(\dfrac{2\gamma}{\gamma - 1} M_1^2 - 1\right)}{M_1^2\left(\dfrac{2\gamma}{\gamma - 1} + \dfrac{\gamma - 1}{2}\right)} \tag{4.10}$$

From Eq. (4.5),

$$\frac{p_2}{p_1} = \frac{1 + \gamma M_1^2}{1 + \gamma M_2^2}$$

Combining with Eq. (4.8) yields

$$\frac{p_2}{p_1} = \frac{2\gamma M_1^2}{\gamma + 1} - \frac{\gamma - 1}{\gamma + 1} \tag{4.11}$$

Finally,

$$\frac{\rho_2}{\rho_1} = \frac{V_1}{V_2}$$

$$= \frac{M_1}{M_2}\sqrt{\frac{T_1}{T_2}}$$

Substituting from Eqs. (4.8) and (4.10), we obtain

$$\frac{\rho_2}{\rho_1} = \frac{M_1}{\sqrt{\dfrac{M_1^2 + \dfrac{2}{\gamma - 1}}{\dfrac{2\gamma}{\gamma - 1} M_1^2 - 1}}} \sqrt{\frac{M_1^2\left(\dfrac{2\gamma}{\gamma - 1} + \dfrac{\gamma - 1}{2}\right)}{\left(1 + \dfrac{\gamma - 1}{2} M_1^2\right)\left(\dfrac{2\gamma}{\gamma - 1} M_1^2 - 1\right)}}$$

$$\frac{\rho_2}{\rho_1} = \frac{V_1}{V_2} = \frac{(\gamma + 1)M_1^2}{(\gamma - 1)M_1^2 + 2} \tag{4.12}$$

Numerical values of M_2, p_2/p_1, T_2/T_1, and ρ_2/ρ_1 are presented in Appendix B for $\gamma = 1.4$, $\gamma = 1.3$, and $\gamma = \frac{5}{3}$.

Changes in stagnation properties across a fixed normal shock can also be related to M_1. For steady flow of a perfect gas with constant specific heats, there is no change in stagnation temperature for the adiabatic shock process. Physically, the increase of static temperature accompanying the compression process is compensated for by a decrease in kinetic energy of the gas, which yields no net change in total or stagnation temperature.

Stagnation pressure variations across a fixed normal shock can be related directly to the entropy rise across the shock. From Eq. (4.9),

$$\frac{s_2 - s_1}{R} = \frac{c_p}{R} \ln \frac{T_2}{T_1} - \ln \frac{p_2}{p_1}$$

Expressing $s_2 - s_1$ in terms of stagnation properties for an adiabatic process, we obtain

$$T_t = T\left(1 + \frac{\gamma - 1}{2} M^2\right)$$

$$p_t = p\left(1 + \frac{\gamma - 1}{2} M^2\right)^{\gamma/(\gamma-1)}$$

so that

$$\frac{s_2 - s_1}{R} = \frac{c_p}{R} \ln \frac{1 + \dfrac{\gamma - 1}{2} M_1^2}{1 + \dfrac{\gamma - 1}{2} M_2^2} - \ln \left\{ \frac{p_{t_2}}{p_{t_1}} \left[\frac{1 + \dfrac{\gamma - 1}{2} M_1^2}{1 + \dfrac{\gamma - 1}{2} M_2^2} \right]^{\gamma/(\gamma-1)} \right\}$$

$$\frac{s_2 - s_1}{R} = -\ln \frac{p_{t_2}}{p_{t_1}} \tag{4.13}$$

But, from the second law $s_2 > s_1$, so that $p_{t_2} < p_{t_1}$ for a fixed normal shock. In particular,

$$\frac{p_{t_2}}{p_{t_1}} = \frac{p_{t_2}}{p_2} \frac{p_2}{p_1} \frac{p_1}{p_{t_1}}$$

$$= \left(1 + \frac{\gamma - 1}{2} M_2^2\right)^{\gamma/(\gamma-1)} \left(\frac{2\gamma M_1^2}{\gamma + 1} - \frac{\gamma - 1}{\gamma + 1}\right) \left(\frac{1}{1 + \dfrac{\gamma - 1}{2} M_1^2}\right)^{\gamma/(\gamma-1)}$$

Substituting for M_2 from Eq. (4.8), we obtain

$$\frac{p_{t_2}}{p_{t_1}} = \left[\frac{\dfrac{\gamma + 1}{2} M_1^2}{1 + \dfrac{\gamma - 1}{2} M_1^2} \right]^{\gamma/(\gamma-1)} \left[\frac{1}{\dfrac{2\gamma}{\gamma + 1} M_1^2 - \dfrac{\gamma - 1}{\gamma + 1}} \right]^{1/(\gamma-1)} \tag{4.14}$$

Values of p_{t_2}/p_{t_1} versus M_1 have been tabulated in Appendix B. For an adiabatic process, stagnation pressure represents a measure of available energy of the flow in a given state. A decrease in stagnation pressure, or increase in entropy, denotes an energy dissipation or loss of available energy.

Example 4.1

An airstream with a velocity of 500 m/s, a static pressure of 50 kPa, and a static temperature of 250 K undergoes a normal shock. Determine the air velocity and the static and stagnation conditions after the wave.

Solution

The Mach number of the airstream, M_1, is given by

$$M_1 = \frac{V_1}{\sqrt{\gamma R T_1}}$$

$$= \frac{500 \text{ m/s}}{\sqrt{1.4(287 \text{ J/kg} \cdot \text{K})250 \text{ K}}}$$

$$= 1.578$$

From Appendix B, Table B.1,

$$\frac{T_2}{T_1} = 1.373, \qquad \frac{p_2}{p_1} = 2.739, \qquad \frac{\rho_2}{\rho_1} = 1.995, \qquad M_2 = 0.675$$

but

$$\frac{V_2}{V_1} = \frac{\rho_1}{\rho_2}$$

so the gas velocity after the wave is

$$V_2 = \frac{500}{1.995} = \underline{250.6 \text{ m/s}}$$

Also,

$$p_2 = 50(2.739) = \underline{137.0 \text{ kPa}}$$

$$T_2 = 250(1.373) = \underline{343.3 \text{ K}}$$

For this fixed normal shock, $T_{t_1} = T_{t_2}$. From Appendix A, Table A.1,

$$\frac{T_1}{T_{t_1}} = 0.6670$$

So

$$T_{t_1} = \frac{250}{0.6670} = \underline{374.8 \text{ K} = T_{t_2}}$$

Also, from Appendix A, Table A.1,

$$\frac{p_1}{p_{t_1}} = 0.2430$$

or

$$p_{t_1} = \frac{50}{0.2430} = 205.8 \text{ kPa}$$

From Appendix B, Table B.1,

$$\frac{p_{t_2}}{p_{t_1}} = 0.9033$$

$$p_{t_2} = 0.9033(205.8) = \underline{185.9 \text{ kPa}}$$

For isentropic flow, the area at which the Mach number is equal to 1 was defined as A^*, with this area being used as a reference. A normal shock, however, is not an isentropic process; so, for example, if a shock occurs in a channel (Figure 4.12a), flow areas downstream of the shock (2 to e) cannot be referenced to the value of A^* for the flow from i to 1. In other words, $A_{i1}^* \neq A_{2e}^*$. It is sometimes convenient to have a relationship between A_{i1}^* and A_{2e}^*. From Figure 4.12b, apply the continuity equation between A_{i1}^* and A_{2e}^*, assuming a perfect gas with constant specific heats.

From Eq. (3.9), with $\dot{m}_{A_{i1}^*} = \dot{m}_{A_{2e}^*}$

$$\frac{p_{t1}A_{i1}^*}{\sqrt{T_{t1}}} f(\gamma, M) = \frac{p_{t2}A_{2e}^*}{\sqrt{T_{t2}}} f(\gamma, M)$$

But $M = 1$ at A_{i1}^* and at A_{2e}^*; also, $T_{t1} = T_{t2}$. Thus

$$p_{t1}A_{i1}^* = p_{t2}A_{2e}^* \tag{4.15}$$

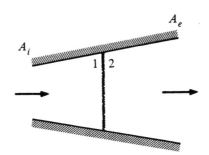

Figure 4.12a

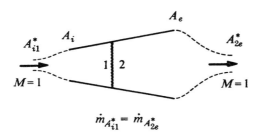

$$\dot{m}_{A_{i1}^*} = \dot{m}_{A_{2e}^*}$$

Figure 4.12b

Example 4.2

An airstream at Mach 2.0, with pressure of 100 kPa and temperature of 270 K, enters a diverging channel, with a ratio of exit area to inlet area of 3.0 (see Figure 4.13). Determine the back pressure necessary to produce a normal shock in the channel at an area equal to twice the inlet area. Assume one-dimensional steady flow, with the air behaving as a perfect gas with constant specific heats; assume isentropic flow except for the normal shock.

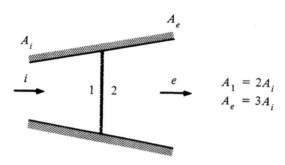

Figure 4.13

Solution

At $M = 2.0$, from Appendix A with $\gamma = 1.4$,

$$\frac{A_i}{A_{i1}^*} = 1.688$$

Therefore,

$$\frac{A_1}{A_{i1}^*} = \frac{A_1}{A_i} \frac{A_i}{A_{i1}^*} = 2(1.688) = 3.376$$

From Appendix A,

$$M_1 = 2.762$$

With the shock Mach number determined, ratios of properties across the shock can be found from Appendix B.

$$\frac{p_{t_2}}{p_{t_1}} = 0.4021 = \frac{A_{i1}^*}{A_{2e}^*}$$

or

$$\frac{A_e}{A_{2e}^*} = \frac{A_e}{A_i} \frac{A_i}{A_{i1}^*} \frac{A_{i1}^*}{A_{2e}^*}$$

$$= (3.0)(1.688)(0.4021)$$

$$= 2.043$$

Flow after the shock is subsonic, so that, from Appendix A, $M_e = 0.299$. We can now solve for the exit pressure, p_e:

$$\frac{p_e}{p_i} = \frac{p_e}{p_{t_2}} \frac{p_{t_2}}{p_{t_1}} \frac{p_{t_1}}{p_i}$$

$$= (0.9399)(0.4021)\frac{1}{0.1278}$$

$$= 2.957$$

or

$$p_e = 100(2.957) = \underline{295.7 \text{ kPa}} = p_{back}$$

With subsonic flow at the channel exit, the channel back pressure is equal to the exit plane pressure.

Example 4.3

A rocket exhaust nozzle has a ratio of exit to throat areas of 4.0. The exhaust gases are generated in a combustion chamber with stagnation pressure equal to 3 MPa and stagnation temperature equal to 1500 K (see Figure 4.14a). Assume the exhaust-gas mixture to behave as a perfect gas with $\gamma = 1.3$ and molecular mass $= 20$. Determine the rocket exhaust velocity for isentropic nozzle flow and for the case where a normal shock is located just inside the nozzle exit plane.

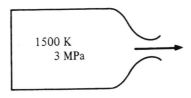

1500 K
3 MPa

Figure 4.14a

Solution

For isentropic flow in the exhaust nozzle, with $A_e/A^* = 4.0$, from Appendix A, Table A.2,

$$M_e = 2.77, \quad \frac{T_e}{T_t} = 0.4643$$

$$T_e = 696.5 \text{ K}$$

$$V_e = 2.77\sqrt{\gamma RT_e}$$

where

$$R = \frac{\bar{R}}{\bar{M}}$$

$$= \frac{8.3143 \text{ kJ/kg-mole} \cdot \text{K}}{20 \text{ kg/kg-mole}}$$

$$= 0.4157 \text{ kJ/kg} \cdot \text{K}$$

$$V_e = 2.77\sqrt{(1.3)(415.7)(696.5)}$$

$$= \underline{1699 \text{ m/s}}$$

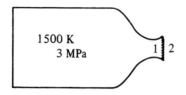

Figure 4.14b

Consider next the case of a normal shock at the nozzle exit plane, as shown in Figure 4.14b. With isentropic flow up to the shock wave, $M_1 = 2.77$. Also, $T_{t2} = T_{t1} = 1500$ K. From Appendix B, Table B.2, $M_2 = 0.4680$. From Appendix A, Table A.2,

$$\frac{T_2}{T_{t2}} = 0.9681 \quad \text{or} \quad T_2 = 0.9681(1500) = 1452 \text{ K}$$

and

$$V_e = 0.468\sqrt{1.3(415.7)1452}$$
$$= 414.6 \text{ m/s}$$

4.4 Moving Normal Shock Waves

Previous sections have dealt with the fixed normal shock wave. However, many physical situations arise in which a normal shock is moving. When an explosion occurs, a shock wave propagates through the atmosphere from the point of the explosion. As a blunt body reenters the atmosphere from space, a shock travels a short distance ahead of the body. When a valve in a gas line is suddenly closed, a shock propagates back through the gas. To treat these cases, it is necessary to extend the procedures already developed for the fixed normal shock wave.

Consider a normal shock moving at constant velocity into still air (Figure 4.15a). Let V_s = absolute shock velocity and V_g = velocity of gases behind the wave; both velocities are measured with respect to a fixed observer. For a fixed observer, the flow is not steady, since conditions at a point are dependent on whether or not the shock has passed over that point.

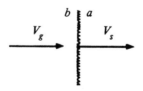

Figure 4.15a

Fixed

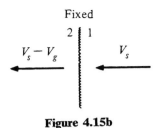

Figure 4.15b

Now consider the same physical situation with an observer moving at the shock-wave velocity, a situation, for instance, with the observer "sitting on the shock wave." The shock is now fixed with respect to the observer (Figure 4.15b). But this is the same case already covered in Sections 4.2 and 4.3. Relations have been derived and results tabulated for the fixed normal shock. To apply these results to the moving shock, consideration must be given to the effect of observer velocity on static and stagnation properties.

Static properties are defined as those measured with an instrument moving at the absolute flow velocity. Thus static properties are independent of the observer velocity, so

$$\frac{p_2}{p_1} = \frac{p_b}{p_a}$$

$$\frac{T_2}{T_1} = \frac{T_b}{T_a}$$

Stagnation properties are measured by bringing the flow to rest. Comparing the situations shown in Figure 4.15, if $T_1 = T_a$ and $p_1 = p_a$, it is evident that $T_{t_1} > T_{t_a}$ and $p_{t_1} > p_{t_a}$, since the gas at state 1 has velocity V_s, and the gas at state a has zero velocity. Thus stagnation properties are dependent on the observer velocity. To calculate the variation of stagnation properties across a moving shock wave, static conditions and velocities must first be determined (see Example 4.4, which follows).

In the following, we will consider two cases dealing with moving normal shock waves. In the first, the shock velocity will be known, and it will be required to find the gas velocity and other properties behind the moving wave. In the second, the gas velocity behind the wave will be known, and it will be required to determine the shock velocity.

Example 4.4

A normal shock moves at a constant velocity of 500 m/s into still air (100 kPa, 0°C). Determine the static and stagnation conditions present in the air after passage of the wave, as well as the gas velocity behind the wave.

Solution

For a fixed observer, the physical situation is shown in Figure 4.16a. With respect to an observer moving with the wave, the situation transforms to that shown in

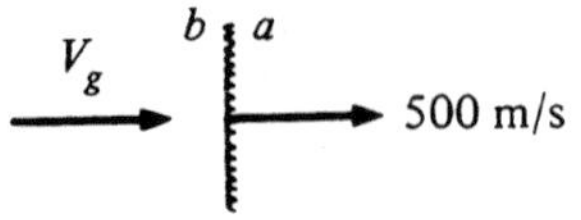

$$\overset{b \;\; a}{\underset{\text{(a)}}{\xrightarrow{\hspace{1cm}} \; \Big| \; \xrightarrow{\hspace{0.5cm}} \text{500 m/s}}}$$

(a)

With respect to an observer moving with the wave:

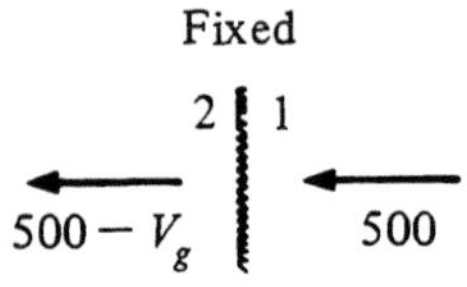

Fixed

$$\overset{2 \;\; 1}{\underset{\text{(b)}}{\underset{500 - V_g}{\xleftarrow{\hspace{1cm}}} \; \Big| \; \underset{500}{\xleftarrow{\hspace{1cm}}}}}$$

(b)

Figure 4.16

Figure 4.16b.

$$M_1 = \frac{500 \text{ m/s}}{\sqrt{\gamma R T_1}}$$

$$= \frac{500 \text{ m/s}}{\sqrt{1.4(287 \text{ J/kg} \cdot \text{K})273 \text{ K}}}$$

$$= 1.510$$

From Appendix B,

$$\frac{T_2}{T_1} = 1.327 \quad \text{or} \quad T_2 = 362.3 \text{ K}$$

$$\frac{p_2}{p_1} = 2.493 \quad \text{or} \quad p_2 = 249.3 \text{ kPa}$$

$$\frac{\rho_2}{\rho_1} = \frac{V_1}{V_2} = 1.879 = \frac{500}{500 - V_g}$$

or

$$V_g = \underline{233.9 \text{ m/s}}$$

Since the velocity of the observer does not affect the static properties,

$$p_b = \underline{249.3 \text{ kPa}}$$
$$T_b = \underline{362.3 \text{ K}}$$

The Mach number of the gas flow behind the wave is given by

$$M_b = \frac{V_g}{\sqrt{\gamma R T_b}}$$

$$= \frac{233.9}{\sqrt{1.4(287)362.3}}$$

$$= 0.613$$

With the Mach number and static properties determined, the stagnation properties of the gas stream can be found from Appendix A. At $M = 0.613$,

$$\frac{T}{T_t} = 0.9301, \qquad \frac{p}{p_t} = 0.7759$$

After passage of the wave, the stagnation pressure is

$$p_{t_b} = \frac{249.3}{0.7759} = 321.3 \text{ kPa}$$

$$T_{t_b} = \frac{362.3}{0.9301} = 389.5 \text{ K}$$

Note that for a fixed observer the stagnation temperature after passage of the
wave is greater than that before passage of the wave. For an observer "sitting on
the wave," however, there is no change of stagnation temperature across the
wave.

Now we shall consider the second case, in which the gas velocity behind
the wave is given, and the shock velocity is to be determined (Figure 4.17) for a
shock moving into a gas at rest. From Eq. (4.12), for a fixed shock,

$$\frac{V_1}{V_2} = \frac{(\gamma + 1)M_1^2}{(\gamma - 1)M_1^2 + 2}$$

where

$$V_1 = V_s, \qquad V_2 = V_s - V_g$$

Substituting yields

$$\frac{V_s}{V_s - V_g} = \frac{(\gamma + 1)\dfrac{V_s^2}{a_1^2}}{(\gamma - 1)\dfrac{V_s^2}{a_1^2} + 2}$$

Cross multiplying gives

$$\frac{(V_s - V_g)(\gamma + 1)V_s}{a_1^2} = \frac{(\gamma - 1)V_s^2}{a_1^2} + 2$$

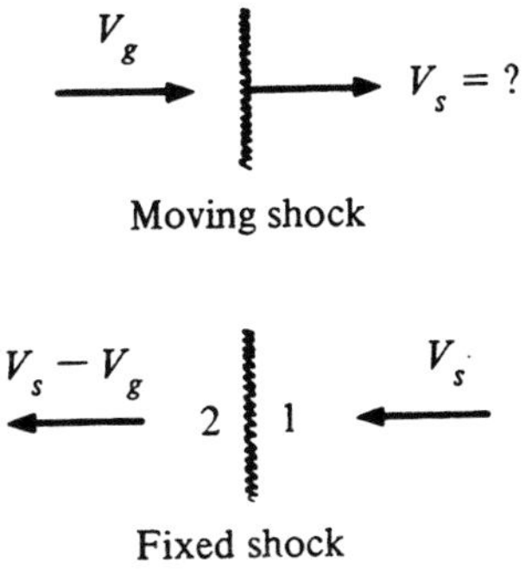

Moving shock

Fixed shock

Figure 4.17

Expanding terms, we obtain

$$\frac{(\gamma + 1)V_s^2}{a_1^2} - \frac{(\gamma + 1)V_gV_s}{a_1^2} = \frac{(\gamma - 1)V_s^2}{a_1^2} + 2$$

$$\frac{2V_s^2}{a_1^2} - \frac{(\gamma + 1)V_gV_s}{a_1^2} - 2 = 0$$

We then solve the quadratic equation for V_s:

$$V_s = \frac{\dfrac{(\gamma + 1)V_g}{a_1^2} \pm \sqrt{\dfrac{(\gamma + 1)^2V_g^2}{a_1^4} + \dfrac{16}{a_1^2}}}{4/a_1^2}$$

$$= \frac{(\gamma + 1)V_g}{4} \pm \sqrt{\frac{(\gamma + 1)^2V_g^2}{16} + a_1^2} \qquad (4.16)$$

Example 4.5

A piston in a tube is suddenly accelerated to a velocity of 50 m/s, which causes a normal shock to move into the air at rest in the tube. Several seconds later, the piston is suddenly accelerated from 50 to 100 m/s, which causes a second shock to move down the tube. Calculate the velocities of the two shock waves for an initial air temperature of 300 K.

Solution

The air next to the piston must move at the same velocity as the piston, since it can neither move through the face of the piston nor move away from the piston and leave a vacuum behind. Therefore, for a fixed observer, the air velocities are as shown in Figure 4.18.

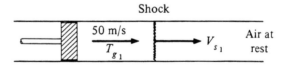

Figure 4.18

From Eq. (4.16),

$$V_{s_1} = \frac{(\gamma + 1)V_g}{4} \pm \sqrt{\frac{(\gamma + 1)^2V_g^2}{16} + a_1^2}$$

$$a_1 = \sqrt{\gamma RT_1} = \sqrt{1.4(287)300} = 347.2 \text{ m/s}$$

$$V_{s_1} = \frac{2.4(50)}{4} \pm \sqrt{\frac{(2.4)^2(50)^2}{16} + 347.2^2}$$

$$= 30 \pm 348.5$$

$$= 378.5 \text{ m/s}$$

$$M_{s_1} = \frac{378.5}{347.2} = 1.090 \quad \text{so} \quad T_{g_1} = 300(1.059) = 317.7 \text{ K}$$

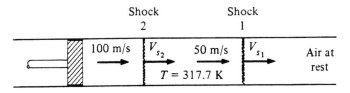

Figure 4.19a

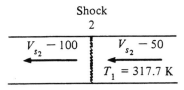

Figure 4.19b

For the second shock, the situation is depicted in Figure 4.19a. For an observer "sitting on the second wave," Figure 4.19b applies. Using Eq. (4.12), we obtain

$$\frac{V_1}{V_2} = \frac{(\gamma + 1)M_1^2}{(\gamma - 1)M_1^2 + 2}$$

where

$$V_1 = V_{s_2} - 50, \qquad V_2 = V_{s_2} - 100, \qquad M_1^2 = \frac{(V_{s_2} - 50)^2}{\gamma R T_1}$$

Substituting yields

$$\frac{V_{s_2} - 50}{V_{s_2} - 100} = \frac{2.4 \dfrac{(V_{s_2} - 50)^2}{1.4 \times 287 \times 317.7}}{0.4 \dfrac{(V_{s_2}-50)^2}{1.4 \times 287 \times 317.7} + 2}$$

Solving the quadratic equation, we obtain

$$\underline{V_{s_2} = 438.54 \text{ m/s}}$$

Thus, the second wave travels at a greater velocity than the first and eventually overtakes it. This result is a demonstration of the principles discussed in Section 4.2: Compression waves are able to overtake and reinforce one another. In this example problem, the second wave travels at a greater velocity because it is both moving into the compressed, higher-temperature gas behind the first wave and also moving into a gas stream already traveling in the same direction with a velocity of 50 m/s.

4.5 Reflected Normal Shock Waves

To complete this study of moving normal shock waves, consider the result of a wave impinging on the end of a tube. Two cases will be studied, a closed tube

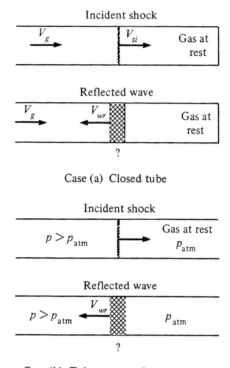

Case (a) Closed tube

Case (b) Tube open to the atmosphere

Figure 4.20

and a tube open to the atmosphere (see Figure 4.20). In both cases it is desired to determine whether the reflected wave is a compression shock wave or a series of weak expansion waves. For case (a), the gas next to the fixed end of the tube must be at rest, with the gas behind the incident shock moving to the right with velocity V_g. For an observer moving with the reflected wave, the physical situation is shown in Figure 4.21, which indicates a decrease in velocity across the reflected wave. Therefore, a normal shock reflects from a closed tube as a normal shock.

For case (b), the boundary condition imposed on the system is the static pressure at the end of the tube. From Figure 4.20, case (b), we see a decrease

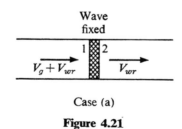

Case (a)

Figure 4.21

in pressure across the reflected wave. A normal shock reflects from an open end of a tube as a series of expansion waves.

Example 4.6

A normal shock wave with pressure ratio of 4.5 impinges on a plane wall (see Figure 4.22). Determine the static pressure ratio for the reflected normal shock wave. The air temperature in front of the incident wave is 20°C.

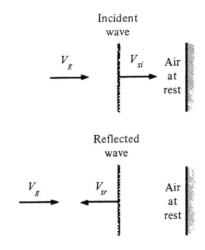

Incident
wave

Reflected
wave

Figure 4.22

Solution

To determine the velocity V_g of the gas behind the incident wave, utilize a

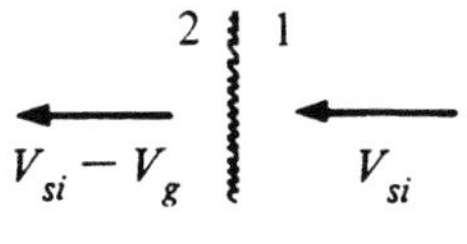

$$2 \mid 1$$

$$V_{si} - V_g \mid V_{si}$$

Figure 4.23

reference system moving with the wave, as shown in Figure 4.23. From Appendix B, for $p_2/p_1 = 4.5$, $M_1 = 2.0$, $\rho_2/\rho_1 = 2.667$, and $\dfrac{T_2}{T_1} = 1.688$:

$$V_{si} = 2.0\sqrt{1.4(287)293} = 686.2 \text{ m/s}$$

$$\frac{V_{si}}{V_{si} - V_g} = \frac{\rho_2}{\rho_1} = 2.667, \qquad V_g = 428.9 \text{ m/s}$$

$$T_2 = 494.6 \text{ K}$$

To find the reflected shock velocity, fix the reflected shock (see Figure 4.24)

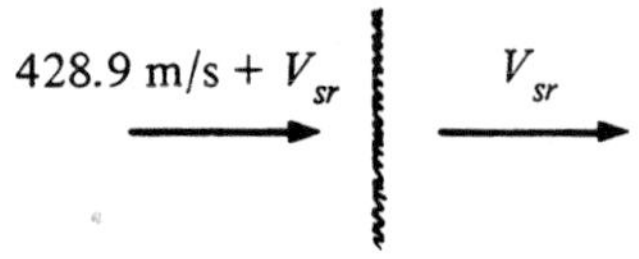

$$428.9 \text{ m/s} + V_{sr} \qquad V_{sr}$$

Figure 4.24

and use Eq. (4.12):

$$\frac{V_1}{V_2} = \frac{(\gamma + 1)M_1^2}{(\gamma - 1)M_1^2 + 2}$$

For this case,

$$V_1 = 428.9 + V_{sr}$$

$$V_2 = V_{sr}$$

$$M_1^2 = \frac{(428.9 + V_{sr})^2}{\gamma R T_1}$$

$$T_1 = 494.6 \text{ K}$$

Solving the quadratic equation for V_{sr} yields 343.2 m/s. For the fixed shock,

$$\frac{V_1}{V_2} = \frac{V_{sr} + 428.9}{V_{sr}} = 2.250 = \frac{p_2}{p_1}$$

From Appendix B, $p_2/p_1 = 3.333$ = static pressure ratio for reflected normal shock.

4.6 Summary

The normal shock represents a sudden, almost discontinuous change in fluid properties, which takes place in the direction of flow. Although the shock process is adiabatic, viscous dissipation and heat conduction effects occurring internal to the wave render the shock wave an irreversible process. Thus, from the second law of thermodynamics, there must be an accompanying rise in entropy across the wave. The equations of continuity, momentum, and energy applied to the fixed shock wave reveal the variations of pressure, temperature, velocity, Mach number, entropy, and so on, across the wave. For an increase of entropy, the Mach number of the flow ahead of the wave must be supersonic with respect to the wave, and the flow behind the wave subsonic, corresponding to compression shock. Possible solutions to the equations of motion corresponding to an expansion shock that involve a decrease of entropy are impossible since they would violate the second law of thermodynamics.

From a more physical standpoint, a series of weak compression waves moving down a tube tend to coalesce, reinforce one another, and form a finite normal shock wave. A series of weak expansion waves, however, tend to spread out, so they are not able to reinforce one another. The formation of a finite expansion shock wave is not possible.

Solutions to the equations of motion for a fixed normal shock wave are presented with M_1 as the independent variable. To handle a shock wave moving with constant velocity, it is necessary to change the frame of reference of the observer. For a fixed observer, the moving wave represents unsteady flow. However, for an observer moving with the same velocity as the wave, the flow is steady and can be handled with the solutions derived for the fixed wave. Thus, the variation of static properties across a moving wave can be determined directly from the equations derived for the fixed-wave case. Stagnation properties, however, are dependent on observer velocity and must be solved for independently.

REFERENCES

1. JOHN, J.E.A., and HABERMAN, W.L., *Introduction to Fluid Mechanics*, 2nd ed., Englewood Cliffs, N.J., Prentice-Hall, Inc., 1980.
2. FOX, R.W., and McDONALD, A.T., *Introduction to Fluid Mechanics*, 2nd ed., New York, John Wiley & Sons, Inc., 1978.
3. ZUCROW, M.J., and HOFFMAN, J.D., *Gas Dynamics*, Vol. 1, New York, John Wiley & Sons, Inc., 1976.
4. MIRONER, A., *Engineering Fluid Mechanics*, New York, McGraw-Hill Book Company, 1979.
5. STREETER, V.L., and WYLIE, E.B., *Fluid Mechanics*, 6th ed., New York, McGraw-Hill Book Company, 1975.

PROBLEMS

1. A helium flow with a velocity of 2500 m/s and static temperature of 200 K undergoes a normal shock. Determine the helium velocity and the static and stagnation temperatures after the wave. Assume the helium to behave as a perfect gas with constant $\gamma = \frac{5}{3}$.

2. A normal shock occurs at the inlet to a supersonic diffuser, as shown in Figure P4.2. A_b/A_a is equal to 3.0. Find M_b, p_b, and the loss in stagnation pressure $(p_t - p_{tb})$. Repeat for a shock at the exit.

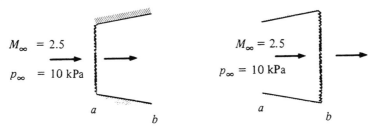

$M_\infty = 2.5$

$p_\infty = 10$ kPa

$M_\infty = 2.5$

$p_\infty = 10$ kPa

a b a b

Figure P4.2

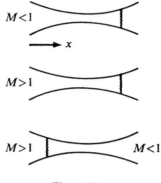

Figure P4.3

3. Sketch p versus x for Figure P4.3, assuming isentropic flow except for normal shocks.

4. Air expands from a storage tank through a converging–diverging nozzle (see Figure P4.4). Under certain conditions it is found that a normal shock exists in the

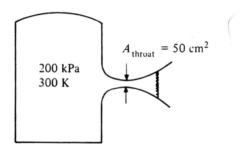

Figure P4.4

diverging section of the nozzle at an area equal to twice the throat area, with the exit area of the nozzle equal to four times the throat area. Assuming isentropic flow except for shock waves, that the air behaves as a perfect gas with constant $\gamma = 1.4$, and that the storage tank pressure and temperature are 200 kPa and 300 K, determine the following:

(a) A^* for flow from inlet to shock
(b) A^* for flow from shock to exit
(c) Mach number at nozzle exit plane
(d) Stagnation pressure at nozzle exit plane
(e) Exit plane static pressure
(f) Exit plane velocity

5. A supersonic flow at Mach 3.0 is to be slowed down via a normal shock in a diverging channel. For the conditions shown in Figure P4.5, find p_2/p_1 and p_e/p_i.

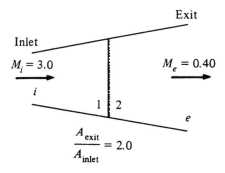

Figure P4.5

6. A projectile moves down a gun barrel with a velocity of 500 m/s (Figure P4.6). Calculate the velocity of the normal shock that would precede the projectile. Assume the pressure in the undisturbed air to be 101 kPa and the temperature to be 25°C. How fast would the projectile have to be moving in order that the shock velocity be twice the projectile velocity?

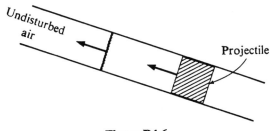

Figure P4.6

7. A body is reentering the earth's atmosphere at a Mach number of 20. Attached to the body is a shock wave, as shown in Figure P4.7. Opposite the nose of the body, the shock can be seen to be normal to the flow direction. Determine the stagnation pressure and temperature to which the nose is subjected. Assume that the air behaves as a perfect gas (neglect dissociation), with constant $\gamma = 1.4$. The ambient pressure and temperature are equal to 1.0 kPa and 220 K.

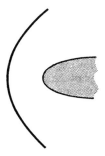

Figure P4.7

8. Determine the back pressure necessary for a normal shock to appear at the exit of a converging–diverging nozzle, as shown in Figure P4.8.

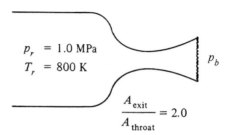

$$p_r = 1.0 \text{ MPa}$$
$$T_r = 800 \text{ K}$$

$$\frac{A_{\text{exit}}}{A_{\text{throat}}} = 2.0$$

p_b

Figure P4.8

9. A normal shock is found to occur in the diverging portion of a converging–diverging nozzle at an area equal to 1.1 times the throat area. If the nozzle has a ratio of exit area to throat area of 2.2, determine the percent of decrease in nozzle exit velocity due to the presence of the shock (compared with the exit velocity of a perfectly expanded isentropic supersonic nozzle flow). Assume the flow is expanded from negligible velocity, that the stagnation temperature of the flow is the same for both cases, and that the working fluid is steam, which behaves as a perfect gas with constant $\gamma = 1.3$.

10. A flow system consists of two converging–diverging nozzles in series (see Figure P4.10a. If the area ratio of each nozzle is 3.0 to 1, find the area ratio A_3/A_1 necessary to produce sonic flow at the second throat, with a shock at A_2. Assume isentropic flow except for the normal shock. Find the percent of loss in stagnation pressure for this flow. At another operating condition, a shock appears at A_3 (Figure P4.10b). Find the percent of loss of stagnation pressure for this condition.

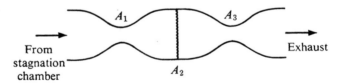

From
stagnation
chamber

A_1 A_3

A_2

Exhaust

Figure P4.10a

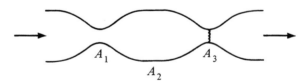

A_1 A_3

A_2

Figure P4.10b

11. For the system shown in Figure P4.11, $M_a = 2.0$, $A_a = 20 \text{ cm}^2$, throat area = 15 cm², shock area = 22 cm², and exit area = 25 cm². With the working fluid

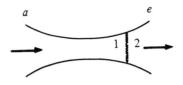

Figure P4.11

behaving as a perfect gas with constant $\gamma = 1.3$, find the following:
(a) Throat Mach number
(b) Exit Mach number
(c) Ratio of exit static pressure to static pressure at a

12. A jet plane uses a diverging passage as a diffuser (Figure P4.12). For a flight Mach number of 1.8, determine the range of back pressures over which a normal shock will appear in the diffuser. Ambient pressure and temperature are 25 kPa and 220 K. Find the mass flow range handled by the diffuser for the determined back pressure ranges: $A_i = 250 \text{ cm}^2$, $A_e = 500 \text{ cm}^2$. Assume isentropic flow except for shocks.

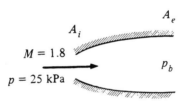

Figure P4.12

13. A normal shock moves into still air with a velocity of 1000 m/s. The still air is at 101 kPa and 20°C. Calculate the following:
(a) Velocity of the air flow behind the wave
(b) Static pressure behind the wave
(c) Stagnation temperature behind the wave

14. A normal shock is observed to move through a constant-area tube into air at rest at 25°C (Figure P4.14). The velocity of the air behind the wave is measured to be 150 m/s. Calculate the shock velocity.

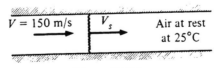

Figure P4.14

15. A piston in a tube is suddenly accelerated to a velocity of 25 m/s, which causes a normal shock to move into helium at rest in the tube. One second later, the piston is suddenly accelerated from 25 to 50 m/s, which causes a second shock to move down the tube. How much time will elapse from the initial acceleration of the piston to the intersection of the two shocks?

16. Air at 100 kPa and 290 K is flowing in a constant-area tube with a velocity of 100 m/s (Figure P4.16). Suddenly the end of the tube is closed, which causes a normal shock to propagate back through the airstream. Find the absolute velocity of this shock.

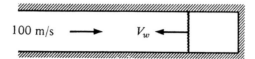

100 m/s $\longrightarrow$ $V_w \longleftarrow$

Figure P4.16

17. A normal shock traveling at 1000 m/s into still air at 0°C and 101 kPa reflects from a plane wall. Determine the velocity of the reflected shock. Compare the pressure ratio across the reflected shock with that across the incident shock. Find the stagnation pressure sensed by a stationary observer behind the reflected wave.

18. Under a certain operating condition, the piston speed in an auto engine is 10 m/s. We will approximate engine knock as the occurrence of a normal shock wave traveling at 1000 m/s downward, as shown in Figure P4.18, into the unburned mixture at 700 kPa and 500 K. Determine the pressure acting on the piston face after the shock reflects from it. Assume the gas has the properties of air and acts as a perfect gas, with γ = constant = 1.4.

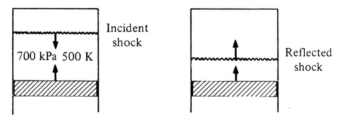

Incident shock

700 kPa 500 K

Reflected shock

Figure P4.18

19. A normal shock moves down a tube with a velocity of 600 m/s into a gas with static p = 50 kPa and static temperature of 300 K. At the end of the tube, a piston is moving with a velocity of 60 m/s, as shown in Figure P4.19. Calculate the velocity

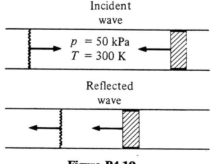

Incident wave

p = 50 kPa
T = 300 K

Reflected wave

Figure P4.19

of the reflected wave and the static pressure behind the reflected wave. Assume all gases have the properties of air.

20. Air enters a converging–diverging diffuser with a Mach number of 2.8, static pressure p_i of 100 kPa, and a static temperature of 20°C. For the flow situation shown in Figure P4.20, find the exit velocity, exit static pressure, and exit stagnation pressure.

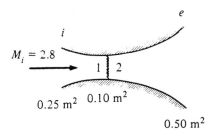

Figure P4.20

21. Write a computer program that will yield values of p_2/p_1, ρ_2/ρ_1, T_2/T_1, and p_{t_2}/p_{t_1} for a fixed normal shock with a working fluid consisting of a perfect gas with constant $\gamma = 1.20$. Use Mach number increments of 0.05 over the range $M = 0$ to $M = 5.0$.

22. Solve Example 4.6 by transforming the reflected wave of Figure 4.22 into a wave moving into a gas at rest and applying Eq. (4.16).

5

APPLICATIONS

5.1 Introduction

Chapters 3 and 4 have covered the subjects of isentropic flow and normal shock waves. Many compressible flow systems involve a combination of these two flows, with the resultant determined by the boundary conditions imposed on the system. It is instructive at this time to consider several applications of the material already studied to gain an appreciation of some of the interactions that may occur in engineering systems. The particular devices that will be discussed include the converging-diverging nozzle, the supersonic wind tunnel, supersonic inlet, and the shock tube. In each case, one-dimensional flow of a perfect gas with constant specific heats will be assumed. It must be noted, however, that in a diffuser, for example, where an adverse pressure gradient exists, rapid boundary layer growth in extreme cases dominate the resultant flow. Two- and three-dimensional effects such as shock-boundary layer interactions are extremely complex; fortunately, they can often, to a first approximation, be neglected.

5.2 Performance of Converging–Diverging Nozzles

Isentropic flow in a converging–diverging nozzle was examined in Section 3.6. It was pointed out that for certain ratios of back pressure to supply pressure, isentropic, one-dimensional solutions to the equations of motion are not possible. However, the normal shock wave was studied in Chapter 4; nozzle flows with normal shocks can now be discussed in detail. Returning to the system illustrated in Figure 3.14, we consider the case in which a fluid stored in a large reservoir is to be discharged through a converging–diverging nozzle. It is desired to find the pressure distribution in the nozzle over a range of values of p_b/p_r, with p_r maintained constant (see Figure 5.1).

With $p_b = p_r$, there is no flow in the nozzle. As p_b is reduced below p_r, subsonic flow is induced through the nozzle, with pressure decreasing to the throat, and then increasing in the diverging portion of the nozzle. When the back pressure is lowered to that of curve 4, sonic flow occurs at the nozzle throat. Further reductions in back pressure can induce no more flow through the nozzle. As the back pressure is reduced below that of curve 4, a normal shock appears in the nozzle just downstream of the throat (curve b) until, for a low enough back pressure, the normal shock positions itself at the nozzle exit plane (curve c). Consider in detail a curve of p versus x with a shock in the nozzle (Figure 5.2). The static pressure decreases in the converging portion of the nozzle, with $M = 1$ at the throat. In the diverging portion, with the flow supersonic, the pressure continues to decrease up to the normal shock. After the shock, flow in the diverging part of the nozzle is subsonic, and the static pressure increases to the exit plane pressure. With subsonic flow at the exit, the exit plane pressure is equal to the back pressure.

As the back pressure is lowered below that of curve c, a shock wave inclined at an angle to the flow appears at the exit plane of the nozzle (Figure

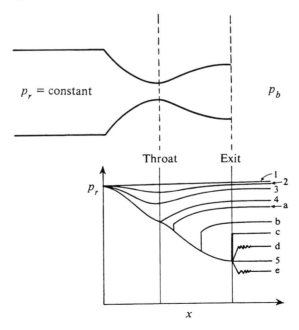

Figure 5.1

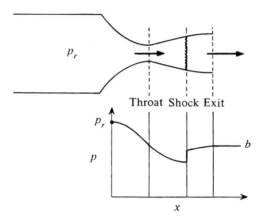

Figure 5.2

5.3). This shock wave, weaker than a normal shock, is called an *oblique shock*. Further reductions in back pressure cause the angle between the shock and the flow to decrease, thus decreasing the shock strength (Figure 5.4), until eventually the isentropic case, curve 5, is reached.

Curve 5 corresponds to the design condition in which the flow is perfectly expanded in the nozzle to the back pressure. For back pressures below that of curve 5, exit plane pressure is greater than the back pressure. A pressure decrease occurs outside the nozzle in the form of expansion waves (Figure 5.5). Oblique shock waves and expansion waves represent flows that are not one dimensional and cannot be treated directly with the methods of Chapters 3 and 4. These flows will be analyzed in detail in Chapters 6 and 7.

It is important to realize that, for all back pressures below that of curve c, the flow adjusts to the back pressure outside the nozzle. Over this range of

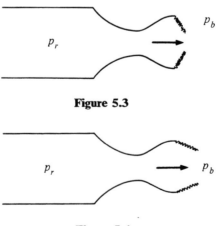

Figure 5.3

Figure 5.4

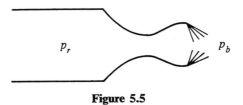

Figure 5.5

back pressures (below curve c), flow inside the nozzle remains unchanged as the back pressure is varied. For example, the exit plane pressure and exit velocity are the same for all back pressures below curve c. If a rocket nozzle is designed to operate isentropically at sea level, the rocket exhaust velocity and exit plane pressure do not change as the rocket moves upward through the atmosphere (assuming constant chamber temperature and pressure).

Figure 5.6 depicts the variation of exit plane pressure with back pressure. For subsonic flow at the exit plane (curves 1 to 4 and a to c) and for the design condition (curve 5), the exit plane pressure is equal to the back pressure. For supersonic flow at the exit plane (curves d, 5, and e), the exit plane pressure is equal to that for the design condition. For back pressures between curves c and 5, the exit plane pressure is less than the back pressure, so the nozzle is termed overexpanded. For back pressures below curve 5, with the exit plane pressure greater than the back pressure, the nozzle is termed underexpanded.

Nozzle design and operation have been studied up to this point by means of a one-dimensional flow analysis. Although this method of analysis is adequate for the solution of many engineering problems, certain limitations become apparent. For example, in the design of a supersonic nozzle, area ratios can be determined for a given supersonic Mach number. But the length of the nozzle or the rate of change of area with axial distance cannot be prescribed from one-dimensional flow considerations. Furthermore, due to the presence of boundary layers on the nozzle walls, the area available to the main flow is

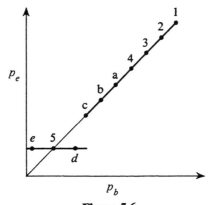

Figure 5.6

somewhat reduced; the areas calculated from a one-dimensional flow analysis may have to be enlarged to account for boundary layers. For an exact and complete analysis of the operation and design of a converging–diverging nozzle, a study of two- and three-dimensional compressible flow is required. However, this should not detract from the one-dimensional analysis under study; good engineering approximations can be obtained for the solution of a wide range of compressible flow problems.

Example 5.1

A converging–diverging nozzle is designed to operate with an exit Mach number of 1.75. The nozzle is supplied from an air reservoir at 5 MPa. Assuming one-dimensional flow, calculate the following:
- (a) Maximum back pressure to choke the nozzle
- (b) Range of back pressures over which a normal shock will appear in the nozzle
- (c) Back pressure for the nozzle to be perfectly expanded to the design Mach number
- (d) Range of back pressures for supersonic flow at the nozzle exit plane.

Solution

The nozzle is designed for $M = 1.75$. From Appendix A, at $M = 1.75$, $A/A^* = 1.386$.

- (a) The nozzle is choked with $M = 1$ at the throat, followed by subsonic flow in the diverging portion of the nozzle. From Appendix A, at $A/A^* = 1.386$, $M = 0.477$ and $p/p_t = 0.8558$. Therefore, the nozzle is choked for all back pressures below 4.279 MPa.
- (b) For a normal shock at the nozzle exit plane (Figure 5.7), $M_1 = 1.75$ and $p_1 = 0.1878(5) = 0.939$ MPa. From Appendix B, at $M_1 = 1.75$, $p_2/p_1 = 3.406$, For a normal shock at the nozzle exit, the back pressure is $3.406(0.939) = 3.198$ MPa. Referring to Figure 5.1, for a shock just downstream of the nozzle throat, the back pressure is 4.279 MPa. A normal shock will appear in the nozzle over the range of back pressures from 3.198 to 4.279 MPa.
- (c) From Appendix A, at $M = 1.75$, $p/p_t = 0.1878$. For a perfectly expanded, supersonic nozzle, the back pressure is 0.939 MPa.
- (d) Referring again to Figure 5.1, supersonic flow will exist at the nozzle exit plane for all back pressures less than 3.198 MPa.

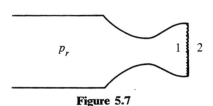

p_r

1 2

Figure 5.7

Example 5.2

A rocket nozzle is designed to operate supersonically with a chamber pressure of 3 MPa and an ambient pressure of 101 kPa. Find the ratio between the thrust at sea level to the thrust in space (0 kPa). Assume a constant chamber pressure, with a chamber temperature of 1600 K. Assume the rocket exhaust gases to behave as a perfect gas with $\gamma = 1.3$ and $R = 0.40 \text{ kJ/kg} \cdot \text{K}$.

Solution

Select a control volume as shown in Figure 5.8 and apply the momentum equation. From the results of Example 1.4,

$$\mathcal{T} = (p_e - p_a)A_e + \rho_e A_e V_e^2$$

The exit plane pressure and exit velocity are the same in space as at sea level.

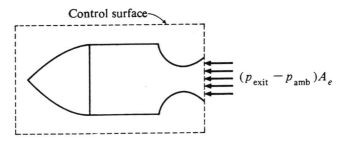

Control surface

$(p_{exit} - p_{amb})A_e$

Figure 5.8

From Appendix A, at $p/p_t = 101/3000 = 0.03367$, $M = 2.81$ and $T/T_t = 0.4578$. The rocket exhaust velocity is $2.81\sqrt{1.3(400)(0.4578 \times 1600)} = 1734$ m/s.

$$\rho_e = \frac{p_e}{RT_e} = \frac{101 \text{ kN/m}^2}{(0.40 \text{ kN} \cdot \text{m/kg} \cdot \text{K})(0.4578 \times 1600 \text{ K})} = 0.3447 \text{ kg/m}^3$$

$$\frac{\mathcal{T}_{sl}}{\mathcal{T}_{space}} = \frac{\rho_e A_e V_e^2}{(p_e - 0)A_e + \rho_e A_e V_e^2} = \frac{(0.3447)1734^2 A_e}{101 \times 10^3 A_e + (0.3447)1734^2 A_e} = \underline{0.911}$$

5.3 Supersonic Wind Tunnels

In a wind tunnel, it is desired to subject a test object to the flow conditions that it will undergo in actual flight. The creation of a uniform supersonic flow in a test section of appreciable size and the maintenance of this flow for a sufficient time to make meaningful measurements on a test object may involve a large consumption of power. This section will be concerned with the use of a supersonic diffuser to recover the fluid pressure after the test section and both improve the operation and reduce the power requirements of the wind tunnel.

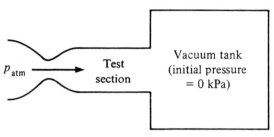

Figure 5.9

A simple, intermittent supersonic wind tunnel can be constructed by allowing air at atmospheric pressure to pass through a converging–diverging nozzle into a vacuum tank (Figure 5.9). With this setup, constant conditions are maintained in the test section until the back pressure in the vacuum tank rises to a value such that a normal shock appears in the test section (corresponding to curve c of Figure 5.1). To provide a longer running time at the design Mach number, a diffuser can be placed at the test section exit. Isentropic flow in the diffuser provides a recovery of static pressure. Ideally, the wind tunnel can now be run until the pressure in the vacuum tank rises to the stagnation pressure after a shock at the design Mach number. This limiting condition is shown in Figure 5.10.

Intermittent wind tunnels present the problem of providing only a relatively short test time before the back pressure rises to a limiting value at which flow in the test section ceases to be supersonic. In a continuous closed-cycle tunnel, test conditions can be maintained almost indefinitely. A compressor is used to boost the air pressure from a somewhat reduced value after the test section to the inlet stagnation pressure p_{t_1} (Figure 5.11). Naturally, the greater the pressure recovery after the test section, the smaller are the compressor power requirements. The use of a normal shock pressure recovery, as shown in Figure 5.10, has the disadvantage of always entailing a loss in stagnation pressure. Even if a perfect diffuser were provided after the shock, causing a rise

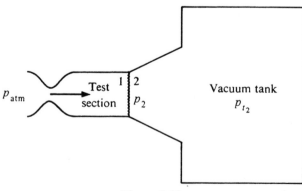

Figure 5.10

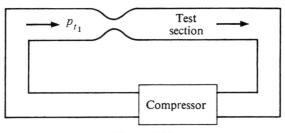

Figure 5.11

in pressure to p_{t_2}, the compressor must still provide the pressure difference $p_{t_1} - p_{t_2}$, which increases as the test section Mach number is increased.

To provide isentropic deceleration and complete pressure recovery after the test section, consider the system shown in Figure 5.12, with a second throat after the test section. In the configuration shown, neglecting friction and boundary layer effects, the wind tunnel could be run at design conditions

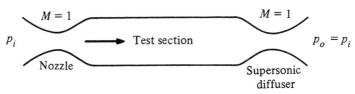

Figure 5.12

indefinitely, with no pressure difference required to maintain the flow. However, difficulties arise during start-up of the system. To initiate flow, a pressure difference must be maintained across the entire system. As the pressure ratio p_o/p_i is decreased from 1.0, the situation for the converging–diverging nozzle is the same as that discussed in Section 5.2. Eventually, the flow condition shown in Figure 5.13 is attained. Now a decrease in overall pressure ratio p_o/p_i causes a shock to appear downstream of the nozzle throat (Figure 5.14a), with a further decrease moving the shock downstream to the nozzle exit (test section) (Figure 5.14b). With a shock in the diverging portion of the nozzle, there is a loss in stagnation pressure in the system. To pass the flow after the shock, the second throat must be at least A_2^*. The worst possible case, involving the maximum loss in stagnation pressure, is that of a normal shock in the test section (Figure 5.14b). For this case, the second throat area must be at least A_2^* with a shock at the test section Mach number. If the second throat area is

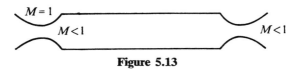

Figure 5.13

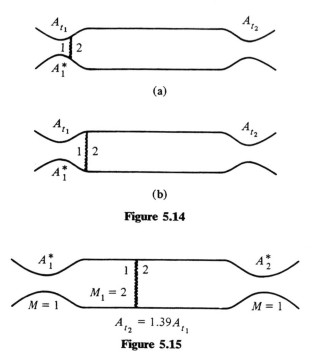

(a)

(b)

Figure 5.14

A_1^* $M_1 = 2$ $M = 1$ A_2^* $M = 1$ $A_{t_2} = 1.39 A_{t_1}$

Figure 5.15

less than this, it cannot pass the required flow, and the shock can never reach the test section and will remain in the diverging part of the nozzle. Under these conditions, supersonic flow can never be established in the test section. For example, if the test-section Mach number is to be 2.0, the ratio A_1^*/A_2^* (from Appendix B) is 0.72. During start-up, then, the second throat must be larger than the first by a factor of $1/0.72$ or 1.39. This condition is illustrated in Figure 5.15. As the ratio p_o/p_i is further lowered, the shock jumps to an area in the diverging part of the diffuser greater than the test section area; the shock is "swallowed" by the diffuser (Figure 5.16).

To maximize the pressure recovery in the diffuser, the pressure ratio p_o/p_i can now be increased, which moves the shock upstream to the diffuser throat, the position at which the shock strength is minimum (Figure 5.17).

Figure 5.17 represents the optimum operating condition for a fixed geometry diffuser. There is still a loss of stagnation pressure across the shock at the diffuser throat, which must be made up for by a compressor.

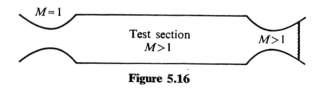

$M = 1$ Test section $M > 1$ $M > 1$

Figure 5.16

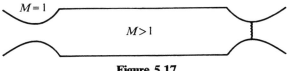

Figure 5.17

An improvement can be obtained with a variable area diffuser throat. After the initial shock has been "swallowed," we reduce the diffuser throat area to that of the nozzle throat, and again raise the overall pressure ratio p_o/p_i so as to bring the shock to the diffuser throat. But at the diffuser throat, the Mach number is 1, so the shock is of vanishing strength. This situation approaches the ideal of Figure 5.12.

To complete this discussion of supersonic wind tunnels, several of the components required for a continuous, cyclic system should be reviewed (see Figure 5.18). Besides the compressor, nozzle, and diffuser, a drier must be

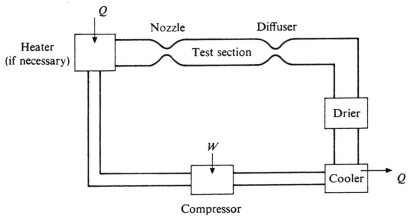

Figure 5.18

provided to remove water vapor from the air so as to prevent its condensation at the low static temperatures encountered in the test section. The condensation of water can lead to shock waves, negate any measurements made in the test section, and possibly damage the test object. At higher Mach numbers, a heater may be required to raise the stagnation temperature of the inlet air and prevent condensation of the air gases (oxygen and nitrogen) in the test section. If this is impractical, a gas with low boiling point (such as helium) may have to be used in place of air. Finally, for a cycle, $\oint \delta Q = \oint \delta W$, so that an amount of energy equal to that put in the air by the compressor and the heater must be rejected in a cooler.

Example 5.3

A continuous supersonic wind tunnel is designed to operate at a test section Mach number of 2.0, with static conditions duplicating those at 20 km (Figure 5.19a).

Figure 5.19a

The test section is to be circular, 25 cm in diameter, with a fixed geometry, and with a supersonic diffuser downstream of the test section. Neglecting friction and boundary layer effects, determine the power requirements of the compressor during start-up and during steady-state operation. Assume an isentropic compressor, with a cooler located between compressor and nozzle (after the compressor), so the compressor inlet temperature can be assumed equal to the test section stagnation temperature.

Solution

During steady-state operation, mass flow through the test section is given by

$$\dot{m} = \rho A V$$

$$= \frac{p}{RT} A M \sqrt{\gamma R T}$$

At 20 km, from Appendix H, $p = 5.5$ kPa and $T = 216.7$ K. Therefore,

$$\dot{m} = \left[\frac{5.5 \text{ kN/m}^2}{(0.287 \text{ kN/kg} \cdot \text{K})(216.7 \text{ K})}\right]\left[\frac{\pi}{4} 0.25^2 \text{ m}^2\right]\left[2.0\sqrt{1.4(287 \text{ N/kg} \cdot \text{K})(216.7 \text{ K})}\right]$$

$$= (0.0884 \text{ kg/m}^3)(0.04909 \text{ m}^2)(590.2 \text{ m/s})$$

$$= 2.56 \text{ kg/s}$$

Test section stagnation temperature during steady-state operation is $216.7/0.5556 = 390.0$ K. For this fixed geometry diffuser, the optimum condition for steady-state operation is a normal shock at the diffuser throat. For this example, the diffuser throat area is $(0.04909 \text{ m}^2)/(0.7209)(1.688) = 0.04034 \text{ m}^2$. From Appendix A, at $A/A^* = 1/0.7209 = 1.387$, $M_1 = 1.751$. From Appendix B, $p_{t_2}/p_{t_1} = 0.8342$. The loss in stagnation pressure must be compensated for by the compressor (Figure 5.19b).

$$W = h_o - h_i = c_p(T_o - T_i)$$

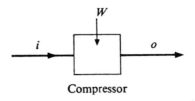

Compressor

Figure 5.19b

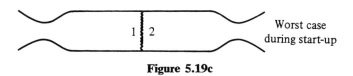

Figure 5.19c

For an isentropic compressor,

$$\frac{T_o}{T_i} = \left(\frac{p_o}{p_i}\right)^{(\gamma-1)/\gamma}$$

$$T_o - T_i = T_i\left[\left(\frac{p_o}{p_i}\right)^{(\gamma-1)/\gamma} - 1\right]$$

$$= 390.0\left[\left(\frac{1}{0.8342}\right)^{0.2857} - 1\right]$$

$$= 20.73 \text{ K}$$

$$W = (1.004 \text{ kJ/kg} \cdot \text{K})(20.73 \text{ K}) = \underline{20.8 \text{ kJ/kg}}$$

$$\text{Power required} = (20.8 \text{ kJ/kg})(2.56 \text{ kg/s}) = \underline{53.2 \text{ kW}}$$

During start-up, the worst possible case (Fig. 5.19c) is that of a shock in the test section, with $M_1 = 2.0$. For this case, $p_{t_2}/p_{t_1} = 0.7209$. Now the isentropic compressor work required is $1.004[(1/0.7209)^{0.2857} - 1]390.0 = 38.4 \text{ kJ/kg}$, or the power required is $38.4(2.56) = \underline{98.2 \text{ kW}}$.

5.4 Converging–Diverging Supersonic Diffuser

In an air-breathing jet propulsion engine, forward thrust is provided by the acceleration of an airstream. In both the turbojet and ramjet, the air is first compressed, energy is added from the combustion of a fuel, and the resultant products are ejected from the exhaust nozzle at high velocity (see Figure 5.20). With the jet engine, the inlet or diffuser must take the incoming air, traveling at high velocity with respect to the engine, and slow it down before entrance to the axial compressor of the turbojet or the combustion zone of the ramjet engine. The amount of static pressure rise achieved during deceleration of the flow in the diffuser is of prime importance to the operation of the jet engine, since the pressure of the air entering the nozzle determines to a large extent the magnitude of the nozzle exhaust velocity. The maximum pressure that can be achieved in the diffuser is the isentropic stagnation pressure. Any loss in available energy (or stagnation pressure) in the diffuser, or for that matter in any other component of the engine, will have a harmful effect on the operation of the engine as a whole. For a supersonic diffuser, it would be highly desirable to provide shock-free isentropic flow.

A first approach is to operate a converging–diverging nozzle in reverse (see Figure 5.21.) At the design Mach number, M_D, for such a diffuser, there is

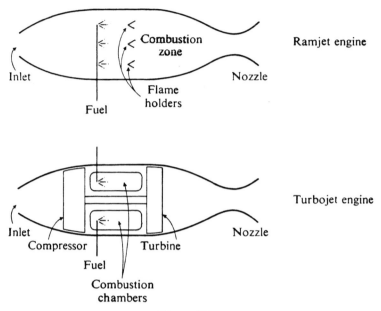

Figure 5.20

no loss in stagnation pressure (neglecting friction). However, off-design performance has to be considered, since the external flow must be accelerated to the
design condition. For example, if a supersonic converging–diverging diffuser is
to be designed for a flight Mach number of 2.0, the ratio A_{inlet}/A_{throat} is 1.688
(see Appendix A). However, for a supersonic flight Mach number less than
design, A/A^* is less than 1.688, which indicates that the throat area is not
large enough to handle this flow. Under these conditions, flow must be
bypassed around the diffuser. A normal shock stands in front of the diffuser,
with subsonic flow after the shock able to sense the presence of the inlet. Now
an appropriate amount of the flow "spills over" or bypasses the inlet (see
Figure 5.22).

As the flight Mach number is increased, the normal shock moves toward
the inlet lip. When the design Mach number is reached during start-up,
however, with a normal shock in front of the diffuser, some of the flow must
still be bypassed, since the throat area of less than A_2^* is still not able to handle

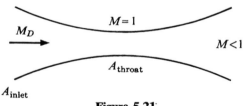

Figure 5.21

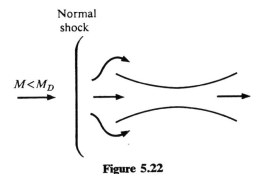

Figure 5.22

the entire subsonic flow after the shock. As the flight Mach number is increased above M_D, the shock moves eventually to the inlet lip. A further increase in M causes the shock to reach a new equilibrium position in the diverging portion of the diffuser; in other words, the shock is "swallowed." Once the shock has been swallowed, a decrease in flight Mach number causes the shock to move back toward the throat, where it reaches an equilibrium position for M equal to M_D. At this position, the shock is of vanishing strength, occurring at $M = 1.0$, so no loss in stagnation pressure occurs at the design condition. In actual operation, it is desirable to operate with the shock slightly past the throat; operation at the design condition is unstable in that a slight decrease in Mach number results in the shock's moving back out in front of the inlet. In this case, the operation of overspeeding to swallow the shock would have to be repeated (see Figure 5.23).

Another method for swallowing the shock is to use a variable throat area. With a shock in front of the diffuser, the throat area would be increased, which

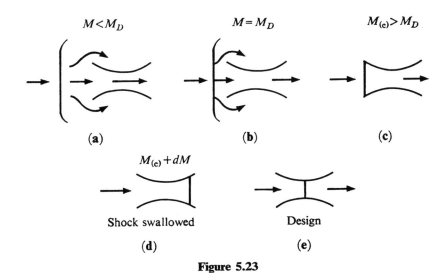

Figure 5.23

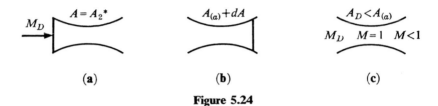

Figure 5.24

would allow more flow to pass through the inlet and consequently bring the shock closer to the inlet lip. To swallow the shock, the throat area would have to be slightly larger than that required to accept the flow with a shock at M_D at the inlet lip, that is, slightly larger than A_2^* with a normal shock at the design Mach number. For M_D equal to 2.0, A_1^*/A_2^* is equal to 0.7209, so an increase of greater than $(1 - 0.7209)/0.7209 = 39$ percent is required to swallow the shock. Once the shock is swallowed, the throat area must be decreased to reach the design condition, as is shown in Figure 5.24.

It can be readily seen that, although the converging–diverging diffuser has favorable operating characteristics at the design condition, off-design operation of such a device involves severe losses. Operation with a normal shock in front of an inlet entails prohibitive losses in stagnation pressure. To swallow this shock, the inlet must be accelerated beyond its design speed, or a variable throat area must be provided. Except for very low supersonic Mach numbers, the amount of overspeeding required to swallow the shock during start-up becomes large enough to be totally impractical. Furthermore, the incorporation of a variable throat area into a diffuser presents many mechanical difficulties. For these reasons, the converging–diverging diffuser is not commonly used; most engines utilize the oblique-shock type to be described in Chapter 6.

Example 5.4

A supersonic converging–diverging diffuser is designed to operate at a Mach number of 1.7. To what Mach number would the inlet have to be accelerated in order to swallow the shock during start-up?

Solution

From Appendix A, at $M = 1.7$, $A/A^* = 1.338$, so the diffuser is designed with $A_{inlet}/A_{throat} = 1.338$. The inlet must be accelerated to a Mach number slightly greater than that required to position the shock at the inlet lip (see Figure 5.25.) For $M = 1.0$ at the diffuser throat, with subsonic flow after a shock at the inlet lip, $A/A^* = 1.338$. From Appendix A, $M_2 = 0.501$. From Appendix B, $M_1 = 2.63$. If the back pressure conditions imposed on the diffuser are such that a Mach number of 1.0 cannot be achieved at the throat, then M_2 will be less than 0.501, and a value of M_1 greater than 2.63 will be required. However, with $M = 1.0$ at the diffuser throat, the diffuser must be accelerated to a Mach number slightly greater than 2.63 to swallow the initial shock during start-up. The

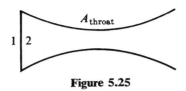

Figure 5.25

impracticality of overspeeding, at least at the flight Mach numbers discussed here and for greater Mach numbers, can be readily seen from this example.

5.5 Shock Tube and One-Dimensional Unsteady Flow

The shock tube is a device in which normal shock waves are generated by the rupture of a diaphragm separating a high-pressure gas from one at low pressure. As such, the shock tube is a useful research tool for investigating not only shock phenomena, but also the behavior of materials and objects when subjected to the extreme conditions of pressure and temperature prevalent in the gas flow behind the wave. Thus the kinetics of a chemical reaction taking place at high temperature can be studied, as well as the performance, for example, of a body during reentry from space back into the earth's atmosphere.

A shock tube is shown in Figure 5.26. After rupture of the diaphragm, the system eventually approaches thermodynamic equilibrium, with the final state in the closed-end tube determined from the first law of thermodynamics. With no external heat transfer, the total internal energy of the gases at the final state is equal to the sum of the internal energies of the gases initially present on either side of the diaphragm. However, of primary interest is not the final equilibrium state of the gases, but the transient shock phenomena occurring immediately after rupture of the diaphragm. Upon rupture of the diaphragm, a normal shock wave moves into the low-pressure side, with a series of expansion waves propagating into the high-pressure gas. Curves of p versus x and T versus x shortly after rupture of the diaphragm are shown in Figure 5.27.

Following the initial normal shock wave down the tube is a *contact surface*,

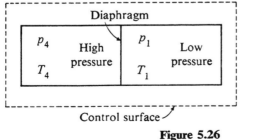

For the control volume shown, $Q = W = 0$

Figure 5.26

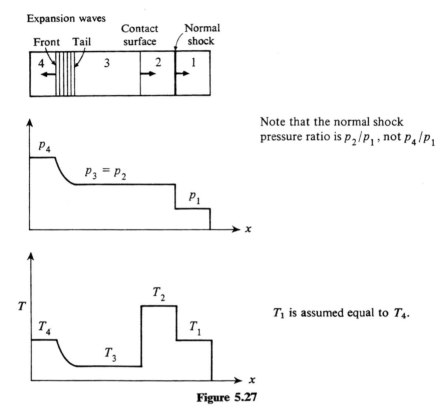

Note that the normal shock pressure ratio is p_2/p_1, not p_4/p_1

T_1 is assumed equal to T_4.

Figure 5.27

a temperature discontinuity separating the gases compressed by the shock from those cooled by the expansion. The gases in regions 2 and 3 move with the same velocity and are at the same pressure, yet there is a density and temperature difference distinguishing these regions. The shock strength and gas velocities are dependent on the initial pressure ratio across the diaphragm, the properties of the gases involved, and the initial temperatures of the gases.

For shock-tube operation, it is of prime interest to develop an expression for shock strength (p_2/p_1) as a function of the initial pressure ratio p_4/p_1 set across the diaphragm. The method we shall use in this analysis will be first to get an expression for p_4/p_3 across the expansion wave system as a function of V_{g_3}, then to develop an expression for the normal shock pressure ratio p_2/p_1 in terms of V_{g_2}, and finally to match the two solutions with $V_{g_2} = V_{g_3}$ and $p_2 = p_3$, thus obtaining a resultant equation relating p_2/p_1 and p_4/p_1 (see Figure 5.28).

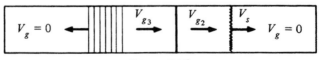

Figure 5.28

Left-moving
wave

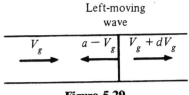

Figure 5.29

First, we shall determine the variation of properties across the expansion waves. Let us consider the general case of an infinitesimal wave moving leftward down a tube into a gas stream moving toward the wave. Assume that the gas behaves as a perfect gas, with constant specific heats (Figure 5.29). The wave moves at the velocity of sound with respect to the gas into which it is moving, so the absolute velocity of the wave will be $a - V_g$. Following our usual procedure, let us fix this wave, as shown in Figure 5.30. From the continuity equation,

$$\rho a = (\rho + d\rho)(a + dV_g)$$

or

$$\frac{d\rho}{\rho} = -\frac{dV_g}{a}$$

For an infinitesimal wave, the flow is isentropic, so that $p/\rho^\gamma = $ constant, or

$$\frac{dp}{p} = \frac{\gamma}{\gamma - 1}\frac{dT}{T}$$

Also, $a^2 = \gamma RT$, or

$$\frac{da}{a} = \frac{1}{2}\frac{dT}{T} \quad \text{and} \quad p = \rho rt$$

Substituting yields

$$\frac{d\rho}{\rho} = \frac{dp}{p} - \frac{dT}{T} = \frac{\gamma}{\gamma - 1}\frac{dT}{T} - \frac{dT}{T} = \frac{2}{\gamma - 1}\frac{da}{a}$$

Thus, we obtain, for the left-moving wave,

$$\frac{2}{\gamma - 1}\frac{da}{a} = -\frac{dV_g}{a}$$

Fixed
wave

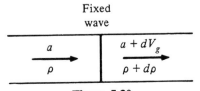

Figure 5.30

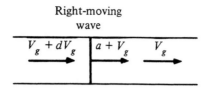

Right-moving
wave

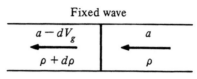

Fixed wave

Figure 5.31

or

$$dV_g + \frac{2}{\gamma - 1} da = 0$$

Integrating gives

$$V_g + \frac{2}{\gamma - 1} a = \text{constant} = C_1 \tag{5.1}$$

If the infinitesimal wave were to be moving to the right, as shown in Figure 5.31, a similar analysis yields

$$-V_g + \frac{2}{\gamma - 1} a = \text{constant} = C_2 \tag{5.2}$$

The constants C_1 and C_2 are called the *Riemann variables*.

For the shock-tube flow of Figure 5.27, Eq. (5.1) can be applied to the region 3 to 4, with $V_{g_4} = 0$:

$$\frac{2}{\gamma - 1} a_4 = \frac{2}{\gamma - 1} a_3 + V_{g_3} \tag{5.3}$$

In the region 3 to 4, the expansion waves do not coalesce, but rather spread out as they move down the tube. Therefore, we can assume the flow from 3 to 4 to be isentropic, with

$$\left(\frac{p_3}{p_4}\right) = \left(\frac{T_3}{T_4}\right)^{\gamma/(\gamma-1)} = \left(\frac{a_3}{a_4}\right)^{2\gamma/(\gamma-1)}$$

Substituting into Eq. (5.3) yields

$$\frac{V_{g_3}}{a_4} = \frac{2}{\gamma - 1}\left(1 - \frac{a_3}{a_4}\right)$$

$$= \frac{2}{\gamma - 1}\left[1 - \left(\frac{p_3}{p_4}\right)^{(\gamma-1)/2\gamma}\right] \tag{5.4}$$

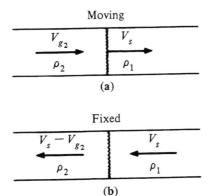

Figure 5.32

We shall now write the equations of motion for a normal shock moving rightward into a gas at rest (Figure 5.32a). To reduce the problem to one of steady flow, first fix the shock (Figure 5.32b). The continuity equation yields

$$\rho_1 V_s = \rho_2(V_s - V_{g2}) \tag{5.5}$$

The momentum equation is

$$p_1 + \rho_1 V_s^2 = p_2 + \rho_2(V_s - V_{g2})^2 \tag{5.6}$$

The energy equation is

$$h_1 + \frac{V_s^2}{2} = h_2 + \frac{(V_s - V_{g2})^2}{2} \tag{5.7}$$

For a perfect gas with constant c_p,

$$h_2 - h_1 = c_p(T_2 - T_1) = \frac{R\gamma}{\gamma - 1}(T_2 - T_1)$$

where $a^2 = \gamma RT$, so that

$$h_2 - h_1 = \frac{a_2^2 - a_1^2}{\gamma - 1} \tag{5.8}$$

Substituting Eq. (5.8) into Eq. (5.7) yields

$$\frac{a_2^2 - a_1^2}{\gamma - 1} = \frac{V_s^2}{2} - \frac{(V_s - V_{g2})^2}{2}$$

$$= V_s V_{g2} - \frac{V_{g2}^2}{2} \tag{5.9}$$

Combining Eqs. (5.5) and (5.6), we obtain

$$p_1 - p_2 = -\rho_1 V_s V_{g2} \tag{5.10}$$

But

$$p_1 = \frac{\rho_1 a_1^2}{\gamma}, \quad p_2 = \frac{\rho_2 a_2^2}{\gamma}$$

so that

$$\rho_1 a_1^2 - \rho_2 a_2^2 = -\gamma \rho_1 V_s V_{g_2}$$

where, from Eq. (5.5),

$$p_2 = \frac{\rho_1 V_s}{V_s - V_{g_2}}$$

Substituting yields

$$a_1^2 - \frac{V_s a_2^2}{V_s - V_{g_2}} = -\gamma V_s V_{g_2} \tag{5.11}$$

Combining Eqs. (5.11) and (5.9) to eliminate a_2^2 gives

$$V_{g_2} = \frac{2}{\gamma + 1}\left(V_s - \frac{a_1^2}{V_s}\right) \tag{5.12}$$

We can now substitute Eq. (5.12) into Eq. (5.10) to obtain

$$\frac{p_2}{p_1} = 1 + \frac{2\gamma}{\gamma + 1}\left[\left(\frac{V_s}{a_1}\right)^2 - 1\right] \tag{5.13}$$

Combining Eqs. (5.12) and (5.13) yields

$$\frac{V_{g_2}}{a_1} = \frac{\dfrac{p_2}{p_1} - 1}{\gamma\sqrt{1 + \dfrac{\gamma + 1}{2\gamma}\left(\dfrac{p_2}{p_1} - 1\right)}} \tag{5.14}$$

To complete the solution to the problem, we will consider the case in which $a_1 = a_4$, and equate Eq. (5.4) to (5.14) with $V_{g_2} = V_{g_3}$ and $p_2 = p_3$. The result is

$$\frac{p_1}{p_4} = \frac{p_1}{p_2}\left[1 - \frac{\gamma - 1}{2\gamma}\frac{\left(\dfrac{p_2}{p_1} - 1\right)}{\sqrt{1 + \dfrac{\gamma + 1}{2\gamma}\left(\dfrac{p_2}{p_1} - 1\right)}}\right]^{2\gamma/(\gamma - 1)} \tag{5.15}$$

Figure 5.33 provides a plot of shock pressure ratio versus diaphragm pressure ratio for a shock tube, with $\gamma = 1.4$ and $T_1 = T_4$.

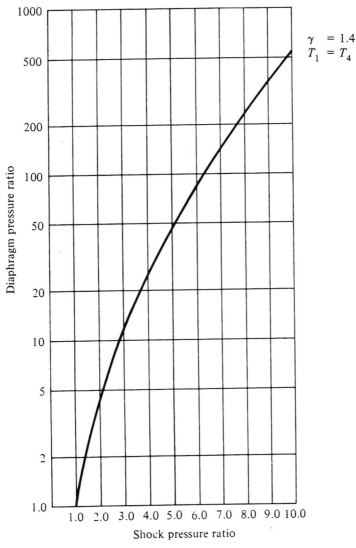

Figure 5.33

Example 5.5

The air pressure on the high-pressure side of the diaphragm in a shock tube is set at 500 kPa, and the air pressure on the low-pressure side is 20 kPa; the initial air temperature on both sides of the diaphragm is 30°C. The diaphragm is suddenly broken, which causes a normal shock to propagate rightward into the low-pressure air and a series of expansion waves to propagate leftward into the high-pressure air, as shown in Figure 5.27. Determine the following:

(a) Velocity of the normal shock

(b) Velocity of the air behind the normal shock
(c) Velocity of the contact surface
(d) Static temperature behind the normal shock
(e) Static temperature behind the contact surface
(f) Velocity of the expansion waves at front and tail of the system of expansion waves

Solution

(a) For $p_4/p_1 = 500/20 = 25$, $p_2/p_1 = 4.05$ [Figure 5.32 or Eq (5.15)]. For a shock pressure ratio of 4.05, from Appendix B, $M_s = 1.90$ and

$$V_s = M_s a_1 = 1.90\sqrt{1.4(287)(303)}$$
$$= 1.90(348.9 \text{ m/s})$$
$$= 662.9 \text{ m/s}$$

(b) From Eq. (5.12),

$$V_{g2} = \frac{2}{\gamma + 1}\left(V_s - \frac{a_1^2}{V_s}\right) = \frac{2}{2.4}\left(662.9 - \frac{348.9^2}{662.9}\right)$$
$$= 399.4 \text{ m/s}$$

(c) The contact surface moves at the same velocity as that of the air behind the normal shock.
(d) For $M_s = 1.90$, $T_2/T_1 = 1.608$ (Appendix B), so $T_2 = 1.608(303 \text{ K}) = 487.2$ K.
(e) For the expansion flow,

$$\frac{p_3}{p_4} = \left(\frac{T_3}{T_4}\right)^{\gamma/(\gamma-1)}$$

or, since $p_2 = p_3$,

$$\frac{4.05(20 \text{ kPa})}{500 \text{ kPa}} = \left(\frac{T_3}{303}\right)^{1.4/0.4} \quad \text{and} \quad \frac{T_3}{303} = 0.5945$$
$$T_3 = 180 \text{ K}$$

(f) The wave at the front of the system moves at the velocity of sound with respect to the air ahead of it, $a_4 = \sqrt{\gamma R T_4} = 349$ m/s = velocity of the front wave. At the tail of the expansion wave system, $a_3 = \sqrt{\gamma R T_3} = 269$ m/s. The absolute velocity of the last wave is $a_3 - V_{g3} = 269 - 399.4 = -130.4$ m/s (the negative sign indicates that the wave is moving rightward). An xt diagram, showing the location of normal shock, contact surface, and expansion waves as a function of time, is given in Figure 5.34.

In a shock tube, it is often desirable to be able to subject a test object to uniform conditions of high pressure and temperature for as long a period of time as possible. Uniform conditions prevail behind the initial shock until the passage of either the contact surface, the shock reflected from the closed end of

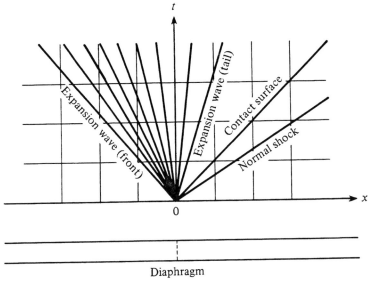

Figure 5.34

the tube, or the front of the expansion waves reflected from the opposite end. Figure 5.35 depicts the location of each of these waves as a function of time from rupture of the diaphragm for a typical shock tube system.

For the system shown, the position at which the test object should be placed in order to subject it to uniform conditions for a maximum time interval

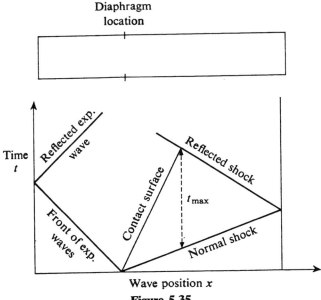

Figure 5.35

is indicated. The test object is exposed to the pressure and temperature behind the initial shock for a period t_{max} from passage of the initial wave to passage over the object of the contact surface. The test time is evidently a function of the shock-tube length, as well as the shock and expansion velocities. An object can be tested under uniform conditions for time periods up to several milliseconds in a shock tube, which is sufficient in many cases to make meaningful measurements on a test object.

REFERENCES

1. GLASS, I.I., "Theory and Performances of Simple Shock Tubes," Institute of Aerophysics, University of Toronto, UTIA Review No. 12 (1958).
2. SHAPIRO, A.H., *The Dynamics and Thermodynamics of Compressible Fluid Flow*, Vol. I. New York, Ronald Press, 1953.
3. HILL, P.G., AND PETERSON, C.R., *Mechanics and Thermodynamics of Propulsion*, Reading, Mass., Addison-Wesley Publishing Co., Inc., 1965.
4. OWCZAREK, J.A., *Fundamentals of Gas Dynamics*. Scranton, Pa., International Textbook Co., 1964.
5. ZUCROW, M.J., AND HOFFMAN, J.D., *Gas Dynamics*, Vol. 1, New York, John Wiley & Sons, Inc., 1976.

PROBLEMS

1. A converging–diverging nozzle has an area ratio of 3.0. The nozzle is supplied from an air reservoir in which the pressure and temperature are maintained at 270 kPa and 35°C, respectively. The nozzle is exhausted to a back pressure of 101 kPa. Find the nozzle exit velocity and nozzle exit-plane static pressure.

2. A supersonic nozzle possessing an area ratio of 3.0 is supplied from a large reservoir and is allowed to exhaust to atmospheric pressure. Determine the range of reservoir pressures over which a normal shock will appear in the nozzle. For what value of reservoir pressure will the nozzle be perfectly expanded, with supersonic flow at the exit plane? Find the minimum reservoir pressure to produce sonic flow at the nozzle throat. Assume isentropic flow except for shocks, with $\gamma = 1.4$.

3. A converging–diverging nozzle with an area ratio of 3.0 exhausts air from a large high-pressure reservoir to a region of back pressure p_b. Under a certain operating condition, a normal shock is observed in the nozzle at an area equal to 2.2 times the throat area. What percent of decrease in back pressure would be necessary to rid the nozzle of the normal shock?

4. Due to variations in fuel flow rate, it is found that the stagnation pressure at the inlet to a jet-engine nozzle varies with time according to $p_t = 200[1 + 0.1 \sin (\pi/4)t]$, with t in seconds and p_t in kilopascals. Determine the resultant variation in nozzle flow rate, nozzle exhaust velocity, and exit-plane static pressure. Assume the nozzle area ratio to be 2.0 to 1, with inlet stagnation temperature of 600 K and negligible inlet velocity. The nozzle exhausts to an ambient pressure of 30 kPa; $\gamma = 1.4$; nozzle exit area is 0.3 m²; $R = 0.3$ kJ/kg·K.

5. Helium enters a converging–diverging nozzle with a negligible velocity; stagnation pressure is 500 kPa and stagnation temperature is 300 K. The nozzle throat area is 50 cm², and the exit area is 300 cm². Determine the range of nozzle back pressures over which a normal shock will appear in the nozzle. Also, find the nozzle exit velocity if the nozzle exhausts into a vacuum.

6. A jet plane uses a diverging passage as a diffuser (Figure P5.6). For a flight Mach number of 1.92, determine the range of back pressures over which a normal shock will appear in the diffuser. Ambient pressure and temperature are 70 kPa and 270 K. Find the mass flow rates handled by the diffuser for the determined back pressure ranges, with $A_{inlet} = 100$ cm² and $A_{exit} = 200$ cm². Assume isentropic flow except for shocks.

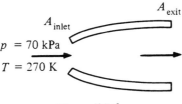

Figure P5.6

7. For the converging–diverging nozzle shown in Figure P5.7, find the range of back pressures for which $p_e > p_{back}$, the range of back pressures for which $p_e < p_{back}$, and the range of back pressures over which the nozzle is choked.

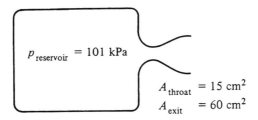

Figure P5.7

8. Nitrogen expands in a converging–diverging nozzle from negligible velocity, a stagnation pressure of 1 MPa, and a stagnation temperature of 1000 K to supersonic velocity in the diverging portion of the nozzle. If the area ratio of the nozzle is 4.0, determine the back pressure necessary for a normal shock to position itself at an area equal to twice the throat area. For this condition, find the nozzle exit velocity.

9. A supersonic wind tunnel is to be constructed as shown in Figure 5.9, with air at atmospheric pressure passing through a converging–diverging nozzle into a constant-area test section and then into a large vacuum tank. The test run is started with a pressure of 0 kPa in the tank. How long can uniform flow conditions be maintained in the test section (i.e., how long will it be before the tank pressure rises to a value such that a shock will appear in the test section)? Assume the test section to be circular, 10 cm in diameter, with a design Mach number of 2.4. The tank

volume is 3 m³, with atmospheric conditions of 101 kPa and 20°C. Assume the air to be brought to rest adiabatically in the tank.

10. Repeat Problem 9 but assume that there is a diffuser of area ratio 2 to 1 between the test section and the tank.

11. A converging–diverging supersonic diffuser is to be used at Mach 3.0. The diffuser is to use a variable throat area so as to swallow the starting shock. What percent of increase in throat area will be necessary? Solve for air and for helium as working fluids.

12. A shock tube is to be used to subject an object to momentary conditions of high pressure and temperature. To provide an adequate measuring time, the tube is to be made long enough so that a period of 100 ms is provided between the time of passage over the body of the initial shock and the time of passage of the shock reflected from the closed end of the tube. The initial pressure ratio across the diaphragm is such as to yield an initial shock with a pressure ratio of 10 to 1, with the object located 3 m from the diaphragm. The initial temperature of the air in the shock tube is 35°C. Determine a suitable length for the low-pressure end of the tube.

13. Air is stored in a tube at 200 kPa and 300 K (Figure P5.13). A diaphragm at the end of the tube is suddenly ruptured, which causes expansion waves to move down the duct. Determine the time required for the first expansion wave to reach the closed end of the tube and the velocity of the air behind the expansion waves.

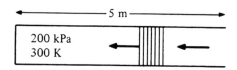

Figure P5.13

14. Write a computer program that will yield values of shock pressure ratio versus diaphragm pressure ratio for a shock tube with helium on both sides of the diaphragm. Determine values of diaphragm pressure ratio for shock pressure ratios from 1.0 to 5.0, using increments of 0.2.

15. A circular tube of length 1.5 m is evacuated to a pressure of 2.5 kPa, with the ambient pressure at 101 kPa. A diaphragm at the end of the tube is ruptured, which causes a normal shock to move down the tube. Determine the velocity of the initial shock that moves down the tube, the velocity and Mach number of the air behind the shock, and the velocity of the shock that reflects from the closed end. Initial air temperature before diaphragm rupture is 300 K. A test object is located midway along the tube. Determine the time that this object is subjected to the pressure and temperature conditions behind the initial shock (before arrival of the reflected shock). Find the static pressure and temperature behind the initial shock.

16. Solve Example 4.4 using Eq. (5.12).

17. The pressure ratio across the diaphragm in a shock tube is set at 10. The diaphragm is ruptured. Determine the velocity of the initial normal shock, the Mach number of the gases behind the shock, and the static pressure and temperature behind the shock for air ($\gamma = 1.4$) as the working fluid and for helium ($\gamma = \frac{5}{3}$) as the working

fluid. Assume the initial temperatures on each side of the diaphragm to be 25°C and the initial pressure in the low-pressure end to be 25 kPa.

18. A normal shock moves down an open-ended tube with a velocity of 1000 m/s (Figure P5.18). The ambient pressure and temperature are 101 kPa and 25°C. Determine the velocity of the first and last expansion waves that move down the tube after reflection of the shock from the open end.

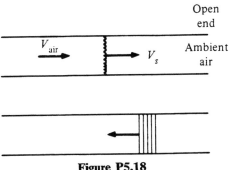

Figure P5.18

19. A shock tube is 10 m long with a 30-cm diameter. The high-pressure section is 4 m long and contains air at 200 kPa; the low-pressure section is 6 m long and contains air at 5 kPa. A test object is placed in the low-pressure section, 3 m from the diaphragm. Both sections initially contain air at 25°C. The diaphragm is suddenly ruptured, which causes a shock to move into the low-pressure section. Determine the following:
    (a) Shock velocity
    (b) Contact surface velocity
    (c) Mach number of air behind shock
    (d) Time between passage of normal shock and contact surface over test object
    (e) Time between passage of normal shock and reflected shock over test object
    (f) Provide an x–t diagram showing location of initial shock, reflected shock, and contact surface as a function of time

20. For the shock tube of Problem 19, determine the location of the test object that would enable the uniform conditions behind the initial shock to be maintained for the maximum amount of time.

6

OBLIQUE SHOCK WAVES

6.1 Introduction

In Chapter 4, the normal shock wave, a compression shock normal to the flow direction, was studied in some detail. However, in a wide variety of physical situations, a compression shock wave occurs that is inclined at an angle to the flow. Such a wave is called an oblique shock.

An oblique shock wave, either straight or curved, can occur in such varied examples as supersonic flow over a thin airfoil, as explained in Chapter 2, or in supersonic flow through an overexpanded nozzle, as mentioned in Chapter 5. The frequency of occurrence of such oblique waves makes the material presented in this chapter essential to an understanding of compressible supersonic flow.

A study of the multidimensional oblique shock wave represents a departure from the one-dimensional flow covered previously, yet in many ways the method of handling the oblique shock parallels that of handling the normal shock. Even though inclined to the flow direction, the oblique shock still represents a sudden, almost discontinuous change in fluid properties, with the shock process itself being adiabatic. In this chapter, attention will be focused on the two-dimensional straight oblique shock wave, a type that might occur during the presence of a wedge in a supersonic stream (Figure 6.1) or during a supersonic compression in a corner (Figure 6.2). As with the normal shock wave, the equations of continuity, momentum, and energy will first be derived. An additional variable is introduced because of the change in flow direction

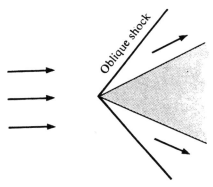

Figure 6.1

across the wave. However, momentum is a vector quantity, so two momentum equations are derivable for this two-dimensional flow. With the additional variable and equation, the analysis of two-dimensional shock flow is somewhat more complex than that for normal shock flow. However, as with the normal shock wave, solutions to the equations of motion will be presented in a form suitable for the working of practical engineering problems.

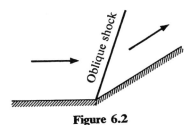

Figure 6.2

6.2 Equations of Motion for a Straight Oblique Shock Wave

When a uniform supersonic stream is forced to undergo a finite change in direction due to the presence of a body in the flow, the stream cannot adjust gradually to the presence of the body; rather, a shock wave or sudden change in flow properties must occur, as described in Chapter 2. A simple case is that of supersonic flow about a two-dimensional wedge with axis aligned parallel to the flow direction. For small wedge angles, the flow adjusts by means of an oblique shock wave, attached to the apex of the wedge. Flow after the shock is uniform, parallel to the wedge surface (as shown in Figure 6.3), with the entire flow having been turned through the wedge half-angle δ.

The equations of continuity, momentum, and energy will now be written for uniform, supersonic flow over a fixed wedge. If one selects the control

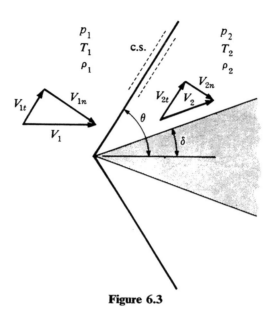

Figure 6.3

volume indicated in Figure 6.3, the continuity equation for steady flow is

$$\iint_s \rho \mathbf{V} \cdot d\mathbf{A} = 0 \tag{1.9}$$

For the case under study, this simplifies to

$$\rho_1 V_{1n} = \rho_2 V_{2n} \tag{6.1}$$

where V_{1n} and V_{2n} are the velocity components normal to the wave. The momentum equation for steady flow is

$$\sum \mathbf{F} = \iint_s \mathbf{V}(\rho \mathbf{V} \cdot d\mathbf{A}) \tag{1.14}$$

Momentum is a vector quantity, so momentum balance equations can be written both in the direction normal to the wave and in the direction tangential to the wave. The normal momentum equation yields

$$p_1 A_1 - p_2 A_2 = \rho_2 A_2 V_{2_n}^2 - \rho_1 A_1 V_{1_n}^2$$

The shock is very thin so that $A_1 = A_2$. Thus

$$p_1 - p_2 = \rho_2 V_{2_n}^2 - \rho_1 V_{1_n}^2 \tag{6.2}$$

In the tangential direction there is no change in pressure, so

$$\iint_s V_t(\rho \mathbf{V} \cdot d\mathbf{A}) = 0$$

or

$$V_{1_t}(\rho_1 A_1 V_{1_n}) = V_{2_t}(\rho_2 A_2 V_{2_n})$$

Canceling, we obtain

$$V_{1_t} = V_{2_t} \qquad\qquad (6.3)$$

where V_{1_t} and V_{2_t} are the velocity components tangential to the wave. The energy equation for adiabatic steady flow simplifies to

$$h_1 + \frac{V_1^2}{2} = h_2 + \frac{V_2^2}{2}$$

Expanding this equation, we obtain

$$h_1 + \frac{V_{1_n}^2 + V_{1_t}^2}{2} = h_2 + \frac{V_{2_n}^2 + V_{2_t}^2}{2}$$

Since $V_{1_t} = V_{2_t}$,

$$h_1 + \frac{V_{1_n}^2}{2} = h_2 + \frac{V_{2_n}^2}{2} \qquad\qquad (6.4)$$

It can be seen that Eqs. (6.1), (6.2), and (6.4) contain only the normal velocity components, and as such are the same as Eqs. (4.1), (4.2), and (4.3) for the normal shock wave. In other words, an oblique shock acts as a normal shock for the component normal to the wave, while the tangential velocity component remains unchanged. The pressure ratio, temperature ratio, and so on, across an oblique shock can be determined by first calculating the component of M_1 normal to the wave and then referring this value to the normal shock tables (Appendix B).

Example 6.1

A uniform supersonic airflow traveling at Mach 2.0 passes over a wedge (Figure 6.4). An oblique shock, making an angle of 40° with the flow direction, is attached to the wedge under these flow conditions. If the static pressure and temperature in the uniform flow are, respectively, 20 kPa and −10°C, determine the static pressure and temperature behind the wave, the Mach number of the flow passing over the wedge, and the wedge half-angle.

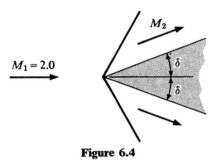

Figure 6.4

Solution

From Figure 6.3, $M_{1_n} = 2.0 \sin 40° = 1.286$. Therefore, from Appendix B,

$$\frac{p_2}{p_1} = 1.763$$

$$\frac{T_2}{T_1} = 1.182$$

$$M_{2_n} = 0.793$$

or

$$p_2 = 35.26 \text{ kPa}, \qquad T_2 = 310.9 \text{ K}$$

For the adiabatic shock process, $T_{t_1} = T_{t_2}$. From Appendix A,

$$\frac{T_1}{T_{t_1}} = 0.5556$$

or

$$T_{t_1} = T_{t_2} = 473.4 \text{ K}$$

From Appendix A, at $T/T_t = 310.9/473.4 = 0.6567$, $M_2 = 1.617$. From Figure 6.3,

$$\sin (\theta - \delta) = \frac{V_{2_n}}{V_2} = \frac{0.793}{1.617} = 0.4904$$

$$\theta - \delta = 29.4°$$

$$\delta = 40 - 29.4°$$

$$\delta = 10.6°$$

Note that the Mach number after an oblique shock wave can be greater than 1 without violating the second law of thermodynamics. The normal component of M_2, however, must still be less than 1.

In most cases, the wave angle is not known, but rather M_1 and δ appear as the independent variables. Therefore, it is more advantageous to be able to express the wave angle θ and M_2 in terms of M_1 and δ. From the geometry of the oblique wave (Figure 6.3),

$$M_{1_n} = M_1 \sin \theta \tag{6.5}$$

$$M_{1_t} = M_1 \cos \theta \tag{6.6}$$

$$M_{2_n} = M_2 \sin (\theta - \delta) \tag{6.7}$$

$$M_{2_t} = M_2 \cos (\theta - \delta) \tag{6.8}$$

It is desired to plot, for example, δ versus θ for a given value of M_1. From Eqs. (6.5) and (6.6), M_{1_n} and M_{1_t} can be found for given values of M_1 and θ. But M_{2_n} can be determined from the normal shock tables at a given M_{1_n} (Appendix B) for a perfect gas with constant specific heats and $\gamma = 1.4$. Since $V_{1_t} = V_{2_t}$ and $V = M\sqrt{\gamma RT}$, it follows that

$$\frac{M_{2_t}}{M_{1_t}} = \sqrt{\frac{T_1}{T_2}}$$

where T_1/T_2 can be found from the normal shock tables at a given M_{1_n}.

With M_{2_n} and M_{2_t}, the two unknowns M_2 and δ are calculable from the two equations (6.7) and (6.8). Now θ can be plotted versus δ for a given value of M_1, and M_2 can be plotted versus δ for given M_1. For $M_1 = 2.0$, the results appear as shown in Figures 6.5 and 6.6. Detailed oblique shock charts are provided in Appendix C, Figures C.1 and C.2, for $\gamma = 1.4$. Several characteristics of the solution to the oblique shock equations can be seen from these charts. For a given M_1 and δ, either two solutions are possible or none at all is. If a solution exists, there may be a weak oblique shock, with M_2 either supersonic or slightly less than 1, or a strong shock, with M_2 subsonic. For the strong shock, the wave makes a large angle θ (close to 90°) with the approach

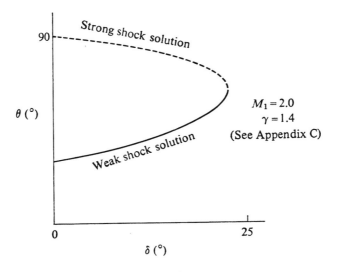

Figure 6.5

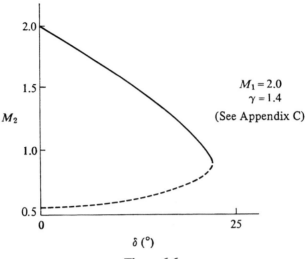

Figure 6.6

flow; for the weak shock, the angle is much less (Figure 6.7). The supersonic flow is turned through the same angle, in both cases, yet the characteristics of the oblique shocks are quite different. The weak shock is accompanied, for example, by a relatively small pressure ratio, the strong shock by a large pressure ratio.

Since the characteristics of the strong and weak oblique shocks are so different, it is essential to have an understanding of which solution will occur in a given flow situation. It is instructive at this point to refer back to a similar discussion already presented relating to the possibility of occurrence of a normal shock. It was shown in Chapter 4 that the normal shock equations are satisfied by the trivial solution $M_2 = M_1$, which corresponds to isentropic flow. This prompts the question, What determines whether isentropic flow or a normal shock will occur in a supersonic flow? The answer, at least for flow through converging–diverging nozzles, was presented in Chapter 5. For low enough back pressures, isentropic flow occurs in the nozzle; for higher back pressures, a normal shock takes place in the diverging section of the nozzle. Thus, for supersonic flow in varying area channels, it is the pressure boundary conditions imposed on the channel that determine the type of solution.

Figure 6.7

The normal shock represents the limiting case of a strong oblique shock ($\delta = 0$), whereas the limiting case of a weak oblique shock ($\delta = 0$) is isentropic flow. Therefore, the result presented for the normal wave can be generalized to the oblique shock. The strong oblique shock occurs when a large back pressure is imposed on a supersonic flow, as might possibly take place during flow through a duct or inlet. When a wedge or airfoil travels through the atmosphere at supersonic velocities with an oblique shock attached to the body, however, only the weak shock solution is found to occur, since in this physical situation, with a uniform pressure after the shock, large pressure differences cannot be supported.

Another characteristic of the oblique shock equations is that, for a great enough turning angle δ, no solution is possible. Under these conditions it is observed that the shock is no longer attached to the wedge, but stands detached, in front of the body (see Figure 6.8). The detached shock is curved, as shown, with the shock strength diminishing progressively from that of a normal shock at the apex of the wedge to that of a Mach wave far from the body. Thus, with a detached shock, the entire range of oblique shock solutions is obtained for the given Mach number M_1. The shape of the wave and the shock-detachment distance are dependent on the Mach number and the body shape. Flow over the body is subsonic in the vicinity of the wedge apex, where the strong oblique shocks occur, and it is supersonic farther back along the wedge, where the weak oblique shocks are present.

A detached oblique shock can also occur with supersonic flow in a concave corner. Again, if the turning angle is too great, a solution cannot be found in Figures C.1 and C.2, so a detached shock forms ahead of the corner (see Figure 6.9). The characteristics of this shock are exactly the same as those of the upper half of the detached shock shown in Figure 6.8. This can be seen if one

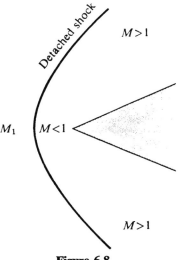

Figure 6.8

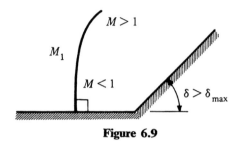

Figure 6.9

replaces the center streamline of Figure 6.8 with a plane wall (the boundary conditions of the flow are not altered, since there can be no flow across a streamline). Thus flow after the shock is subsonic near the wall and supersonic farther out in the flow.

Example 6.2

Uniform flow at $M = 2.0$ passes over a wedge of 10° half-angle. Using the oblique shock charts, find M_2, p_2/p_1, T_2/T_1, p_{t_2}/p_{t_1}, and also the half-angle above which the shock will become detached. Assume $\gamma = 1.4$.

Solution

From Figure C.1, at $M = 2.0$ and $\delta = 10°$, the weak solution yields $\theta = 39.3°$. (For supersonic flow over an isolated wedge, the weak shock solution is found to occur, as described previously.) From Figure C.2, $M_2 = 1.64$. The normal component M_{1_n} can be found from Eq. (6.5): $M_{1_n} = M_1 \sin \theta = 1.27$. From the normal shock tables (Appendix B), at $M_{1_n} = 1.27$, $p_2/p_1 = 1.715$, $T_2/T_1 = 1.17$, and $p_{t_2}/p_{t_1} = 0.984$. From Figure C.1, it can be seen that δ_{max} for $M = 2.0$ is 23°.

Example 6.3

A supersonic two-dimensional inlet is to be designed to operate at Mach 3.0. Two possibilities will be considered, as shown in Figure 6.10. In one, the compression

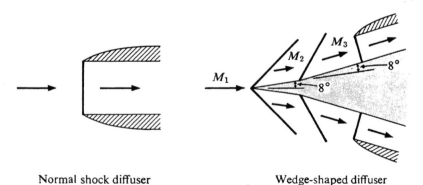

Normal shock diffuser   Wedge-shaped diffuser

Figure 6.10

and slowing down of the flow take place through one normal shock; in the other, a wedge-shaped diffuser, the deceleration occurs through two weak oblique shocks, followed by a normal shock. The wedge turning angles are each 8°. Compare the loss in stagnation pressure for the two cases shown in Figure 6.10.

Solution

For the normal shock diffuser, the ratio p_{t_2}/p_{t_1} can be found from Appendix B. Thus $p_{t_2}/p_{t_1} = \underline{0.328}$. For the wedge-shaped diffuser, M_2 and M_3, as well as the wave angles, can be found from Figures C.1 and C.2. Thus $M_2 = 2.60$ and $M_3 = 2.255$. The wave angles are, respectively, 25.6° and 29.0°. From Eq. (6.5), $M_{1n} = 3.0 \sin 25.6 = 1.30$, and $M_{2_n} = 2.60 \sin 29.0° = 1.26$. From Appendix B, $p_{t_2}/p_{t_1} = 0.979$ and $p_{t_3}/p_{t_2} = 0.986$. For $M_3 = 2.255$, $p_{t_4}/p_{t_3} = 0.603$, so that

$$p_{t_4}/p_{t_1} = (p_{t_4}/p_{t_3})(p_{t_3}/p_{t_2})(p_{t_2}/p_{t_1}) = \underline{0.582}$$

Therefore, the overall stagnation pressure ratio is 0.582. The advantage of diffusing through several oblique shocks rather than one normal shock can be seen. The greater the number of oblique shocks, the less the overall loss in stagnation pressure. Theoretically, if the flow is allowed to pass through an extremely large number of oblique shocks, each turning the flow through a very small angle, the inlet flow should approach that of an isentropic compression. The oblique shock diffuser will be discussed in detail in Chapter 8.

6.3 Oblique Shock Reflections

When a weak, two-dimensional oblique shock impinges on a plane wall, the presence of a reflected wave is required to straighten the flow, since the boundary condition at the wall imposes the restriction that there can be no flow across the wall surface (see Figure 6.11.).

Flow after the incident wave is deflected toward the wall. Hence, a reflected oblique shock wave must be present to deflect the flow back through the same angle and restore the flow direction parallel to the wall. The reflected shock is weaker than the incident shock, since $M_2 < M_1$.

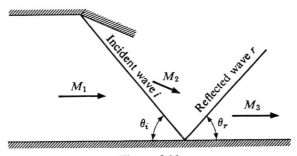

Figure 6.11

Example 6.4

For $M_1 = 2.0$ and $\theta_i = 40°$, determine θ_r, M_2, and M_3. Refer to Figure 6.11.

Solution

From Figure C.1, for $M_1 = 2.0$ and $\theta = 40°$, the deflection angle δ is equal to 10.6°. This corresponds to the angle through which the flow is turned after the incident wave and also the angle through which the flow is turned back after the reflected wave. From Figure C.2, for $M_1 = 2.0$ and $\delta = 10.6°$, M_2 is equal to 1.62. From the same chart, for $M_2 = 1.62$ and $\delta = 10.6°$, M_3 is equal to 1.24. From Figure C.1, for $M_2 = 1.62$ and $\delta = 10.6°$, the shock wave angle θ is 51.2°, which is the angle between the flow direction in region 2 and the reflected wave. From geometrical consideration, $\theta_r = 51.2° - 10.6° = \underline{40.6°}$.

If M_2 is low enough, a simple shock reflection may be impossible. That is, for a given M_2, the required turning angle may be great enough so that no solution exists from Figures C.1 and C.2. In this case, a Mach reflection occurs (see Figure 6.12).

A curved, strong oblique shock forms in the stream, extending from O to the wall at W. At W, the shock must be normal to the wall to prevent the possibility of any flow deflection, and hence satisfy the wall boundary condition at this point. Flow after the curved shock OW is subsonic, this flow adjusting smoothly to the presence of the wall. A weak oblique shock OR also appears, with flow after this shock supersonic. The combination of supersonic and subsonic flow after the waves makes an analysis of the Mach reflection extremely difficult, and certainly beyond the scope of this text.

In a real fluid, the problem of oblique shock reflections is complicated by the presence of a boundary layer on the wall. The interaction of the incident oblique shock with the wall boundary layer may have a pronounced effect on the resultant flow. It must be emphasized, then, that the analysis presented here of oblique shock reflections is an approximate one, which neglects real fluid effects.

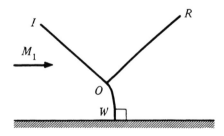

Figure 6.12

6.4 Conical Shock Waves

Supersonic flow about a right circular cone, while considerably more complex than that about a wedge, bears many similarities to wedge flow. For a cone at zero angle of attack with the oncoming stream, a conical shock is attached to the apex of the cone for small cone angles. It is interesting to compare the resultant wedge and cone flows (see Figure 6.13.) For a wedge, straight parallel flow exists before the oblique shock and after the shock. For the three-dimensional semi-infinite cone, this is no longer possible. Streamlines after the conical shock must be curved in order that the three-dimensional continuity equation be satisfied. For axisymmetric flow about a semi-infinite cone, with no characeristic length along the cone surface, conditions after the shock are dependent only on the conical coordinate ω. That is, along each line of constant ω, the flow pressure, velocity, and so on, are constant. This indicates that the pressure on the surface of the cone after the shock is constant, independent of distance from the cone apex. At each point on the conical wave, the oblique shock equations already presented are valid. Conical flow behind the wave is isentropic, with the static pressure increasing to the cone surface pressure. A solution for the conical shock thus requires fitting the isentropic compression behind the shock to the shock equations already derived. The resultant nonlinear differential equations were solved numerically by Taylor and Maccoll[1,2], with solutions tabulated by Kopal[3] and more recently by Sims[4]. Results[5] are shown in Figures C.3, C.4, and C.5 of Appendix C, which depict the variation of shock wave angle, surface pressure coefficient, and surface Mach number with cone semivertex angle and Mach number. Whereas the conical flow equations yield two shock solutions, the only one observed on an isolated conical body is the weak shock. As with wedge flow, for large enough cone angles there is no solution; the shock stands detached from the cone.

If we compare again the wedge and cone solutions, it can be seen from Figures C.3, C.4, and C.5 that, for a given body half-angle and M_1, the shock

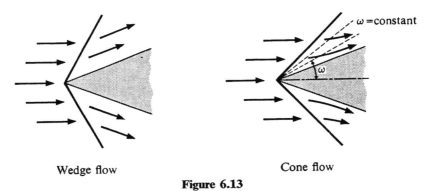

Wedge flow   Cone flow

Figure 6.13

on the wedge is inclined at a greater angle to the flow direction than the shock on the cone; this indicates that a stronger compression takes place across the wedge oblique shock. In other words, the wedge presents a greater flow disturbance than the cone. Again, this results from three-dimensional effects. From a physical standpoint, the flow is unable to pass around the side of the two-dimensional wedge since it extends to infinity in the third dimension. Flow can pass around the sides of the three-dimensional cone, however, so the cone presents less overall disruption to the supersonic flow.

Example 6.5

Uniform supersonic flow at Mach 2.0 and $p = 20$ kPa passes over a cone of semivertex angle of 10° aligned parallel to the flow direction. Determine the shock wave angle, Mach number of the flow along the cone surface, and the surface pressure.

Solution

From Figure C.3, the shock wave angle is 31.2°. From Figure C.5, the Mach number along the cone surface is 1.85. From Figure C4, the surface pressure coefficient is $0.104 = (p_c - p_1)/(\frac{1}{2}\gamma p_1 M_1^2)$ or

$$p_c = [(0.104)2.8 + 1]p_1 = 1.29p_1 = \underline{25.8 \text{ kPa}}.$$

6.5 Summary

Chapter 6 has presented an analysis of the oblique shock wave. Emphasis has been placed on demonstrating how the oblique shock can be treated as a normal shock for the velocity component perpendicular to the wave. With this approach, the oblique shock and resultant two-dimensional flow can be handled with the one-dimensional equations already derived for the normal shock in Chapter 4.

The student should now appreciate the importance of recognizing the boundary conditions in a given flow situation. It is the boundary conditions, for example, that determine whether the strong or weak oblique shock solution is to occur; again, it is the boundary conditions that determine the characteristics of the flow after impingement of an oblique shock on a plane wall.

REFERENCES

Specific References

1. TAYLOR, G.I., and MACCOLL, J.W., "The Air Pressure on a Cone Moving at High Speed," *Proceedings of the Royal Society A*, Vol. 139 (1933), pp. 278–311.
2. MACCOLL, J.W., "The Conical Shock Wave Formed by a Cone Moving at High Speed," *Proceedings of the Royal Society A*, Vol. 159 (1937), pp. 459–472.

3. KOPAL, Z., *Tables of Supersonic Flow around Cones*, M.I.T. Department of Electrical Engineering, Center of Analysis, Report No. 1, 1947.
4. SIMS, J., "Tables for Supersonic Flow around Right Circular Cones at Zero Angle of Attack," NASA SP-3004, 1964.
5. Ames Research Staff, *Equations, Tables and Charts for Compressible Flow*, NACA Report 1135, 1953.

General References

6. SHAPIRO, A.H., *The Dynamics and Thermodynamics of Compressible Fluid Flow*, Vols. 1 and 2, New York, Ronald Press, 1953.
7. OWCZAREK, J.A., *Fundamentals of Gas Dynamics*. Scranton, Pa., International Textbook Co., 1964.
8. CHEERS, F., *Elements of Compressible Flow*, New York, John Wiley & Sons, Inc., 1963.
9. ZUCROW, M.J., and HOFFMAN, J.D., *Gas Dynamics*, Vol. 1, New York, John Wiley & Sons, Inc., 1976.

PROBLEMS

1. Uniform air flow at Mach 3 passes into a concave corner of angle 15°, as shown in Figure P6.1. The pressure and temperature in the supersonic flow are, respectively, 72 kPa and 290 K. Determine the tangential and normal components of velocity and Mach number upstream and downstream of the wave. Also find the static and stagnation pressure ratios across the wave. How great would the corner angle have to be before the shock would detach from the corner?

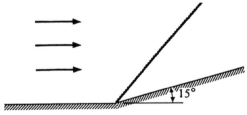

Figure P6.1

2. In a helium wind tunnel, flow at Mach 4.0 passes over a wedge of unknown half-angle aligned symmetrically with the flow. An oblique shock is observed attached to the wedge, making an angle of 30° with the flow direction. Determine the half-angle of the wedge and the ratios of stagnation pressure and stagnation temperature across the wave.

3. A wedge is to be used as an instrument to determine the Mach number of a supersonic airstream; that is, with the wedge axis aligned to the flow, the wave angle of the attached oblique shock is measured; this permits a determination of the incident Mach number. If the total included angle of such a wedge is 45°, give the Mach number range over which such an instrument would be effective.

4. The leading edge of a supersonic wing is wedge shaped, with a total included angle

$M_\infty = 2.5$

—— 2.0 m ——

Figure P6.4

of 10° (Figure P6.4). If the wing is flying at zero angle of attack, determine the lift and drag force on the wing per meter of span. Repeat for an angle of attack of 3°. Assume the wing is traveling at Mach 2.5.

5. An oblique shock wave is incident on a solid boundary, as shown in Figure P6.5. The boundary is to be turned through such an angle that there will be no reflected wave. Determine the angle θ.

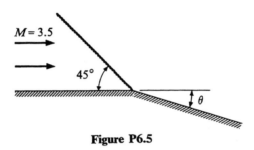

$M = 3.5$

45°

θ

Figure P6.5

6. Explain in physical terms why the angle of incidence and the angle of reflection of a reflected oblique shock are not equal.

7. A converging–diverging nozzle is designed to provide flow at Mach 2.0. With the nozzle exhausting to a back pressure of 80 kPa, however, and a reservoir pressure of 280 kPa, the nozzle is overexpanded, with oblique shocks at the exit (Figure P6.7). Determine the flow direction and flow Mach number in region R with air the working fluid.

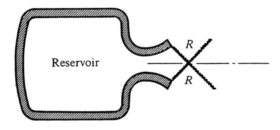

Reservoir

R

R

Figure P6.7

8. (a) Oblique shock waves appear at the exit of a supersonic nozzle, as shown in Figure P6.8. If the nozzle back pressure is 101 kPa, determine the nozzle inlet stagnation pressure. Nozzle throat area is 50 cm², and nozzle exit area is 120 cm². (b) Find the velocity at the nozzle exit plane. (c) Find the mass flow rate through the nozzle.

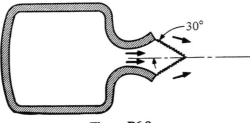

Figure P6.8

9. A supersonic flow leaves a two-dimensional nozzle in parallel, horizontal flow (region A) with a Mach number of 2.6 and static pressure (in region A) of 50 kPa. The pressure of the atmosphere into which the jet discharges is 101 kPa. Find the pressures in regions B and C of Figure P6.9.

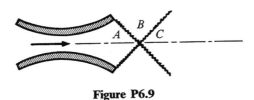

Figure P6.9

10. For the two-dimensional diffuser shown in Figure P6.10, find V_i and p_{t_i}.

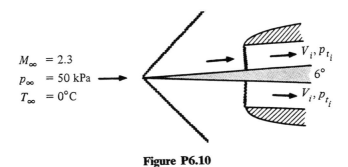

$$M_\infty = 2.3$$
$$p_\infty = 50 \text{ kPa}$$
$$T_\infty = 0°C$$

Figure P6.10

11. A two-dimensional supersonic inlet is to be designed to operate at Mach 2.4. Deceleration is to occur through a series of oblique shocks followed by a normal shock, as shown in Figure 6.10. Determine the loss of stagnation pressure for the cases of two, three, and four oblique shocks. Assume the wedge turning angles are each 6°.

12. Two oblique shocks intersect as shown in Figure P6.12. Determine the flow conditions after the intersection, with $\gamma = 1.4$.

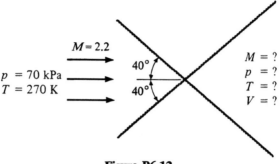

M = 2.2

p = 70 kPa
T = 270 K

40°

40°

M = ?
p = ?
T = ?
V = ?

Figure P6.12

13. A supersonic diffuser contains a conical spike of semivertex angle 5°; the spike is aligned with the flow (Figure P6.13). Determine the Mach number of the flow along the cone surface and the static pressure at the surface of the cone. Altitude = 5 km.

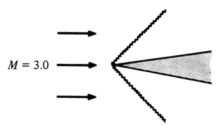

M = 3.0

Figure P6.13

14. For air flow at Mach 3.6 over a cone aligned with the flow, determine the cone semivertex angle above which the conical shock will be detached. Repeat for a wedge aligned with the flow.

15. For the two-dimensional case shown in Figure P6.15, determine M_3 and p_3. $\gamma = 1.4$.

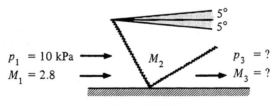

5°
5°

p_1 = 10 kPa
M_1 = 2.8

M_2

p_3 = ?
M_3 = ?

Figure P6.15

7

PRANDTL MEYER FLOW

7.1 Introduction

When a supersonic compression takes place at a concave corner, an oblique shock has been shown to occur at the corner. When supersonic flow passes over a convex corner, it is evident that some sort of supersonic expansion must take place. Previous results indicate that an expansion shock is impossible. However, a means must be available for the supersonic flow of Figure 7.1 to negotiate the corner. Chapter 7 will present an analysis of the mechanism of two-dimensional, supersonic expansion flow, as might occur, for example,

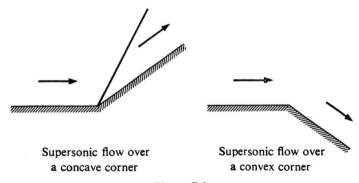

Supersonic flow over
a concave corner

Supersonic flow over
a convex corner

Figure 7.1

during supersonic flow over a convex corner or at the exit of an underexpanded supersonic nozzle.

7.2 Thermodynamic Considerations

Two-dimensional, supersonic flow is to be turned through a finite angle at a convex corner. The mechanism of the resultant flow is of interest. Consider first the possibility of an oblique adiabatic shock occurring at the corner. Figure 7.2 shows the velocity vectors normal and tangential to such a wave. For this two-dimensional flow, uniform conditions prevail upstream and downstream of the wave. The equations of motion are exactly the same as those presented in Chapter 6 for the oblique compression shock. Again, with no pressure gradient in the direction tangential to the wave, the tangential momentum equation yields $V_{1_t} = V_{2_t}$. From geometrical considerations, then, it follows that V_{2_n} must be greater than V_{1_n}. The normal momentum equation, Eq. (6.2), yields

$$p_1 + \rho_1 V_{1_n}^2 = p_2 + \rho_2 V_{2_n}^2$$

Combining this with the continuity equation, Eq. (6.1), we obtain

$$p_2 - p_1 = \rho_1 V_{1_n}(V_{1_n} - V_{2_n})$$

Since $V_{2_n} > V_{1_n}$, it follows that $p_2 < p_1$, indicating that the resultant flow must be an expansion.

It has been shown that an oblique shock reduces to a normal shock for the velocity component normal to the wave, with the tangential component remaining unchanged. The ratios of pressure, temperature, and density across an oblique shock are functions of M_{1_n} alone (see Section 6.2). The entropy change across an oblique shock can be written, then, in terms of M_{1_n}, the resultant

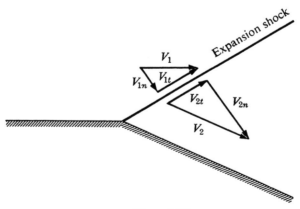

Figure 7.2

variation of ΔS with M_{1_n} being exactly the same as that shown in Figure 4.11 for the normal shock. Hence, an oblique expansion shock $(V_{2_n} > V_{1_n})$, just as a normal expansion shock, would involve a decrease in entropy during an adiabatic process. This violates the second law of thermodynamics and is impossible. Therefore, the expansion shock, with sudden changes in flow properties, cannot occur at the convex corner. Instead, a more gradual type of supersonic expansion must take place.

7.3 Gradual Compressions and Expansions

When a supersonic stream undergoes a compression due to a finite, sudden change of direction at a concave corner, an oblique shock occurs at the corner. However, if the flow is allowed to change direction in a more gradual fashion, the compression can approach an isentropic process. Allowing supersonic flow to pass through several weak oblique shocks rather than one strong shock has been shown to reduce the resultant loss in stagnation pressure (or entropy rise) for a given change in flow direction (see Figure 7.3). In the limit, as the number of oblique shocks gets larger and larger, with each shock turning the flow through a smaller and smaller angle, the oblique shocks approach the Mach waves discussed in Chapter 2. The Mach wave, brought about by the presence of an infinitesimal disturbance in a supersonic flow, here corresponds to an oblique shock of vanishing strength, with infinitesimally small changes of velocity, flow direction, entropy, and so on, taking place across the wave (see Figure 7.4).

The wave angle θ is given by Equation (2.8), $\mu = \sin^{-1} 1/M$. Note that, from the oblique shock charts, Appendix C, for an oblique shock of vanishing strength ($\delta = 0$), $\mu = \sin^{-1} 1/M$; for example, at $M_1 = 2.0$, $\delta = 0$ and $\mu = 30°$. So, by employing a smooth turn, with the resultant oblique shocks approaching Mach waves, a continuous compression is achieved in the vicinity of the wall with vanishingly small entropy rise (see Figure 7.5).

Away from the wall, however, the compression waves converge (Figure 7.6), coalescing to form a finite oblique shock wave. The characteristics of this shock are the same as those already discussed in Chapter 6 for an oblique shock wave of given M_1 and turning angle δ. In fact, far enough away from the wall, flow about the smooth turn cannot be distinguished from the flow about a sharp, concave corner of angle δ. It is important to note that here, again, the

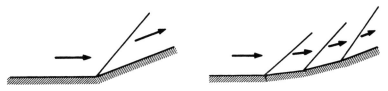

Figure 7.3

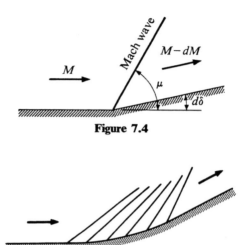

Figure 7.4

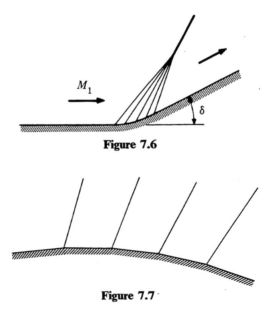

Figure 7.5

weak compression waves, each involving only an infinitesimal entropy rise, are able to reinforce one another to form a compression shock wave, with the resultant shock process involving a finite increase of entropy.

Now consider a supersonic expansion through a series of infinitesimally small convex turns (see Figure 7.7). Mach waves are generated at each corner, with each wave inclined at an angle to the flow direction. For this expansion flow, unlike the compressive flow discussed previously, waves do not coalesce but rather spread out. The divergent waves cannot reinforce one another; the

Figure 7.6

Figure 7.7

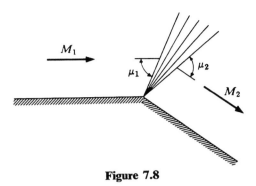

Figure 7.8

oblique expansion shock is physically impossible. Flow between each of the waves in Figure 7.7 is uniform, so the length of the wall between waves has no effect on the variation of flow properties. Thus the lengths of the wall segments can be made vanishingly small, without affecting the overall variation of flow properties across the expansion. By thus reducing the wall segments, the series of convex turns becomes a sharp corner (see Figure 7.8.) The resultant series of expansion waves, centered at the corner, is called a Prandtl Meyer expansion fan.

7.4 Flow Equations for a Prandtl Meyer Expansion Fan

It has been shown that supersonic expansion flow around a convex corner involves a smooth, gradual change in flow properties. The Prandtl Meyer fan consists of a series of Mach waves, centered at the convex corner. The initial wave is inclined to the approach flow at an angle $\mu_1 = \sin^{-1} 1/M_1$; the final wave is inclined to the downstream flow at an angle $\mu_2 = \sin^{-1} 1/M_2$, with the component of velocity normal to the wave at each point in the flow equal to the local velocity of sound (see Section 2.4). Flow conditions along each Mach wave are uniform; the variation of pressure, velocity and so on, through the expansion is only a function of angular position.

The equations for two-dimensional Prandtl Meyer flow will now be presented so that, just as with oblique shock flow, the variation of flow properties can be determined for a given flow turning angle. A perfect gas with constant specific heats will be assumed in the following analysis.

Consider first a single Mach wave, expanding the supersonic flow through an angle of magnitude dv. With no pressure gradient in the tangential direction, again, there is no change of the tangential velocity component across the wave. Equating the expressions for V_t upstream and downstream of the Mach

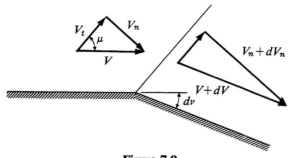

Figure 7.9

wave (see Figure 7.9),

$$V \cos \mu = (V + dV) \cos (\mu + d\nu)$$
$$= (V + dV)(\cos \mu \cos d\nu - \sin \mu \sin d\nu) \qquad (7.1)$$

Since $d\nu$ is a very small angle,

$$\cos d\nu \approx 1$$
$$\sin d\nu \approx d\nu$$

Therefore,

$$V \cos \mu = (V + dV)(\cos \mu - \sin \mu \, d\nu)$$

Expanding yields

$$V \cos \mu = V \cos \mu + \cos \mu \, dV - V \sin \mu \, d\nu - dV \sin \mu \, d\nu$$

The last term, containing the product of two differentials, can be dropped in comparison with the other terms of the equation. Simplifying, we obtain

$$\frac{dV}{V} = \tan \mu \, d\nu$$

Since $\sin \mu = 1/M$, it follows that

$$\tan \mu = \frac{1}{\sqrt{M^2 - 1}}$$

so

$$\frac{dV}{V} = \frac{1}{\sqrt{M^2 - 1}} \, d\nu \qquad (7.2)$$

To solve for M as a function of ν, V must be expressed in terms of M. For a perfect gas with constant specific heats, we can write,

$$V = M\sqrt{\gamma RT}$$

Taking logs and differentiating, we obtain

$$\frac{dV}{V} = \frac{dM}{M} + \frac{1}{2}\frac{dT}{T} \tag{7.3}$$

But, for this adiabatic flow, there is no change in stagnation temperature.

$$T_t = \text{constant} = T\left(1 + \frac{\gamma - 1}{2}M^2\right)$$

Taking logs and differentiating, we find

$$\frac{dT}{T} + \frac{(\gamma - 1)M\,dM}{1 + \frac{\gamma - 1}{2}M^2} = 0 \tag{7.4}$$

If Eqs. (7.3) and (7.4) are combined,

$$\frac{dV}{V} = \frac{dM}{M} - \frac{\frac{1}{2}(\gamma - 1)M\,dM}{1 + \frac{\gamma - 1}{2}M^2} \tag{7.5}$$

$$\frac{dV}{V} = \frac{dM}{M}\left[\frac{1}{1 + \frac{\gamma - 1}{2}M^2}\right] \tag{7.6}$$

Substituting Eq. (7.6) into Eq. (7.2),

$$d\nu = \frac{dM}{M}\frac{\sqrt{M^2 - 1}}{1 + \frac{\gamma - 1}{2}M^2} \tag{7.7}$$

To determine the change of Mach number associated with a finite turning angle, the above Eq. (7.7) can be integrated.

$$\Delta\nu = \nu_2 - \nu_1$$

$$= \int_{M_1}^{M_2}\frac{\sqrt{M^2 - 1}}{M}\frac{dM}{1 + \frac{\gamma - 1}{2}M^2}$$

$$= \left[\sqrt{\frac{\gamma + 1}{\gamma - 1}}\tan^{-1}\sqrt{\frac{\gamma - 1}{\gamma + 1}(M^2 - 1)} - \tan^{-1}\sqrt{M^2 - 1}\right]_{M_1}^{M_2} \tag{7.8}$$

For the purpose of tabulating this result, it is convenient to define a reference state 1, so that

$$\nu - \nu_{\text{ref}} = \left[\sqrt{\frac{\gamma + 1}{\gamma - 1}}\tan^{-1}\sqrt{\frac{\gamma - 1}{\gamma + 1}(M^2 - 1)} - \tan^{-1}\sqrt{M^2 - 1}\right]_{M_{\text{ref}}}^{M}$$

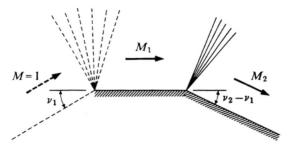

Figure 7.10

Let the reference state be $\nu = 0$ at $M = 1$. Now

$$\nu = \left[\sqrt{\frac{\gamma + 1}{\gamma - 1}} \tan^{-1} \sqrt{\frac{\gamma - 1}{\gamma + 1} (M^2 - 1)} - \tan^{-1}\sqrt{M^2 - 1} \right] \quad (7.9)$$

The symbol ν represents the angle through which a stream, initially at Mach 1, must be expanded to reach a supersonic Mach number M. Values of ν have been tabulated in Appendix D, for Mach numbers from 1.0 to 5.0 for $\gamma = 1.4$. Also presented are values of the wave angle μ, with both ν and μ expressed in degrees.

To determine the angle through which a flow would have to be turned to expand from M_1 to M_2, with M_1 not equal to 1, it is necessary only to subtract the value of ν_1 at M_1 from the value of ν_2 at M_2, where ν_1 and ν_2 are found in Appendix D (see Figure 7.10).

The variation of pressure, temperature, and other thermodynamic properties through the expansion can be found from the usual thermodynamic relations for isentropic flow, presented in Chapter 3. For this isentropic process, with no change in stagnation pressure,

$$\frac{p_2}{p_1} = \left[\frac{1 + \frac{\gamma - 1}{2} M_1^2}{1 + \frac{\gamma - 1}{2} M_2^2} \right]^{\gamma/(\gamma-1)}$$

Since the expansion is adiabatic with no change of stagnation temperature,

$$\frac{T_2}{T_1} = \frac{1 + \frac{\gamma - 1}{2} M_1^2}{1 + \frac{\gamma - 1}{2} M_2^2}$$

Example 7.1

A uniform supersonic flow at Mach 2.0, with static pressure of 75 kPa and a temperature of 250 K, expands around a 10° convex corner. Determine the

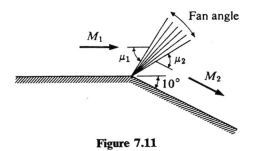

Figure 7.11

downstream Mach number M_2, pressure p_2, temperature T_2, and the fan angle. See Figure 7.11.

Solution

From Appendix D, $\nu_1 = 26.380°$. But

$$\nu_2 = \nu_1 + 10° = 36.380°$$

So again, from Appendix D,

$$M_2 = 2.385$$

Since

$$\frac{p_2}{p_1} = \frac{p_2/p_{t_2}}{p_1/p_{t_1}} \qquad \text{where } p_{t_2} = p_{t_1}$$

and, from Appendix A, $\gamma = 1.4$,

$$\frac{p_2}{p_{t_2}} = 0.07003$$

$$\frac{p_1}{p_{t_1}} = 0.1278$$

then

$$\frac{p_2}{p_1} = \frac{0.07003}{0.1278} = 0.5480$$

or

$$p_2 = 41.10 \text{ kPa}$$

From Appendix A, $\gamma = 1.4$,

$$\frac{T_2}{T_{t_2}} = 0.4678$$

$$\frac{T_1}{T_{t_1}} = 0.5556$$

or

$$T_2 = 210.5 \text{ K}$$

$$\text{fan angle} = \nu_2 - \nu_1 + \mu_1 - \mu_2$$
$$= 10° + 30° - 24.79°$$
$$= \underline{15.21°}$$

Example 7.2

Flow in Example 7.1 is expanded through a second convex turn of angle 10° (see Figure 7.12). Determine the downstream Mach number M_3 and the angle of the second fan.

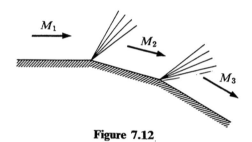

Figure 7.12

Solution

The initial wave of the second fan must be parallel to the final wave of the first fan. Again, the distance between waves can have no effect on the resultant flow, since the flow between the waves is uniform. Therefore, the variation of properties is the same whether the flow is expanded through two 10° turns or one 20° turn.

$$\nu_3 = \nu_2 + 10°$$
$$= 46.38°$$

or

$$M_3 = 2.831$$

Alternatively,

$$\nu_3 = \nu_1 + 20°$$
$$= 46.38°$$
$$M_3 = 2.831$$

Note the difference between Prandtl Meyer flow and oblique shock flow. In the latter, there was an advantage to be derived from compressing through several weak shocks rather than one strong shock. The fan angle for the second fan is $10° + 24.79° - 20.68° = \underline{14.11°}$.

Example 7.3

An underexpanded, two-dimensional, supersonic nozzle exhausts into a region where $p = 100$ kPa (Figure 7.13). Flow at the nozzle exit plane is uniform, with $p = 200$ kPa and $M = 2.0$. Determine the flow direction and Mach number after the initial expansion.

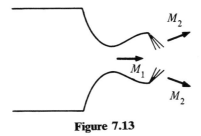

Figure 7.13

Solution

From Appendix A, at $M_1 = 2.0$,

$$\frac{p_1}{p_t} = 0.1278$$

Since $p_{t_2} = p_{t_1}$ for this isentropic expansion,

$$\frac{p_2}{p_{t_2}} = (0.1278)\left(\frac{100}{200}\right) = 0.0639$$

or

$$M_2 = 2.444$$

From Appendix D,

$$\nu_1 = 26.380°$$

$$\nu_2 = 37.803°$$

or the flow has been turned through $\nu_2 - \nu_1 = \underline{11.42°}$.

7.5 Prandtl Meyer Flow in a Smooth Compression

It was shown in Section 7.3 that, at a smooth compressive turn in supersonic flow, Mach waves emanate from the wall, coalescing farther out in the stream to form an oblique shock wave. In the region from the wall out to the point of coalescence of the waves (see Figure 7.6), the flow is isentropic and possesses the same characteristics as Prandtl Meyer flow. Therefore, the equations derived for Prandtl Meyer flow can be applied to the isentropic flow region at a concave corner, even though a compression takes place at the corner. Naturally, the turning angle $\Delta\nu$ will here be negative, corresponding to a decrease in

Mach number. The extent of the isentropic flow region at a concave corner depends on the curvature of the wall. For a sharp turn, the region that can be treated as Prandtl Meyer flow is negligible; for a gradual turn, with a large radius of curvature, a much greater region has the characteristics of Prandtl Meyer flow.

7.6 Maximum Turning Angle for Prandtl Meyer Flow

From Eq. (7.9), it can be seen that, as $M \to \infty$, or as the static pressure $p_2 \to 0$ (see Figure 7.14), the turning angle approaches a finite value of 130.4°. This result has significance, for example, in a determination of the shape of the exhaust plume of an underexpanded nozzle discharging into the vacuum of space. To prevent the impingement of rocket exhaust gases on a part of a spacecraft, the designer must have a knowledge of the shape of the rocket-nozzle exhaust plume; modification of a spacecraft geometrical design may be required to prevent possible damage from the hot exhaust gases. Furthermore, the axial thrust of a rocket depends on the direction of the exhaust velocity vectors.

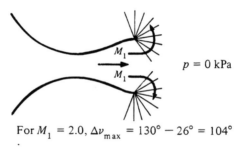

For $M_1 = 2.0$, $\Delta\nu_{max} = 130° - 26° = 104°$

Figure 7.14

The actual magnitude of the maximum turning angle presented here has only academic interest, in that effects such as liquefaction of air gases and other departures from perfect gas flow would occur long before the ultimate pressure could be attained. However, the result does indicate the presence of a maximum turning angle for a supersonic expansion.

7.7 Reflections

When a Prandtl Meyer expansion flow impinges on a plane wall, as shown in Figure 7.15, sufficient waves must be generated to maintain the wall boundary condition; that is, at the wall surface, the flow must be parallel to the wall. Each Mach wave of the initial Prandtl Meyer fan, then, must reflect as an expansion Mach wave. The resultant wave interactions present complexities

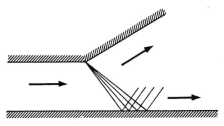

Figure 7.15

that render an exact analysis of the flow extremely difficult; however, the general nature of the flow can be recognized. An application is the expansion that takes place at the exit of an underexpanded, two-dimensional nozzle. Since, from symmetry, there can be no flow across the center streamline, this streamline can be replaced by a plane wall. The resultant flow situation is shown in Figure 7.16.

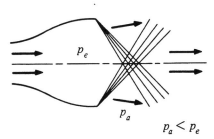

Figure 7.16

7.8 Summary

Just as Chapter 6 presented an analysis of supersonic, compressive oblique flow, so Chapter 7 has provided a study of supersonic, oblique expansion flow. A discussion of expansion and compression supersonic flows around convex and concave corners bears many similarities to the previous discussion of wave motion in constant-area tubes. In Chapter 4, it was shown that weak compression waves moving down a tube overtake one another and eventually coalesce to form a finite normal shock; expansion waves move farther apart and, hence, cannot reinforce one another. At a convex turn, the expansion waves diverge from the corner and, again, no reinforcement is possible; at a concave turn, the compression waves converge to form a finite oblique shock. It was shown, in each case, that the expansion shock represents a violation of the second law of thermodynamics.

So, at a concave corner in supersonic flow, an isentropic expansion takes place. This flow consists of a large number of expansion Mach waves; across each wave the changes in flow properties are infinitesimally small. The resultant flow is analyzed by using the equations of continuity, momentum, and energy for a perfect gas with constant specific heats. The result of this analysis has been presented in tabular form, showing the change in Mach number occurring for a given flow turning angle. The use of the isentropic relationships then permits an evaluation of the change of pressure and temperature taking place through the expansion.

REFERENCES

1. CHEERS, F., *Elements of Compressible Flow*, New York, John Wiley & Sons, Inc., 1963.
2. LIEPMANN, H.W., AND ROSHKO, A., *Elements of Gas Dynamics*, New York, John Wiley & Sons, Inc., 1957.
3. SHAPIRO, A.H., *The Dynamics and Thermodynamics of Compressible Fluid Flow*, Volume 1, New York, Ronald Press, 1963.
4. ZUCROW, M.J., AND HOFFMAN, J.D., *Gas Dynamics*, Vol. 1, New York, John Wiley & Sons, Inc., 1976.

PROBLEMS

1. A uniform supersonic flow of air at Mach 2.6, with stagnation pressure of 5 MPa and stagnation temperature of 1000 K, expands around a 20° convex corner. Determine the downstream Mach number, the stagnation pressure and temperature, and the static pressure and temperature.
2. A reservoir containing air at 2 MPa is connected to ambient air at 101 kPa through a converging–diverging nozzle designed to produce flow at Mach 2.0, with axial flow at the nozzle exit plane (Figure P7.2). Under these conditions, the nozzle is underexpanded, with a Prandtl Meyer expansion fan at the exit. Find the flow direction after the initial expansion fan. How does this turning angle affect the net axial thrust forces exerted by the fluid on the nozzle?

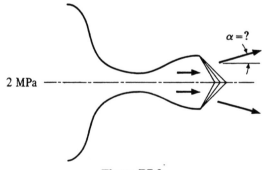

Figure P7.2

3. Develop a computer program that will yield values of ν versus M for Prandtl Meyer flow for $\gamma = 1.3$ over the range $M = 1.0$ to $M = 5.0$, using Mach number increments of 0.1. Repeat for μ versus M for $\gamma = 1.3$.

4. A uniform supersonic flow of a perfect gas with $\gamma = 1.3$ and Mach number 3.0 expands around a 5° convex corner. Determine the downstream Mach number, ratio of downstream to upstream velocity, and ratio of downstream to upstream stagnation temperature.

5. For flow at Mach 2.5 and $\gamma = 1.4$ over the symmetrical protrusion shown in Figure P7.5, find M_2, M_3, M_4, T_2, T_3, and T_4.

$T_1 = 300$ K
$M_1 = 2.5$

Figure P7.5

6. A two-dimensional, flat plate is inclined at a positive angle of attack in a supersonic air stream of Mach 2.0 (Figure P7.6). Below the plate, an oblique shock wave starts at the leading edge, making an angle of 42° with the stream direction. On the upper side, an expansion occurs at the leading edge.
   (a) Find the angle of attack of the plate.
   (b) What is the pressure on the lower surface of the plate?
   (c) What is the pressure on the upper surface of the plate?

$M = 2.0$
$p = 50$ kPa

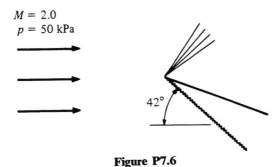

Figure P7.6

7. A two-dimensional supersonic wing has the profile shown in Figure P7.7. At zero angle of attack, determine the drag force on the wing per unit length of span at Mach 2 and at Mach 4. Repeat for the lift force. Take the maximum thickness of the airfoil to be 0.2 m.

$p_\infty = 20$ kPa

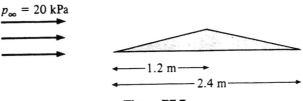

—1.2 m—
—2.4 m—

Figure P7.7

8. In Problem 7, a compression occurs at the trailing edge, with the resultant flows in regions (a) and (b) parallel (Figure P7.8). Is there any difference in pressure, velocity, or entropy between regions (a) and (b)? Discuss.

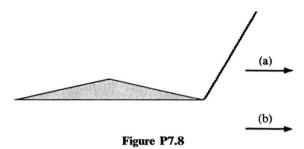

(a)

(b)

Figure P7.8

9. A reservoir containing air at 10 MPa is discharged through a converging–diverging nozzle of area ratio 3.0. An expansion fan is observed at the exit, with the flow immediately downstream of the fan turned through an angle of 10°. Determine the pressure of the region into which the nozzle is exhausting, if the air can be assumed to behave as a perfect gas with constant $\gamma = 1.4$.

8

APPLICATIONS II

8.1 Introduction

Chapters 6 and 7 have presented cases of two-dimensional flow: the oblique shock wave and the Prandtl Meyer expansion. To analyze supersonic flow over a finite body in which changes in flow direction must occur, it is necessary to apply the characteristics of Prandtl Meyer flow and oblique shock flow to a physical situation. It is appropriate, then, at this point in the text, to present several applications in which the two-dimensional flows discussed in the previous chapters are present.

Several such examples have already been alluded to in the text. The oblique shock diffuser and the expansion or compression flows that take place at the exit of an underexpanded or overexpanded supersonic nozzle have been introduced in previous chapters. Along with a detailed study of these two examples, Chapter 8 will present an analysis of the supersonic airfoil, involving both oblique shocks and Prandtl Meyer flow, and the plug nozzle, a relatively new development in jet propulsion systems, which essentially replaces the diverging portion of a conventional supersonic nozzle with a Prandtl Meyer expansion, so as to increase the off-design performance of the nozzle.

8.2 Supersonic Oblique Shock Diffuser

For a turbojet or ramjet traveling at high velocity, it is necessary to provide an inlet, or diffuser, that will perform the function of slowing down the incoming

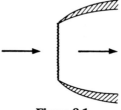

Figure 8.1

air with a minimum loss of stagnation pressure. The use of a converging–diverging passage as an inlet for supersonic flow was studied in Chapter 5. Because such an internal deceleration device can operate isentropically only at the design speed, this type of diffuser was found to be impractical during start-up and when operating off design. In fact, without provisions for varying the throat area or overspeeding, the design condition could not be attained.

To eliminate the starting problem involved with the converging–diverging passage, the internal throat must be removed. Thus a possible design is the normal shock diffuser, the deceleration taking place through a normal shock followed by subsonic diffusion in a diverging passage (see Figure 8.1). The disadvantage of this setup is the large loss in stagnation pressure incurred by the normal shock. Only at Mach numbers close to 1 would this design be practicable.

In Chapter 6, the advantage of decelerating through several oblique shocks rather than one normal shock was demonstrated (Example 6.3). The oblique shock, spike-type diffuser takes advantage of this, and hence represents a practical device for decelerating a supersonic flow. The operation of a single oblique shock inlet at design speed is depicted in Figure 8.2. External deceleration is accomplished through an oblique shock attached to the spike. Further deceleration takes place through a normal shock at the engine cowl inlet, with subsonic deceleration occurring internally. Even though a normal shock occurs in this system, the flight Mach number has been reduced by the oblique shock, thus reducing the normal shock strength and resultant stagnation pressure loss.

Theoretically, the greater the number of oblique shocks, the less the resultant total loss in stagnation pressure. For example, a two-shock inlet is

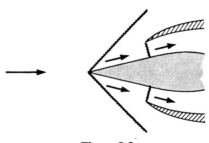

Figure 8.2

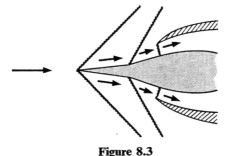

Figure 8.3

shown in Figure 8.3. Note, however, that along the surface of the spike, the boundary layer increases in thickness. The adverse pressure gradient created by the second shock may be sufficient to cause flow separation, with resultant loss of available energy. The greater the number of shocks, then, the greater the tendency toward flow separation. It is necessary to effect a compromise in supersonic diffuser design between the increased total pressure recovery achieved by increasing the number of oblique shocks through which the flow must be diffused and the increased tendency toward separation brought about by the shocks. For this reason, with flight Mach numbers up to 2.0, a single shock diffuser is generally employed, whereas multiple shock inlets are required for higher flight Mach numbers.

Several different modes of operation of the spike diffuser may occur, depending on the downstream engine conditions such as nozzle opening, turbine speed, and fuel flow rate. This is in contrast to the converging–diverging inlet, where operation was dependent on the inlet geometry. The spike diffuser's modes of operation are termed subcritical, critical, and supercritical, depending on the location of the normal shock. Critical operation occurs with the normal shock at the cowl inlet, as shown in Figure 8.4, with the engine operating at design speed. If the flow resistance downstream of the inlet is increased, with the engine still at the design flight Mach number, the normal shock moves ahead of the inlet, with some of the subsonic flow after the shock able to spill over or bypass the inlet (see Figure 8.4). For this condition, the inlet is not handling maximum flow rate; furthermore, the pressure recovery is unfavorable, since at least some of the inlet air passes through a normal shock at the design Mach number. If the downstream resistance is reduced below that for critical operation, the normal shock reaches an equilibrium position inside the diffuser. For this condition, the inlet is still handling maximum mass flow, yet the pressure recovery is less than that for critical operation, since the normal shock occurs at a higher Mach number in the diverging passage.

A turbojet engine must be able to operate efficiently both at other-than-design speeds and also at different angles of attack. An engine operating at the critical mode may be pushed over into the undesirable subcritical mode by a small change of speed or angle of attack. For this reason, in actual operation it

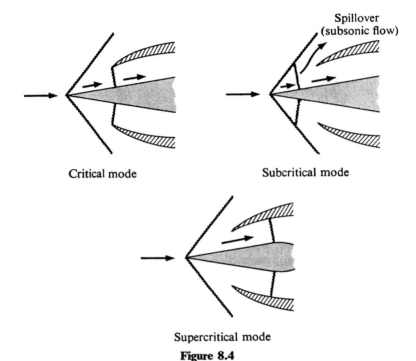

Critical mode   Subcritical mode

Supercritical mode

Figure 8.4

is more practical to operate in the supercritical mode. While not providing quite as good a pressure recovery as critical operation, the supercritical mode still yields maximum engine-mass flow and also furnishes a safety margin so that a small decrease in engine speed will not cause a transition to the subcritical mode. Thus the supercritical mode provides a more stable engine operation.

Example 8.1

Compare the loss in total pressure incurred by a one-shock spike diffuser (two dimensional) with that incurred by a two-shock diffuser operating at Mach 2. Repeat at Mach 4.0 (see Figure 8.5). Assume that each oblique shock turns the flow through an angle of 10°.

Solution

From Figure C.1, at $M = 2.0$ and $\delta = 10°$, the weak solution yields $\theta = 39.3°$. Thus $M_{1_n} = M_1 \sin \theta = 1.27$. From Figure C.2, $M_2 = 1.64$. For the one-shock diffuser,

$$\frac{p_{t_3}}{p_{t_1}} = \frac{p_{t_3}}{p_{t_2}} \cdot \frac{p_{t_2}}{p_{t_1}}$$

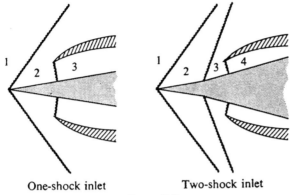

One-shock inlet   Two-shock inlet

Figure 8.5

From Appendix B, $p_{t_2}/p_{t_1} = 0.984$ and $p_{t_3}/p_{t_2} = 0.880$, so

$$\left(\frac{p_{t_3}}{p_{t_1}}\right)_{\text{one shock}} = (0.880)(0.984) = \underline{0.866}$$

For the two-shock inlet, $M_2 = 1.64$. At $M_2 = 1.64$ and $\delta = 10°$, $\theta = 49.4$ and $M_3 = 1.28$. $M_{2n} = 1.64 \sin 49.4° = 1.25$.

$$\left(\frac{p_{t_4}}{p_{t_1}}\right)_{\text{two shocks}} = \frac{p_{t_4}\,p_{t_3}\,p_{t_2}}{p_{t_3}\,p_{t_2}\,p_{t_1}} = (0.983)(0.987)(0.984) = \underline{0.955}$$

From Figure C.1, at $M = 4.0$ and $\delta = 10°$, the weak solution yields $22.2°$. Thus $M_{1_n} \sin \theta = 1.51$. From Figure C.2, $M_2 = 3.29$. For the one oblique shock diffuser,

$$\left(\frac{p_{t_3}}{p_{t_1}}\right)_{\text{one shock}} = \frac{p_{t_3}\,p_{t_2}}{p_{t_2}\,p_{t_1}} = (0.256)(0.927) = \underline{0.237}$$

For the two-shock inlet, $M_2 = 3.29$. At $M_2 = 3.29$ and $\delta = 10°$, $\theta = 25.5°$, so $M_{2_n} = 3.29 \sin 25.5° = 1.42$ and $M_3 = 2.74$. From Appendix B, $p_{t_3}/p_{t_2} = 0.953$ and $p_{t_4}/p_{t_3} = 0.4097$. For this case,

$$\left(\frac{p_{t_4}}{p_{t_1}}\right)_{\text{two shocks}} = \frac{p_{t_4}\,p_{t_3}\,p_{t_2}}{p_{t_3}\,p_{t_2}\,p_{t_1}} = (0.410)(0.953)(0.984) = \underline{0.384}$$

It can be seen that, whereas the improvement in total pressure ratio gained by using a two-shock inlet over a one-shock inlet is only about 10 percent at Mach 2, this improvement amounts to 62 percent at Mach 4. Thus, at flight Mach numbers of 2.0 and below, the use of an inlet with one oblique shock is satisfactory; at flight Mach numbers of 4.0, an inlet with two oblique shocks (or more) is necessary.

Example 8.2

A two-dimensional, spike-type inlet is operating in the supercritical mode at a flight Mach number of 2.0. The flow cross-sectional area at the cowl inlet is 0.1 m^2; the cross-sectional area at the location where the normal shock occurs in the

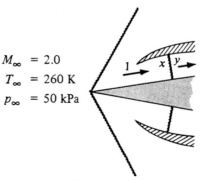

$M_\infty = 2.0$

$T_\infty = 260$ K

$p_\infty = 50$ kPa

Figure 8.6

diverging passage is $0.12 \, \text{m}^2$ (see Figure 8.6). Calculate the mass flow rate and total pressure ratio p_{t_y}/p_{t_∞}. Neglect friction. The spike half-angle is 10°.

Solution

From Figure C.1, $\theta = 39.3°$. From Figure C.2, $M_1 = 1.64$. For isentropic flow from 1 to x,

$$\frac{A_x}{A_x^*} = \frac{A_x}{A_1} \frac{A_1}{A^*}$$

From Appendix A,

$$\frac{A_x}{A_x^*} = \left(\frac{0.12}{0.10}\right) 1.284 = 1.54$$

Therefore,

$$M_x = 1.89$$

Next,

$$\frac{p_{t_y}}{p_{t_\infty}} = \frac{p_{t_y}}{p_{t_x}} \frac{p_{t_1}}{p_{t_\infty}}$$

From Appendix B,

$$\frac{p_{t_y}}{p_{t_x}} = 0.772$$

From Example 8.1,

$$\frac{p_{t_1}}{p_{t_\infty}} = 0.984$$

Therefore,

$$\frac{p_{t_y}}{p_{t_\infty}} = (0.772)(0.984) = \underline{0.760}$$

$$p_{t_\infty} = \frac{50}{p_\infty/p_{t_\infty}} = \frac{50}{0.1278} = 391 \text{ kPa}$$

or

$$p_{t_1} = 0.984(391) = 385 \text{ kPa}$$

Since

$$p_1 = \frac{p_1}{p_{t_1}} p_{t_1}$$

where $p_1/p_{t_1} = 0.222$ at $M_1 = 1.64$, then

$$p_1 = (0.222)(385) = 85.5 \text{ kPa}$$

Since no change in stagnation temperature occurs across a normal shock,

$$T_1 = \frac{T_1}{T_{t_1}} T_{t_\infty} = (0.650)\left(\frac{260}{0.556}\right) = 304 \text{ K}$$

$$\dot{m} = \rho_1 A_1 V_1 = \left(\frac{p_1}{RT_1}\right) A_1 M_1 \sqrt{\gamma R T_1}$$

$$= \left[\frac{85.5}{287(304)}\right][0.10]\left[1.64\sqrt{1.4(287)304}\right]$$

$$= \underline{56.2 \text{ kg/s}}$$

8.3 Exit Flow for Underexpanded and Overexpanded Supersonic Nozzles

In Chapters 3 and 5, the flow through a converging–diverging nozzle was studied in detail. The variation in flow patterns inside the nozzle obtained by changing the back pressure, with a constant reservoir pressure, was discussed. It was shown that, over a certain range of back pressures, the flow was unable to adjust to the prescribed back pressure inside the nozzle, but rather adjusted externally in the form of compression waves or expansion waves. Having studied oblique waves in Chapter 6 and Prandtl Meyer flow in Chapter 7, we can now discuss in detail the wave pattern occurring at the exit of an underexpanded or overexpanded nozzle.

Consider, first, flow at the exit plane of an underexpanded, two-dimensional nozzle (see Figure 8.7). Since the expansion inside the nozzle was insufficient to reach the back pressure, expansion fans form at the nozzle exit plane. As is shown in Figure 8.7, flow at the exit plane is assumed to be uniform and parallel, with $p_1 > p_b$. For this case, from symmetry, there can be no flow across the centerline of the jet. Thus the boundary conditions along the centerline are the same as those at a plane wall in nonviscous flow, and the normal velocity component must be equal to zero. The pressure is reduced to the prescribed value of back pressure in region 2 by the expansion fans. However, the flow in region 2 is turned away from the exhaust-jet centerline. To maintain the zero normal-velocity component along the centerline, the flow

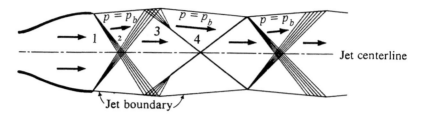

Underexpanded supersonic nozzle

Figure 8.7

must be turned back toward the horizontal. Thus the intersection of the expansion fans centered at the nozzle exit yields another set of expansion waves, just as did the reflection of the expansion fan from a plane wall discussed in Section 7.7. The second expansion, however, produces a pressure in region 3 less than the back pressure, so the expansion waves reflect from the external air as oblique shocks. These compression waves produce a static pressure in region 4 equal to the back pressure, but again turn the flow away from the centerline. The intersection of the oblique shocks from either side of the jet then requires another set of oblique shocks to turn the flow back toward the horizontal, with the shocks reflecting from the external air as expansion waves. The process thus goes through a complete cycle and continues to repeat itself. The flow pattern discussed appears as a series of diamonds, often visible at the exit of high-speed rocket nozzles. Theoretically, the wave pattern should extend to infinity. Actually, however, mixing of the jet with ambient air along the jet boundaries eventually causes the wave pattern to die out.

Flow at the exit of an overexpanded nozzle is shown in Figure 8.8. Since the exit-plane pressure is less than the back pressure, oblique shock waves form at the nozzle exit. The intersection of these shocks at the centerline yields a second set of oblique shocks, which in turn reflect from the ambient air as expansion waves. Thus, except for being out of phase with the wave pattern from the underexpanded nozzle, the jet flow of the overexpanded nozzle exhibits the same characteristics as the underexpanded nozzle.

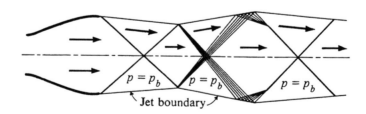

Overexpanded supersonic nozzle

Figure 8.8

Example 8.3

A supersonic nozzle is designed to operate at Mach 2.0. Under a certain operating condition, however, an oblique shock making a 45° angle with the flow direction is observed at the nozzle exit plane. What percent of increase in stagnation pressure would be necessary to eliminate this shock and maintain supersonic flow at the nozzle exit? (See Figure 8.9.)

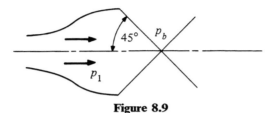

Figure 8.9

Solution

From Appendix A, for $M = 2.0$, $p_1/p_{t_1} = 0.128$. For this case, then, p_1 is equal to $0.128 \times p_{t_1}$. The component of M_1 normal to the oblique wave is $M_1 \sin 45° = 1.41$. From Appendix B, $p_b/p_1 = 2.15$. Therefore, with the oblique shock, the ratio $p_b/p_{t_1} = p_b/p_1 \times p_1/p_{t_1} = 2.15 \times 0.128 = 0.276$. With the shock, p_{t_1} is equal to $(1/0.276)p_b = 3.62p_b$. For supersonic exit flow with no shocks (perfectly expanded case), $p_t = (1/0.128)p_b = 7.81p_b$. Thus, an increase of $(7.81 - 3.62)/3.62 = 116$ percent in stagnation pressure is required.

8.4 Plug Nozzle

The expansion of a high-pressure gas through a converging–diverging nozzle has been shown, at least over a reasonably wide range of operating pressure ratios, to be independent of back pressure. The adjustment to the back pressure in these cases occurs outside the nozzle in the form of expansion waves and oblique shock waves, as shown in the previous section. Hence, the actual expansion of the gases in the nozzle is controlled solely by the nozzle walls. For example, in an overexpanded nozzle, the gases continue to expand to pressures well below the ambient pressure. The thrust developed by a nozzle is dependent on the nozzle exhaust velocity and the pressure at the nozzle exit plane (see Example 1.4). In a jet propulsion device, for example, an exit-plane pressure greater than ambient gives a positive contribution to the thrust of the device, whereas an exit-plane pressure less than ambient gives a negative thrust component. Consider a rocket moving at constant velocity, and write the momentum equation for a fluid control volume as shown in Figure 8.10. If one assumes one-dimensional flow at the nozzle exit,

$$T - (p_e - p_a)A_e = \dot{m}V_e$$
$$T = (p_e - p_a)A_e + \dot{m}V_e \tag{8.1}$$

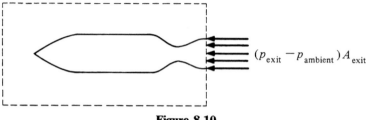

$(p_{exit} - p_{ambient})A_{exit}$

Figure 8.10

When a supersonic nozzle is operating in the under- or overexpanded regimes, with flow in the nozzle independent of back pressure, the exit velocity is unaffected by back pressure. Thus, over this range of back pressures, Eq. (8.1) shows that the greater thrusts are developed in the underexpanded case, and the lesser in the overexpanded case. A plot of thrust versus back pressure for a converging–diverging nozzle is shown in Figure 8.11.

For back pressures greater than the upper limit indicated, a normal shock appears in the diverging portion of the nozzle, the exit velocity becoming subsonic, and this analysis no longer applies.

The plug nozzle is a device that is intended to allow the flow to be directed or controlled by the ambient pressure rather than by the nozzle walls. In this nozzle, the supersonic flow is not confined within solid walls, but is exposed to the ambient pressure. Plug nozzle operation at the design pressure ratio is depicted in Figure 8.12. Figure 8.12a shows the expansion wave pattern and part b shows the streamlines at the nozzle exit. The annular flow first expands internally up to Mach 1 at the throat. The remainder of the expansion to the back pressure occurs with the flow exposed to ambient pressure. Since the throat pressure is considerably higher than the back pressure, a Prandtl Meyer expansion fan is attached to the throat cowling as shown. The plug is designed

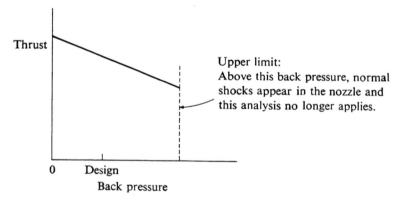

Thrust

Upper limit:
Above this back pressure, normal shocks appear in the nozzle and this analysis no longer applies.

0   Design
    Back pressure

Design refers to a perfectly expanded nozzle.

Figure 8.11

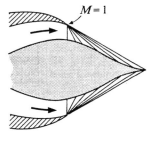
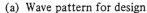

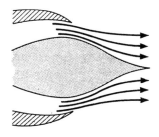

(a) Wave pattern for design   (b) Streamlines for design

Figure 8.12

so that, at the design pressure ratio, the final expansion wave intersects the plug apex. Thus, under this operating condition, the pressure at the plug wall decreases continuously from throat pressure to ambient pressure, just as with the converging–diverging perfectly expanded nozzle.

To produce a maximum axial thrust, it is necessary for the exit flow to have an axial direction. Therefore, the flow at the throat cowling must be directed toward the axis so that the turning produced by the expansion fan will yield axial flow at the plug apex.

For the underexpanded case, the operation of the plug nozzle (Figure 8.13) is similar to that of the converging–diverging nozzle (see Section 5.2). The pressure along the plug is the same as for the design case, just as the static pressure along the converging–diverging nozzle wall is the same as for the perfectly expanded case. With a lower back pressure than that for the design case depicted in Figure 8.12, the flow continues to expand after the apex pressure, yielding a nonaxial jet velocity component, just as with the underexpanded supersonic converging–diverging nozzle.

The major improvement to be derived from the plug nozzle occurs with the overexpanded mode of operation. This is significant, in that a rocket nozzle, for example, accelerating from sea level up to design speed and altitude, must pass through the overexpanded regime. With the ambient pressure greater than the design back pressure, the flow expands along the plug

Underexpanded   Overexpanded

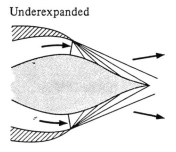

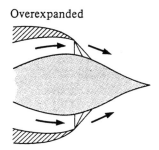

Figure 8.13   **Figure 8.14**

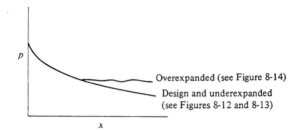

Figure 8.15

only up to the design back pressure. The final wave of the expansion fan centered at the cowling intersects the plug at a point upstream of the apex. As shown in Figure 8.14, the outer boundaries of the exhaust jet are directed inward. Further weak compression and expansion waves occur downstream of the point of impingement of the final wave from the fan; the strength and location of these waves are dependent on the plug contour. Thus the expansion along the plug is controlled by the back pressure, whereas the converging–diverging nozzle expansion is controlled by nozzle geometry. A plot of pressure along the plug surface versus x is given in Figure 8.15. The pressure along the plug surface does not decrease below ambient, so there is not a negative thrust term due to pressure difference. As a result, the plug nozzle provides improved thrust over the converging–diverging nozzle for the overexpanded case (see Figure 8.16).

It would appear desirable to design the plug so as to provide for isentropic expansion flow along its surface. However, it has been shown[1] that a simple cone of up to 40° half-angle can be used without undue loss in thrust performance. Thus the plug nozzle has the further advantage over the converging–diverging nozzle of being short and compact. One major problem with the plug nozzle, however, is that of designing a plug to withstand the high temperatures that exist, for example, in the exhaust gases of a rocket engine. Either cooling of the plug or allowance for its ablation is necessary.

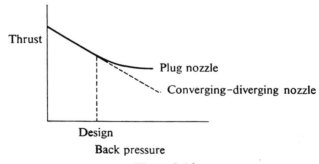

Figure 8.16

Example 8.4

A rocket nozzle is designed to operate with a ratio of chamber pressure to ambient pressure of 50. Compare the performance of a plug nozzle with that of a converging–diverging nozzle for two cases where the nozzle is operating overexpanded; $p_e/p_a = 40$ and $p_e/p_a = 20$. Compare on the basis of thrust coefficient; $C_T = $ thrust$/(p_c \times A_{\text{throat}})$. Assume $\gamma = 1.4$ and in both cases neglect the effect of nonaxial exit velocity components.

Solution

For the design case, with p/p_t equal to 0.02, $M_e = 3.208$ (see Appendix A), and $T_e/T_t = 0.327$,

$$C_T = \frac{\dot{m}_{\text{throat}} V_e}{p_c A_{\text{th}}}$$

$$= \frac{p_{\text{th}}}{RT_{\text{th}}} \frac{A_{\text{th}} V_{\text{th}} V_e}{p_c A_{\text{th}}}$$

$$= \left[\frac{0.5283 p_c}{R(0.8333 T_c)} \right] \left[\frac{\sqrt{1.4(R)0.8333 T_c}}{p_c} \right] 3.21 \sqrt{1.4(R)0.327 T_c}$$

$$= 1.49$$

For the converging–diverging nozzle operating off design,

$$C_T = 1.49 + \frac{A_e(p_e - p_a)}{A_{\text{th}} p_c}$$

where $A/A^* = 5.17$ at $M = 3.21$. For $p_e/p_a = 40$,

$$C_T = 1.49 + 5.17(0.02 - 0.025) = \underline{1.46}$$

For $p_e/p_a = 20$,

$$C_T = 1.49 + 5.17(0.02 - 0.05) = \underline{1.33}$$

Flow in the plug nozzle does not continue to expand below ambient pressure, so there is no pressure term in the expression for thrust.

$$\frac{p_c}{p_a} = 40, \qquad M_e = 3.057, \quad \text{and} \quad \frac{T_e}{T_c} = 0.3485$$

For the plug nozzle,

$$C_T = \frac{\dot{m}_{\text{throat}} V_e}{p_c A_{\text{th}}}$$

$$= \left[\frac{0.5283 p_c}{R(0.8333 T_c)} \right] \left[\frac{A_{\text{th}} \sqrt{1.4R(0.8333 T_c)}}{p_c A_{\text{th}}} \right] 3.057 \sqrt{1.4R(0.3485 T_c)}$$

$$= 1.46 \text{ at } \frac{p_c}{p_a} = 40$$

For $p_c/p_a = 20$,

$$M_e = 2.602 \quad \text{and} \quad \frac{T_e}{T_c} = 0.4248$$

For the plug nozzle,

$$C_T = 1.37 \quad \text{at} \quad \frac{p_c}{p_a} = 20$$

The advantage of using a plug nozzle when operating in the overexpanded regime can be seen from the example.

8.5 Supersonic Airfoils

The design of an airfoil should be such as to provide a lift force normal to the undisturbed flow accompanied by low drag force in the direction of the undisturbed flow. The shape of a wing section to be used in low-speed, incompressible flow is the well-known teardrop or streamlined profile. This shape is predicated on incompressible aerodynamics, where, for example, drag is composed of skin friction on the airfoil surface and pressure or profile drag, due to the effects of flow separation at the rear of the airfoil. In supersonic flow, however, the design must be completely modified owing to the occurrence of shocks. For example, if a streamlined profile with a rounded blunt nose were used in supersonic flow, either an attached shock of relatively high strength would occur at the nose, or, if θ were great enough, a detached shock (Figure 8.17) would take place in front of the airfoil. In both cases, the high pressures after the shock wave produce excessive drag forces on the airfoil. To minimize wave drag, or drag due to the presence of shocks, the supersonic airfoil must have a pointed nose and also be as thin as possible. The ideal case is a flat plate airfoil, possessing zero thickness.

Consider a two-dimensional flat plate at an angle of attack to the approach flow as shown in Figure 8.18. Flow over the upper surface is turned through an

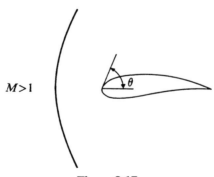

Figure 8.17

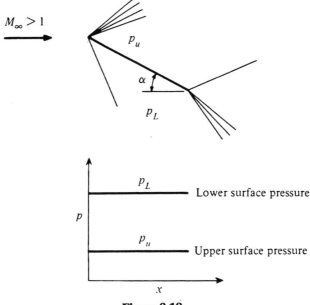

Figure 8.18

expansion fan centered at the nose; flow over the lower surface is compressed through an oblique shock attached to the nose. The difference in pressure between the upper and lower surfaces causes a net upward force, or lift, on the airfoil. This lift is accompanied by a drag force, opposing the motion of the airfoil. Expressions for the lift and drag are given by

$$L = (p_L - p_u)c \cos \alpha$$
$$D = (p_L - p_u)c \sin \alpha$$

where c is chord length.

Example 8.5

Compute the lift and drag coefficients,

$$C_L = \frac{L}{\frac{1}{2}\rho_\infty V_\infty^2 c} \quad \text{and} \quad C_D = \frac{D}{\frac{1}{2}\rho_\infty V_\infty^2 c}$$

for a flat plate airfoil with $M_\infty = 2.5$, $c = 1$ m, and $\alpha = 10°$.

Solution

First, find the static pressure on the lower surface behind the oblique shock. From Figure C.1, for $M_1 = 2.5$ and $\delta = 10°$, $\theta_w = 31.9°$. Therefore, $M_{1_n} = 2.5 \sin 31.9° = 1.32$. From Appendix B, $p_L/p_\infty = 1.87$. For the Prandtl Meyer expansion fan, at $M_\infty = 2.5$, $\nu_\infty = 39.1°$, so that $\nu_u = 39.1° + 10° = 49.1°$. From

Appendix D, $M_u = 2.97$. For isentropic flow, p_t is constant, so

$$\frac{p_u}{p_\infty} = \frac{p_u}{p_t}\frac{p_t}{p_\infty} = \frac{0.0285}{0.0585} = 0.487$$

$$C_L = \frac{L}{\frac{1}{2}\rho_\infty V^2 c} = \frac{L}{\frac{1}{2}\gamma p_\infty M_\infty^2 c} = \frac{(p_L - p_u)\cos\alpha}{\frac{1}{2}(1.4)p_\infty(2.5)^2} = \frac{1.38\cos 10°}{4.375} = \underline{0.311}$$

$$C_D = \frac{D}{\frac{1}{2}\gamma p_\infty M_\infty^2 c} = \underline{0.055}$$

The flat plate is, however, an idealization; structurally, such an airfoil is unsound. Again, for a supersonic airfoil, the requirement exists for a thin airfoil with pointed nose. The curved, symmetrical airfoil represents one possibility. For small angles of attack, oblique shocks are attached to the nose, the stronger shock occurring on the lower surface since the flow turning angle must be greater on this surface (see Figure 8.19). Due to the continuous curvature of the airfoil, flow over the airfoil continually changes direction, and a gradual expansion occurs over the upper and lower surfaces. Expansion waves are produced as shown in Figure 8.19. If the angle of attack becomes too great, or if the nose half-angle is too large, the oblique shocks may detach from the nose, yielding excessive drag.

Another airfoil shape for supersonic flow is the diamond profile, shown in Figure 8.20. Flow over the upper surface is first expanded through a fan

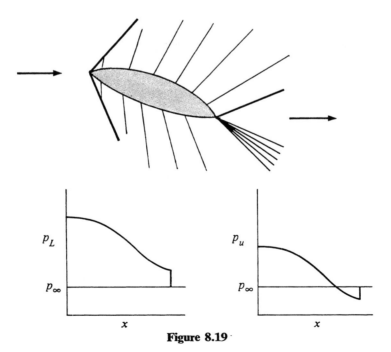

Figure 8.19

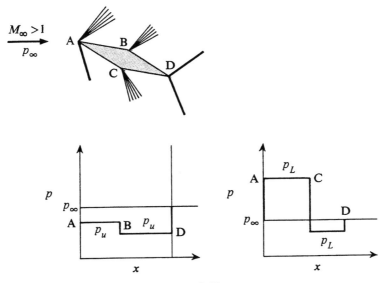

Figure 8.20

centered at A, and then is turned through another expansion fan at B. If the angle of attack is small enough, or if the airfoil is thick enough, flow over the upper surface may first be compressed through an oblique shock attached at A (see Figure 8.21). Flow over the lower surface is turned through an oblique shock at A, and then through an expansion fan at C. As shown by the pressure distribution, higher pressures over the lower surfaces yield a lift force; higher pressures at the front surfaces cause a drag force.

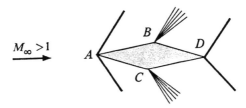

Figure 8.21

Example 8.6

For the two-dimensional airfoil shown in Figure 8.22, compute the lift and drag coefficients for an angle of attack of 10°.

Solution

On the upper surface, supersonic flow is first expanded through a Prandtl Meyer fan. From Appendix D,

$$v_\infty = 49.757$$

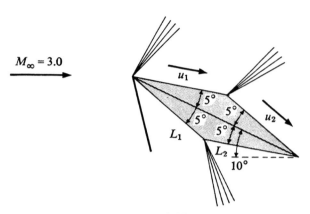

Figure 8.22

so that

$$\nu_{u_1} = 54.757$$

and

$$\nu_{u_2} = 64.757$$

Therefore,

$$M_{u_1} = 3.27$$
$$M_{u_2} = 3.92$$

Now, from Appendix A,

$$\frac{p_{u_1}}{p_\infty} = \frac{p_{u_1}}{p_{t_\infty}} \times \frac{p_{t_\infty}}{p_\infty}$$

$$= \frac{0.01826}{0.02722} = 0.671$$

$$\frac{p_{u_2}}{p_\infty} = \frac{p_{u_2}}{p_{t_\infty}} \frac{p_t}{p_\infty}$$

$$= \frac{0.007332}{0.02722} = 0.269$$

Flow on the lower surface is first compressed through an oblique shock, which turns the flow by an angle of 15°. For $M_\infty = 3.0$ and $\delta = 15°$, $\theta_w = 32.2°$ and $M_{L_1} = 2.25$, or $M_{\infty_n} = 3 \sin 32.2° = 1.60$ so that $p_{L_1}/p_\infty = 2.82$. Flow on the lower surface then expands through a fan for which

$$\nu_{L_1} = 33.018$$
$$\nu_{L_2} = \nu_{L_1} + 10° = 43.018 \quad \text{so that} \quad M_{L_2} = 2.67$$
$$\frac{p_{L_2}}{p_\infty} = \frac{p_{L_2}/p_{L_{2t}}}{p_{L_1}/p_{L_{1t}}} \frac{p_{L_1}}{p_\infty} = \frac{0.04498}{0.08648} 2.82 = 1.47$$

$$\text{lift} = \left(p_{L_1}\frac{c/2}{\cos 5°}\right)\cos 15° + p_{L_2}\frac{c/2}{\cos 5°}\cos 5°$$

$$-p_{u_1}\frac{c/2}{\cos 5°}\cos 5° - p_{u_2}\frac{c/2}{\cos 5°}\cos 15°$$

$$= p_\infty c\left[\frac{2.82}{2}\frac{0.966}{0.996} + \frac{1.47}{2} - \frac{0.671}{2} - \frac{0.269}{2}\frac{0.966}{0.996}\right]$$

$$= p_\infty c\, 1.64$$

or

$$C_L = \frac{L}{\frac{1}{2}\gamma p_\infty M_\infty^2 c} = \underline{0.260}$$

$$\text{drag} = \left(p_{L_1}\frac{c}{\cos 5}\right)\sin 15° + \left(p_{L_2}\frac{c/2}{\cos 5°}\right)\sin 5°$$

$$-p_{u_1}\frac{c/2}{\cos 5°}\sin 5° - p_{u_2}\frac{c/2}{\cos 5°}\sin 15°$$

$$= \frac{c}{2}p_\infty\left[\frac{2.82}{0.996}0.259 + \frac{1.47}{0.996}0.0872 - \frac{0.671}{0.996}0.0872 - \frac{0.269}{0.996}0.259\right]$$

$$= \frac{c}{2}p_\infty[0.733 + 0.129 - 0.0587 - 0.0700]$$

$$= 0.733\frac{cp_\infty}{2}$$

$$C_D = \frac{D}{\frac{1}{2}\gamma p_\infty M_\infty^2 c} = \underline{0.058}$$

It can be seen, then, that subsonic airfoil design cannot be applied to supersonic airfoils. The shape of a supersonic wing is determined mainly by the effect of shock waves and thus requires a thin, pointed profile. The conventional teardrop shape, used for subsonic flow, would have extremely poor lift and drag characteristics in supersonic flow.

REFERENCES

Specific References

1. BERMAN, K., and CRIMP, F.W., "Performance of Plug-Type Rocket Exhaust Nozzles," *ARS Journal*, Vol. 31 (January 1961), pp. 18–23.

General References

2. HILL, P.G., and PETERSON, C.R., *Mechanics and Thermodynamics of Propulsion*, Reading, Mass., Addison-Wesley Publishing Co., Inc., 1965, Chapter 7.
3. ZUCROW, M.J., and HOFFMAN, J.D., *Gas Dynamics*, Vol. 1, New York, John Wiley & Sons, Inc., 1976.

4. BERMAN, K., and NEUFFER, B., "Plug Nozzle Flexibility," *Astronautics*, Vol. 5 (September 1960), pp. 30–31.

5. RAO, G., "Recent Developments in Rocket Nozzle Configurations," *ARS Journal*, Vol. 31 (November 1961), pp. 1488–1494.

6. BERMAN, K., "The Plug Nozzle: A New Approach to Engine Design," *Astronautics*, Vol. 5 (April 1960), pp. 22–24.

7. DONOVAN, A., and LAWRENCE, H. (eds.), *Aerodynamic Components of Aircraft at High Speeds*, Princeton, N.J., Princeton University Press, 1957.

PROBLEMS

1. A supersonic inlet (Figure P8.1) is to be designed to handle air at Mach 1.75 with static pressure and temperature of 50 kPa and 250 K. Determine the diffuser inlet area A_i if the device is to handle 10 kg/s of air.

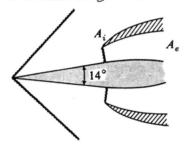

Figure P8.1

2. The diffuser in Problem 1 is to further decelerate flow after the normal shock so that the velocity entering the compressor is not to exceed 25 m/s. Assuming isentropic flow after the shock, determine the area A_e required. For this condition, find the static pressure p_e.

3. A converging nozzle is supplied from a large air reservoir maintained at 600 K and 2 MPa. If the nozzle back pressure is 101 kPa, determine the pressure and Mach number that exist at the nozzle exit plane. Since the nozzle is operating in the underexpanded regime, expansion waves occur at the nozzle exit. Determine the flow direction after the initial expansion fans and the flow Mach number. Also determine the fan angles.

4. A converging–diverging nozzle is designed to provide flow at Mach 2.2. With the nozzle exhausting to a back pressure of 101 kPa, however, and a reservoir pressure of 350 kPa, the nozzle is overexpanded, with oblique shocks appearing at the exit. Determine the flow direction, static pressure, and Mach number in regions R_1, R_2, and R_3 of Figure P8.4.

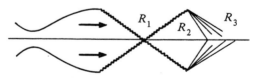

Figure P8.4

5. Determine the flow direction in R_1 and R_3 if the reservoir pressure were increased to 2 MPa.

6. A plug nozzle is designed to produce Mach 2.5 flow in the axial direction at the plug apex. Flow at the throat cowling must therefore be directed toward the axis. Determine the flow direction at the throat cowling to produce axial flow at the apex. Assume $\gamma = 1.4$.

7. Compute the drag coefficient for a symmetric, diamond-shaped airfoil (Figure P8.7) flying at Mach 3.5 at 10 km at zero angle of attack.

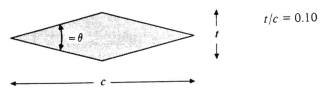

$t/c = 0.10$

Figure P8.7

8. Compute the lift and drag coefficients for the airfoil described in Problem 7 for an angle of attack of 5°.

9. Compare the lift to drag ratio of the diamond airfoil with that of a flat-plate airfoil at an angle of attack of 5°.

10. Consider a flat-plate supersonic airfoil with a flap, as shown in Figure P8.10. For a flap angle of 5°, an angle of attack 10°, and a flight Mach number of 2.2, find the lift and drag coefficients of the airfoil.

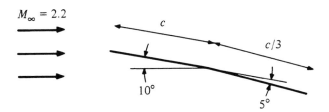

Figure P8.10

11. Compute the lift and drag coefficients for the supersonic, symmetric airfoil shown at an angle of attack of 5° in Figure P8.11.

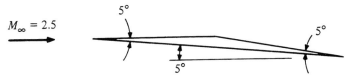

Figure P8.11

12. A jet plane is flying at 150 m above ground level at a Mach number of 2.5, as shown in Figure P8.12. The airfoil is symmetric and diamond shaped, with chord length of 4 m. As the plane passes over, a ground observer hears the "sonic boom" caused by the shock waves. Find the time between the two "booms," one from the shock at the leading edge and one from the shock at the trailing edge. Ambient pressure and temperature are 100 kPa and 20°C.

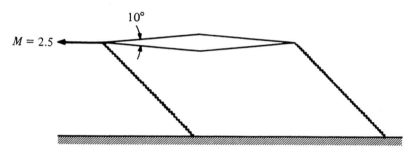

Figure P8.12

9

FLOW WITH FRICTION

9.1 Introduction

In previous chapters, compressible flow in ducts was analyzed for the case in which changes in flow properties were brought about solely by area change. In a real flow situation, however, frictional forces are present and may have a decisive effect on the resultant flow characteristics. Naturally, the inclusion of friction terms in the equations of motion makes the resultant analysis far more complex. For this reason, to study the effect of friction on compressible flow in ducts, certain restrictions will be placed on the flow. The first part of this chapter is concerned with compressible flow with friction in constant-area, insulated ducts, which eliminate the effects of area change and heat addition. In a practical sense, these restrictions limit the applicability of the resultant analysis; however, certain problems such as flow in short ducts can be handled and, furthermore, an insight is provided into the general effects of friction on a compressible flow.

The second part of the chapter will deal with flow with friction in constant-area ducts, in which the fluid temperature is assumed constant. The latter case approximates the flow of a gas through a long, uninsulated pipeline. Thus, these two cases cover a wide range of frictional flows and are consequently of great significance.

The third section of this chapter will present a procedure for handling flows with friction in variable-area channels, such as rocket or turbine nozzles.

The analyses presented in this chapter again will be simplified by assuming one-dimensional, steady flow of a perfect gas with constant specific heats.

9.2 Fanno Line Flow

Consider one-dimensional steady flow of a perfect gas with constant specific heats through a constant-area channel (see Figure 9.1). For the case of adiabatic flow with no external work, termed *Fanno line flow*, the energy equation (see Example 1.6) can be written as

$$h + \frac{V^2}{2} = \text{constant} = h_t \tag{9.1}$$

The continuity equation for constant area steady flow is given by

$$\rho V = \text{constant} \tag{9.2}$$

It is instructive to develop a Ts or hs diagram for this flow. From Eq. (1.22),

$$T\,ds = dh - \frac{dp}{\rho} = du - \frac{p}{\rho^2}\,d\rho$$

so that, for a perfect gas,

$$ds = \frac{du}{T} - R\frac{d\rho}{\rho} \tag{9.3}$$

Assuming constant specific heats, with state 1 a reference state in the flow,

$$s - s_1 = c_v \ln\frac{T}{T_1} - R \ln\frac{\rho}{\rho_1}$$

Using the continuity equation

$$s - s_1 = c_v \ln\frac{T}{T_1} + R \ln\frac{V}{V_1} \tag{9.4}$$

From the energy equation,

$$V = \sqrt{2(h_t - h)}$$
$$= \sqrt{2c_p(T_t - T)} \tag{9.5}$$

so that

$$\frac{s - s_1}{c_v} = \ln\frac{T}{T_1} + \frac{\gamma - 1}{2}\ln\frac{T_t - T}{T_t - T_1} \tag{9.6}$$

$$= \ln T + \frac{\gamma - 1}{2}\ln(T_t - T) + \text{constant} \tag{9.7}$$

$$Q = 0$$
$$A = \text{constant}$$

Figure 9.1

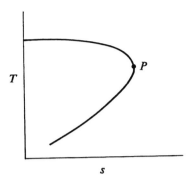

Figure 9.2   Fanno line

Figure 9.2, a plot of Eq. (9.7), depicts the Fanno line on Ts coordinates. Note that for a perfect gas with constant specific heats, $\Delta h = c_p\,\Delta T$, so the Ts and hs diagrams are similar.

The curve shown in Figure 9.2 represents the locus of states that can be obtained under the assumptions of Fanno flow for a fixed mass flow and total enthalpy. Consider the point of tangency P, where $ds/dT = 0$. To determine the characteristics of this point, differentiate Eq. (9.7):

$$\frac{d}{dT}\left(\frac{s}{c_v}\right) = \frac{1}{T} - \frac{\gamma - 1}{2(T_t - T)} = 0 \qquad \text{at } P$$

But

$$c_p(T_t - T) = \frac{V^2}{2}$$

so that

$$\frac{V^2}{c_p} = (\gamma - 1)T$$

or

$$V^2 = \gamma RT$$
$$V^2 = a^2 \qquad \text{for } ds = 0 \tag{9.8}$$

so that $M = 1$ at P.

According to the energy equation, Eq. (9.1), higher velocities are associated with lower enthalpies or temperatures, so the section of the Fanno line on Ts coordinates that lies above P corresponds to subsonic flow, and the section below P to supersonic flow. The Fanno line becomes a most useful tool in describing the variations in properties for this frictional compressible flow. Consider subsonic adiabatic flow in a constant-area tube. The flow is irreversible because of friction, so for this adiabatic case, $ds > 0$. In other words, the entropy increases in the flow direction. Returning to the Ts diagram (see

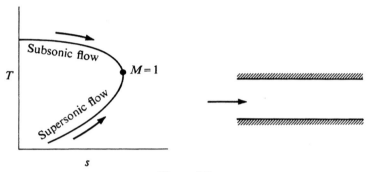

Figure 9.3

Figure 9.3), we see that for a given mass flow the state of the fluid continually moves to the right, corresponding to an entropy rise. Thus, for subsonic flow with friction, the Mach number increases to 1. For supersonic flow, the entropy must again increase, so the flow Mach number here decreases to 1. Suppose now that the duct is long enough for a flow initially subsonic to reach Mach 1, and an additional length is added, as shown in Figure 9.4. The flow Mach number for the given mass flow cannot go past 1 without decreasing entropy; this is impossible from the second law. Hence the additional length brings about a reduction in mass flow; the flow jumps to another Fanno Line (see Figure 9.5). Essentially, the duct is choked due to friction. Corresponding to a given inlet subsonic Mach number, there is a certain maximum duct length L_{max} beyond which a flow reduction occurs. Now suppose the inlet flow is supersonic and the duct length is made greater than L_{max} to produce Mach 1. With the supersonic flow unable to sense changes in duct length occurring ahead of it, the flow adjusts to the additional length by means of a normal shock rather than a flow reduction. The location of the shock in the duct is determined by the back pressure imposed on the duct. (This will be covered in detail later in this chapter.)

From practical considerations, it is necessary to determine the change of properties with actual duct length. This requires the use of the momentum equation, with a term accounting for the frictional forces acting on the control volume. Select a control volume as shown in Figure 9.6. The forces acting on

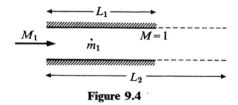

Figure 9.4

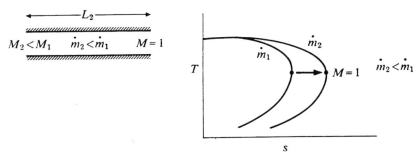

Figure 9.5

the control volume are shown in Figure 9.7, where

$$\tau_f = \text{shear stress due to wall friction}$$
$$A_s = \text{surface area over which the friction acts}$$
$$A = \text{cross-sectional area of duct}$$

Applying the momentum equation for steady flow,

$$\sum F_x = \iint_s V_x(\rho \mathbf{V} \cdot d\mathbf{A})$$

$$pA - (p + dp)A - \tau_f A_s = (\rho AV)(V + dV) - (\rho AV)V \qquad \textbf{(9.9)}$$
$$-A\,dp - \tau_f A_s = \rho AV\,dV$$

Define hydraulic diameter:

$$D_h = \frac{4A}{\text{perimeter}} \qquad \textbf{(9.10)}$$

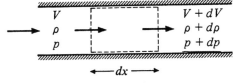

Figure 9.6

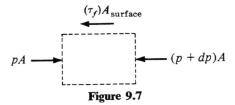

Figure 9.7

For a circular duct,

$$D_h = \frac{4(\pi D^2/4)}{\pi D} = D$$

For a square duct,

$$D_h = \frac{4S^2}{4S} = S \qquad \text{where } S = \text{length of one side}$$

Substituting yields

$$-A\,dp - \tau_f(dx)\frac{4A}{D_h} = \rho AV\,dV \tag{9.11}$$

As was done for incompressible flow[1], define a friction coefficient $f = 4\tau_f/(\frac{1}{2}\rho V^2)$, where f is dependent on the flow Reynolds number and wall roughness. Now,

$$-dp - \frac{1}{2}\rho V^2 f\frac{dx}{D} = \rho V\,dV \tag{9.12}$$

It is desirable to integrate Eq. (9.12) to obtain, for example, an expression for Mach number and pressure change over a given duct length. Divide by p:

$$\frac{dp}{p} + \frac{1}{2}\gamma M^2 f\frac{dx}{D} + \gamma M^2 \frac{dV}{V} = 0 \tag{9.13}$$

To get M in terms of x, eliminate dV/V and dp/p. From continuity, $d\rho/\rho + dV/V = 0$. Also, using the perfect gas law and the definition of Mach number,

$$\frac{dp}{p} = \frac{d\rho}{\rho} + \frac{dT}{T}$$

$$\frac{dM}{M} = \frac{dV}{V} - \frac{1}{2}\frac{dT}{T}$$

Substituting into Eq. (9.13), we obtain

$$\frac{dT}{T} - \frac{1}{2}\frac{dT}{T} - \frac{dM}{M} + \frac{1}{2}\gamma M^2 f\frac{dx}{D} + \gamma M^2\frac{dM}{M} + \frac{1}{2}\gamma M^2\frac{dT}{T} = 0$$

But, for this adiabatic flow, $T_t = $ constant so

$$T\left(1 + \frac{\gamma - 1}{2}M^2\right) = \text{constant}$$

or

$$\frac{dT}{T} + \frac{(\gamma - 1)M^2\dfrac{dM}{M}}{1 + \dfrac{\gamma - 1}{2}M^2} = 0 \tag{9.14}$$

Substituting yields

$$\frac{(\gamma - 1)M^2 \dfrac{dM}{M}}{1 + \dfrac{\gamma - 1}{2}M^2}\left[\frac{1}{2} + \frac{\gamma M^2}{2}\right] - (\gamma M^2 - 1)\frac{dM}{M} = \frac{1}{2}\gamma M^2 f \frac{dx}{D}$$

Combining terms, we obtain

$$\frac{f\,dx}{D} = \frac{2dM}{M}\left[\frac{1 - M^2}{\left(1 + \dfrac{\gamma - 1}{2}M^2\right)\gamma M^2}\right] \tag{9.15}$$

This equation must then be integrated to determine M as a function of duct length. It is convenient for Fanno flow to choose, as a reference point, $M = 1$ and $L = L_{max}$. For the lower limit of integration, select $x = 0$ at $M = M$. Thus,

$$\frac{fL_{max}}{D} = \int_{M=M}^{1}\frac{2(1 - M^2)\dfrac{dM}{M}}{\left(1 + \dfrac{\gamma - 1}{2}M^2\right)\gamma M^2} \tag{9.16}$$

Values of the integral have been tabulated in Appendix E as fL_{max}/D versus M.

Likewise, if it is required to find p versus M or $f\,dx/D$, dV/V can be eliminated from Eq. (9.13):

$$\frac{dp}{p} + \frac{1}{2}\gamma M^2 f\frac{dx}{D} + \gamma M^2 \frac{dM}{M} - \frac{1}{2}\frac{\gamma M^2(\gamma - 1)M^2\dfrac{dM}{M}}{1 + \dfrac{\gamma - 1}{2}M^2} = 0$$

Substituting for $f\,dx/D$, we obtain

$$\frac{dp}{p} + \frac{1 - M^2}{1 + \dfrac{\gamma - 1}{2}M^2}\frac{dM}{M} + \gamma M^2\frac{dM}{M} - \frac{\dfrac{1}{2}\gamma M^2(\gamma - 1)M^2\dfrac{dM}{M}}{1 + \dfrac{\gamma - 1}{2}M^2} = 0$$

Collecting terms yields

$$\frac{dp}{p} = -\frac{dM}{M}\left[\frac{1 + (\gamma - 1)M^2}{1 + \dfrac{\gamma - 1}{2}M^2}\right] \tag{9.17}$$

Integrating between the limits p and p^*, where $*$ denotes the value of a property at Mach 1, gives

$$\int_{p}^{p^*}\frac{dp}{p} = -\int_{M}^{1}\frac{dM}{M}\left[\frac{1 + (\gamma - 1)M^2}{1 + \dfrac{\gamma - 1}{2}M^2}\right] \tag{9.18}$$

Thus a relationship is obtained for p/p^* versus M. Tabulated values are provided in Appendix E. With pressure and the term fL_{max}/D now calculated as functions of M, the pressure variation in a duct can be found in terms of duct length. Likewise, from Eq. (9.14), T/T^* can be found in terms of M, as can V/V^* from Eq. (9.13). With p and M, the variation of stagnation pressure p_t is determinable.

Some qualitative conclusions can be drawn concerning the variation of these properties for Fanno flow. For subsonic flow, as shown by the Ts diagram, static temperature decreases while Mach number increases to 1. For supersonic flow, the opposite occurs; that is, temperature increases while Mach number decreases to 1. For subsonic flow, with dM positive, Eq. (9.17) shows that the static pressure decreases. For supersonic flow, however, the static pressure increases. For adiabatic flow with friction, whether the Mach number be subsonic or supersonic, the stagnation pressure decreases, since for both cases there is a loss of available energy (see Section 4.3).

Example 9.1

Flow enters a constant-area, insulated duct (Figure 9.8) with a Mach number of 0.60, static pressure of 150 kPa, and static temperature of 300 K. Assume a duct length of 45 cm, duct diameter of 3 cm, and a friction coefficient of 0.02. Determine the Mach number, static pressure, and static temperature at the duct outlet.

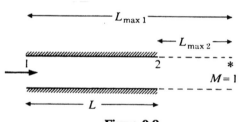

Figure 9.8

Solution

From Appendix E, at $M_1 = 0.60$, $(fL_{max}/D)_1 = 0.49081$, where L_{max} is the duct length required to reach Mach 1. The actual fL/D for the duct, however, is 0.30. Therefore, as can be seen from Figure 9.8,

$$\left(\frac{fL_{max}}{D}\right)_2 = \left(\frac{fL_{max}}{D}\right)_1 - \frac{fL}{D}$$

or

$$\left(\frac{fL_{max}}{D}\right)_2 = 0.49081 - 0.30 = 0.19081$$

Thus, from Appendix E, $M_2 = 0.709$. Also,

$$\frac{p_2}{p_1} = \frac{p_2/p^*}{p_1/p^*} = \frac{1.4728}{1.7634} = 0.835$$

$$\frac{T_2}{T_1} = \frac{T_2/T^*}{T_1/T^*} = \frac{1.0904}{1.1194} = 0.974$$

or $p_2 = 125.3$ kPa and $T_2 = 292.2$ K.

9.3 Flow Through a Nozzle and Constant-Area Duct in Series

Very often a situation occurs where a duct is fed by a nozzle, with the back pressure and nozzle stagnation pressure the known quantities. Consider, for example, a duct supplied by a converging nozzle, with flow provided by a reservoir at pressure p_r (see Figure 9.9). Assuming isentropic nozzle flow, with Fanno flow in the duct, the system pressure distribution, p versus x, can be determined for various back pressures (p_r is maintained constant). As p_b is lowered below p_r, curves such as (l) and (m) are obtained, with pressure decreasing in both nozzle and duct. Finally, when the back pressure is decreased to that of curve (n), Mach number 1 occurs at the duct exit (note that the Mach number at the nozzle exit is still less than 1). Further decreases in back pressure cannot be sensed by the reservoir; for all back pressures below that of curve (n) the mass flow rate remains the same as that of curve (n); $\dot{m}$ is

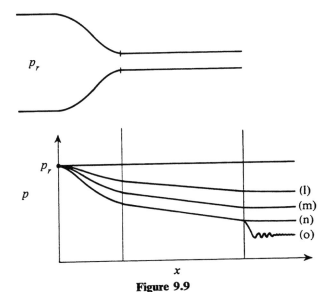

Figure 9.9

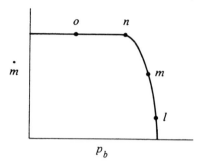

Figure 9.10

plotted versus p_b in Figure 9.10. The system here is choked by the duct, not the converging nozzle. The maximum mass flow that can be passed by this system is less for the same reservoir pressure than that for a converging nozzle with no duct.

Example 9.2

A constant-area duct, 20 cm in length by 2 cm in diameter, is connected to a reservoir through a converging nozzle, as shown in Figure 9.11a. For a reservoir pressure and temperature of 1 MPa and 500 K, determine the maximum air flow rate in kilograms per second through the system and the range of back pressures over which this flow is realized. Repeat these calculations for a converging nozzle with no duct. Assume f equal to 0.032.

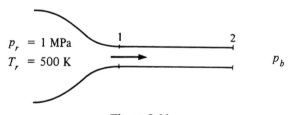

p_r = 1 MPa
T_r = 500 K

p_b

Figure 9.11a

Solution

For maximum mass flow through the nozzle-duct system, M_2 is equal to 1. For this condition, the actual fL/D of the duct becomes equal to $(fL_{max}/D)_1$, so that, from Appendix E, $M_1 = 0.652$ at $(fL_{max}/D)_1 = 0.32$. For isentropic nozzle flow, from Appendix A, at $M_1 = 0.652$, $(p/p_t)_1 = 0.7515$ and $(T/T_t)_1 = 0.9217$, or, for

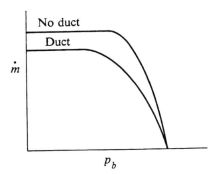

Figure 9.11b

a choked duct, p_1 = 751.5 kPa and T_1 = 460.9 K. Now,

$$\dot{m}_{max} = \left(\frac{p}{RT}\right)_1 A_1 M_1 \sqrt{\gamma R T_1}$$

$$= \left[\frac{(751.5 \text{ kN/m}^2)}{(0.2870 \text{ kJ/kg} \cdot \text{K})(460.9 \text{ K})}\right]\left[\frac{\pi}{4}(4 \times 10^{-4})\text{m}^2\right]$$

$$\times [0.652\sqrt{1.4(287 \text{ J/kg} \cdot \text{K})460.9 \text{ K}}]$$

$$= (5.681 \text{ kg/m}^3)(0.0003142 \text{ m}^2)(280.6 \text{ m/s})$$

$$= \underline{0.5009 \text{ kg/s}}$$

Also, p_2/p_1 = p^*/p_1 = 1/1.6130, or p^* = 465.9 kPa, so the system is choked over the range of back pressures from 0 to 465.9 kPa. If the duct were to be removed, choking would occur with Mach 1 at the nozzle exit. For this condition, p_1 = 0.5283(1000 kPA) = 528.3 kPa and T_1 = 0.8333(500 K) = 416.7 K, so the maximum mass flow is 0.5679 kg/s. For this case, the system is choked over the back pressure range from 0 to 528.3 kPa. Results are shown in Figure 9.11b.

When a duct is connected to a reservoir through a converging–diverging nozzle, the situation becomes somewhat more complex. Consider first the case of subsonic flow in both nozzle and duct. A typical pressure distribution is shown in Figure 9.12. Depending on the duct length, the minimum pressure point, or point of maximum Mach number, can occur at the nozzle throat or duct exit. If the duct is long enough (see dashed curve), the system reaches Mach 1 first at the duct exit; in this case, the nozzle is not choked. Once Mach 1 is reached, no further increase in mass flow rate can occur by reduction of the system back pressure. Supersonic flow in this system is impossible with the converging–diverging nozzle unchoked. Generally, however, the duct length required to cause choking is very long. For this reason, the more important case is that in which the system is choked at the nozzle throat, and supersonic flow can occur in the duct (see Figure 9.13).

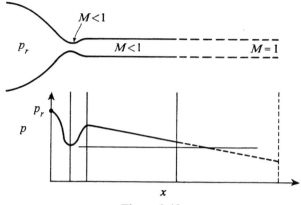

Figure 9.12

With supersonic flow at the nozzle exit, there is the possibility of shocks in the duct. Note, however, that once the back pressure is just low enough to produce Mach 1 at the nozzle throat, the system is choked, with no further increase in mass flow possible. Unlike the case previously discussed, in which a converging nozzle exhausted into the duct, and mass flow was affected by duct length, here, once the throat velocity reaches the velocity of sound, the mass flow rate is unaffected by duct length. Now the system is choked by the nozzle, not the duct.

Let us consider the flow pattern obtained with supersonic flow at the duct inlet. First, suppose the duct length L is less than L_{max}, corresponding to the given inlet supersonic Mach number M_1. The change in flow pattern is to be described as the back pressure is increased from 0 kPa. A back pressure of 0 kPa, or a very low back pressure implies the existence of expansion waves at the duct exit. This means that the exit Mach number must be either supersonic or 1. Since L is less than L_{max}, supersonic flow occurs at the duct exit, with p_{exit} greater than p_b (see curve a, Figure 9.14). When p_b is raised to a value corresponding to curve b, the exit-plane pressure is equal to the back pressure. A further increase in back pressure yields oblique shock waves at the duct exit (curve c), until eventually, for a back pressure equal to that of curve d, a normal shock stands at the duct exit. It can be seen that the flow described is exactly the same as that obtained at the exit of a converging–diverging nozzle

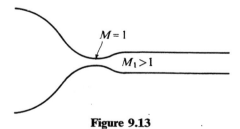

Figure 9.13

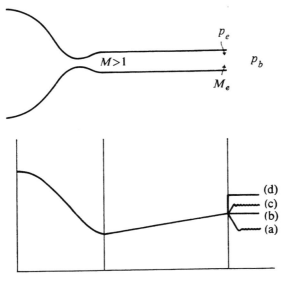

Figure 9.14

(see Section 5.2). Increases in back pressure over that of curve d cause the shock to move into the duct. For a high enough back pressure, the shock moves into the nozzle, thus eliminating supersonic flow in the duct.

Example 9.3

A converging–diverging nozzle, with area ratio of 2 to 1, is supplied by a reservoir containing air at 500 kPa. The nozzle exhausts into a constant-area duct of length-to-diameter ratio of 10 and friction coefficient f of 0.02. Determine the range of system back pressures over which a normal shock appears in the duct. Assume isentropic flow in the nozzle and Fanno flow in the duct.

Solution

From Appendix A, at $A/A^* = 2.0$, $M = 2.197$. From Appendix E, $fL_{max}/D = 0.3601$. For the duct under consideration, $fL/D = 0.20$ so that $L < L_{max}$. Calculations must be made for two limiting cases, one with shock at the duct inlet (Figure 9.15), the other with shock at the duct outlet. For case a, $M_1 = 2.197$. From Appendix B, for a normal shock, $M_2 = 0.5475$. From Appendix E, $(fL_{max}/D)_2 = 0.7427$. Thus

$$\left(\frac{fL_{max}}{D}\right)_0 = 0.7427 - \frac{fL}{D} = 0.5427$$

so that $M_0 = 0.5875$. For this case,

$$p_b = p_0 = \frac{p_0}{p^*}\frac{p^*}{p_2}\frac{p_2}{p_1}\frac{p_1}{p_r}500 = (1.803)\left(\frac{1}{1.944}\right)(5.465)(0.0940)500$$

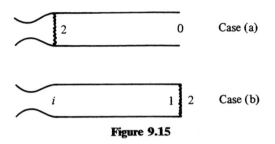

Figure 9.15

or

$$p_b = 238.2 \text{ kPa}$$

For case b, $M_i = 2.197$ so that $(fL_{max}/D)_1 = 0.3601 - 0.20 = 0.1601$, or from Appendix E, $M_1 = 1.566$. From Appendix B, for a normal shock, $p_2/p_1 = 2.695$. For this case,

$$p_b = p_2 = \frac{p_2}{p_1}\frac{p_1}{p^*}\frac{p^*}{p_i}\frac{p_i}{p_r} 100 = (2.695)\left(\frac{0.5730}{0.3557}\right)(0.0940)500 = 204.0 \text{ kPa}$$

Thus a shock will appear in the duct over the back pressure range from 204.0 to 238.2 kPa.

For L greater than L_{max}, the duct length is greater than that required to reach Mach 1 for supersonic duct flow. For very low back pressures, however, there must be sonic or supersonic flow at the duct exit to provide expansion waves outside the duct. As was discussed previously, whereas for subsonic duct

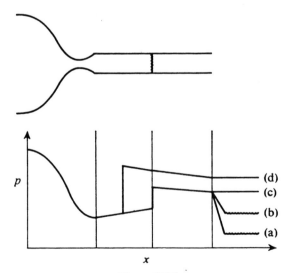

Figure 9.16

flow a reduction of flow occurs for $L > L_{max}$, for supersonic flow, no such flow reduction can occur; instead, a normal shock appears in the duct. The resultant flow is shown in Figure 9.16. For a back pressure of 0 kPa and for very low back pressures, it is evident that the back pressure is less than the exit-plane pressure, so expansion waves must occur at the duct exit, with the exit plane Mach number equal to 1 (flow after the shock cannot reach supersonic velocities without violating the second law of thermodynamics). For curves a and b, therefore, a normal shock occurs in the duct, with sonic flow at the duct exit and expansion waves outside the duct. For curve c, the exit plane pressure is equal to the back pressure. As the back pressure is raised above curve c, the normal shock moves toward the duct inlet, with the exit Mach number subsonic and the back pressure equal to the exit-plane pressure. Again, for high enough back pressures, the shock moves into the nozzle, eliminating supersonic flow in the duct.

Example 9.4

A converging–diverging nozzle, with an area ratio of 2 to 1, is supplied by a reservoir containing air at 500 kPa. The nozzle exhausts into a constant-area duct of length-to-diameter ratio of 25 and friction coefficient of 0.02. Determine the normal shock location and exit-plane pressure for a back pressure of 0 kPa. Refer to Figure 9.17.

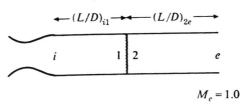

$$\longleftarrow (L/D)_{i1} \longrightarrow \longleftarrow (L/D)_{2e} \longrightarrow$$

$$i \qquad 1 | 2 \qquad e$$

$$M_c = 1.0$$

Figure 9.17

Solution

For this case, $M_i = 2.197$ and $M_e = 1.0$, with $fL/D = 0.02(25) = 0.50$ which is greater than fL_{max}/D at $M = 2.197$. The solution involves trial and error, in that a shock location, or L/D_{i1}, must be tried so that M_e can be made equal to 1. For example, assume $L/D_{i1} = 7.5$, so that $fL/D_{i1} = 0.15$. This yields a value for fL_{max}/D_1 of $0.3601 - 0.15 = 0.2101$. From Appendix E, $M_1 = 1.707$. For a normal shock at $M_1 = 1.707$, $M_2 = 0.639$. From Appendix E, $(fL_{max}/D)_2 = 0.3564$. But, since $M_e = 1$, $(fL/D)_{2-e}$ should equal $(fL_{max}/D)_2$. $(fL_{max}/D)_{2-e} = (0.02)17.5 = 0.35$ for the assumed shock location. A more exact answer yields $L/D_{i1} = 7.7$, with $M_1 = 1.695$ and $M_2 = 0.642$. For this case,

$$p_e = p^* = \frac{p^*}{p_2}\frac{p_2}{p_1}\frac{p_1}{p^*}\frac{p^*}{p_i}\frac{p_i}{p_r} 500$$

$$= \left(\frac{1}{1.640}\right)(3.185)\left(\frac{0.5150}{0.3557}\right)(0.0940)(500) = \underline{132.2 \text{ kPa}}$$

9.4 Isothermal Flow

During gas flow in long, uninsulated ducts, the gas is able to achieve temperature equilibrium with the surroundings. In other words, the flow is isothermal.
This flow possesses different characteristics from Fanno flow, involving heat
transfer as well as friction. Consider a control volume as shown in Figure 9.18,
with isothermal flow in a constant-area duct. The continuity, momentum, and
energy equations are

$$\frac{d\rho}{\rho} + \frac{dV}{V} = 0 \tag{9.19}$$

$$dp + \rho V\,dV + \frac{f\,dx}{D}\rho V^2 = 0 \tag{9.20}$$

or

$$\frac{dp}{p} + \frac{\gamma M^2}{2}\frac{f\,dx}{D} + \gamma M^2\frac{dV}{V} = 0 \tag{9.21}$$

$$d\dot{q} = \dot{m}c_p\,dT_t \tag{9.22}$$

or

$$dq = c_p\,dT_t \tag{9.23}$$

Also, the perfect gas law is

$$\frac{dp}{p} = \frac{d\rho}{\rho} + \frac{dT}{T} = \frac{d\rho}{\rho} \quad \text{for} \quad \frac{dT}{T} = 0 \tag{9.24}$$

and the definition of Mach number is

$$M = \frac{V}{a}$$

or

$$\frac{dM}{M} = \frac{dV}{V} \quad \text{for} \quad a = \text{constant} \tag{9.25}$$

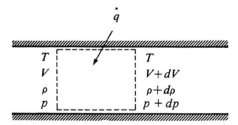

Figure 9.18

Combining Eqs. (9.19), (9.24), and (9.25) yields

$$\frac{dp}{p} = -\frac{dM}{M} \tag{9.26}$$

Substituting Eq. (9.26) into Eq. (9.21), we get

$$\frac{dM}{M} = \frac{f\,dx}{D}\frac{(\gamma M^2/2)}{1 - \gamma M^2} \tag{9.27}$$

Equation (9.27) shows that the critical Mach number for isothermal flow is not Mach 1, but $M = 1/\sqrt{\gamma}$. For example, for $M < 1/\sqrt{\gamma}$, Eq. (9.27) indicates that M increases with x, whereas for $M > 1/\sqrt{\gamma}$, M decreases with x.

To determine the heat transfer necessary to maintain isothermal flow, refer to Eq. (9.23). Since

$$T_t = T\left(1 + \frac{\gamma - 1}{2}M^2\right)$$

$$\frac{dT_t}{T_t} = \frac{(\gamma - 1)M\,dM}{1 + \frac{\gamma - 1}{2}M^2}$$

for isothermal flow with $dT/T = 0$. Substituting for dM from Eq. (9.27), we obtain

$$dq = c_p\,dT_t$$

$$= \frac{c_p T_t(\gamma - 1)M^4\dfrac{\gamma}{2}}{\left(1 + \dfrac{\gamma - 1}{2}M^2\right)(1 - \gamma M^2)} \cdot \frac{f\,dx}{D} \tag{9.28}$$

As $M \to 1/\sqrt{\gamma}$ it can be seen that $q \to \infty$, indicating that an infinite rate of heat transfer is necessary to maintain the isothermal flow. It is apparent, then, that it is physically impossible to isothermally accelerate flow in a constant-area duct from low Mach number up to $M = 1/\sqrt{\gamma}$. The assumption of isothermal flow is valid, however, and can be realized physically for low Mach number flow. For such flow, it is desirable to obtain an expression for pressure drop and Mach number variation as a function of x. Integrating Eq. (9.27) yields

$$\int_{M_1}^{M_2}\left(\frac{dM}{M}\right)\frac{1 - \gamma M^2}{(\gamma M^2/2)} = \frac{fL}{D}$$

$$\frac{fL}{D} = \frac{1 - \gamma M_1^2}{\gamma M_1^2} - \frac{1 - \gamma M_2^2}{\gamma M_2^2} + \ln\frac{M_1^2}{M_2^2} \tag{9.29}$$

Also, $p_2/p_2 = M_1/M_2$ from Eq. (9.26), so pressure ratio can be determined from Mach number ratio.

Example 9.5

Methane enters a gas pipeline with a velocity of 15 m/s and static pressure of 500 kPa. The pipeline is 200 m long and 8 cm in diameter, with the gas temperature essentially constant at 288 K over the pipe length. Determine the pressure drop. For methane, $\gamma = 1.32$ and molecular mass $= 16$. Assume $f = 0.025$.

Solution

The inlet Mach number is

$$M_1 = \frac{15}{\sqrt{\gamma RT}} = \frac{15}{\sqrt{1.32\left(\frac{8314}{16}\right)288}} = 0.0337$$

and

$$\frac{fL}{D} = 0.025\frac{200}{0.08} = 62.5$$

From Eq. (9.29),

$$62.5 = 666 - \frac{1 - \gamma M_2^2}{M_2^2} + \ln\frac{(0.0337)^2}{M_2^2}$$

As a first trial, assume the term $\ln (0.0337)^2/M_2^2$ is small, so that $(1 - \gamma M_2^2)/M_2^2 = 603.5$. Solving yields $M_2 = 0.0407$. The validity of the assumption, in this case, can be realized.

$$\frac{p_2}{p_1} = \frac{M_1}{M_2} = \frac{0.0337}{0.0407} = 0.828$$

so

$$p_1 - p_2 = 500\left(1 - \frac{p_2}{p_1}\right) = \underline{86 \text{ kPa}}$$

9.5 Adiabatic Flow with Friction and Area Change

If the cross-sectional area is allowed to vary, for adiabatic flow with friction the analysis becomes considerably more complex than Fanno flow. Consider a control volume as shown in Figure 9.19.

The pressure on the side walls of the control volume is assumed to be $p + dp/2$ (see Section 3.2). The continuity equation is

$$\frac{d\rho}{\rho} + \frac{dA}{A} + \frac{dV}{V} = 0 \tag{9.30}$$

The momentum equation in the x direction is

$$\sum F_x = \iint_s V_x(\rho \mathbf{V} \cdot d\mathbf{A}) \quad \text{for steady flow}$$

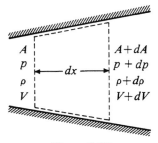

Figure 9.19

or

$$pA + \left(p + \frac{dp}{2}\right) dA - (p + dp)(A + dA) - \tau_f A_s = \rho A \dot{V} \, dV \quad (9.31)$$

Substituting for τ_f from Eqs. (9.11) and (9.12) and simplifying and dropping second-order terms yields

$$-A \, dp - \frac{1}{2} \rho V^2 \frac{f A_s}{4} = \rho A V \, dV$$

For a circular duct,

$$dp + \frac{1}{2} \rho V^2 \frac{f \, dx}{D} + \rho V \, dV = 0$$

Following the same analysis as was carried out for Fanno flow (see Section 9.2), divide by p and substitute for dp/p from the perfect gas law:

$$\frac{d\rho}{\rho} + \frac{dT}{T} + \gamma M^2 \frac{dV}{V} + \frac{1}{2} \gamma M^2 \frac{f \, dx}{D} = 0$$

Eliminate $d\rho/\rho$ from continuity:

$$-\frac{dA}{A} - \frac{dV}{V} + \frac{dT}{T} + \gamma M^2 \frac{dV}{V} + \frac{1}{2} \gamma M^2 \frac{f \, dx}{D} = 0 \quad (9.32)$$

or, since

$$\frac{dV}{V} = \frac{dM}{M} + \frac{1}{2} \frac{dT}{T}$$

by definition of Mach number, then

$$-\frac{dA}{A} - \frac{dM}{M} + \frac{1}{2} \frac{dT}{T} + \gamma M^2 \frac{dM}{M} + \frac{\gamma M^2}{2} \frac{dT}{T} + \frac{1}{2} \gamma M^2 \frac{f \, dx}{D} = 0$$

Substituting for dT/T from Eq. (9.14) for this adiabatic flow, we obtain

$$\frac{1}{2} \gamma M^2 \frac{f \, dx}{D} = \frac{dA}{A} + \frac{dM}{M} \frac{1 - M^2}{1 + \frac{\gamma - 1}{2} M^2} \quad (9.33)$$

or

$$\frac{1}{2}\gamma M^2 \frac{f}{D} = \frac{1}{A}\frac{dA}{dx} + \frac{dM}{dx}\frac{(1 - M^2)/M}{1 + \frac{\gamma - 1}{2}M^2} \qquad (9.34)$$

The direct integration of this equation, except for certain specialized cases, is not possible; instead, some sort of numerical integration is required. However, let us discuss some special cases. The case $dA/dx = 0$, or that of constant-area flow, has been discussed in detail; this is Fanno flow. Another possibility is to let $dM/dx = 0$, eliminating the right-hand term in Eq. (9.34). The resultant equation for a duct of circular cross section is

$$\frac{1}{2}\gamma M^2 \frac{f}{D} = \frac{1}{A}\frac{dA}{dx} = \frac{2}{D}\frac{dD}{dx} \qquad (9.35)$$

Thus, if it is required to design a duct to provide constant Mach number flow with friction, Eq. (9.35) gives a relationship for the diameter variation with x. Solving yields

$$\frac{dD}{dx} = \frac{\gamma M^2}{4}f \qquad (9.36)$$

(see Figure 9.20).

A case of great interest is nozzle flow with friction. Although direct integration of Eq. (9.34) is not possible, several statements based on this equation can be made concerning nozzle flow. It is to be remembered that, for isentropic flow ($f = 0$) in a converging–diverging nozzle, $M = 1$ corresponds to $dA/dx = 0$, or to the minimum area. For flow with friction, from Eq. (9.34), Mach 1 occurs for dA/dx positive, or in the diverging part of the nozzle (see Figure 9.21). At the throat, $dA/dx = 0$, but, for $M < 1$, dM/dx is positive. With the throat area no longer suitable as a reference area, and with direct integration of Eq. (9.34) not possible, an exact solution of nozzle flow with friction becomes extremely difficult.

However, the engineer must still obtain at least an approximate solution to

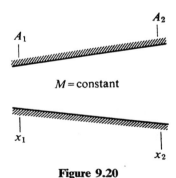

Figure 9.20

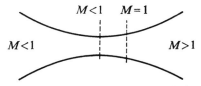

Figure 9.21   Nozzle flow with friction

the problem of nozzle flow with friction. One procedure is to define a nozzle velocity coefficient C_V, with C_V equal to the ratio of actual nozzle exit velocity to the velocity that could be obtained in an isentropic expansion over the same overall pressure ratio. Alternatively, a coefficient C_h can be defined, relating the actual enthalpy drop in the nozzle to the enthalpy drop for an isentropic nozzle:

$$C_h = \frac{\Delta h_{\text{actual}}}{\Delta h_{\text{isentropic}}}$$

If the nozzle inlet velocity is negligible,

$$\Delta h = \frac{V_{\text{exit}}^2}{2} \quad \text{for adiabatic flow}$$

so that

$$C_h = \left(\frac{V_{\text{exit actual}}}{V_{\text{exit isentropic}}}\right)^2$$

or

$$C_h = C_V^2$$

In a well-designed nozzle, operating at or close to design pressure ratio, values of C_V obtained are greater than 0.95, for Reynolds numbers based on nozzle diameter greater than 10^6 (in the turbulent flow regime). With a favorable pressure gradient associated with nozzle expansion flows, the only losses are due to the small amount of skin friction at the nozzle walls.

Example 9.6

Flow is expanded in a jet engine exhaust nozzle from a stagnation pressure and temperature of 300 kPa and 1000 K to an exhaust pressure of 100 kPa. For a velocity coefficient of 0.95, determine the actual nozzle exit velocity. Assume $\gamma = 1.40$ and $R = 250$ J/kg·K.

Solution

First calculate the exit velocity for an isentropic expansion. For $p/p_t = 0.3333$, $M = 1.358$ and $T/T_t = 0.7306$ (from Appendix A). Therefore, $V_{e_{\text{isen}}} = 1.358\sqrt{1.4(250)730.6} = 686.7$ m/s and $V_{e_{\text{actual}}} = 0.95(686.7) = \underline{652.4 \text{ m/s}}$.

9.6 Summary

Chapter 9 has presented a study of compressible flows with friction. In each case, the analysis was started by referring to the continuity, momentum, and energy equations, with a term now included in the momentum equation to account for frictional effects acting on the control volume. Three basic flows were considered: first, adiabatic, constant-area or Fanno flow; second, constant-area, constant-temperature or isothermal flow; and, finally, varying-area, adiabatic flow. The phenomenon of choking was shown again to occur for frictional flows, even without area change.

The student should now appreciate the simplicity of the approach taken to analyze compressible flows. Whether for isentropic flow, flow with friction, or, as we shall see in later chapters, flows with heat addition and applied electrical and magnetic fields, emphasis is on the use of the continuity, momentum, and energy equations and the organization of the resultant solutions into corresponding tables using Mach number as the independent variable.

REFERENCES

Special References

1. JOHN, J.E.A., AND HABERMAN, W.L., *Introduction to Fluid Mechanics*, 2nd ed., Englewood Cliffs, N.J., Prentice-Hall, Inc., 1980, p. 157.

General References

2. FOX, R.W., AND MCDONALD, A.T., *Introduction to Fluid Mechanics*, 2nd ed., New York, John Wiley & Sons, Inc., 1978.
3. SHAPIRO, A., *The Dynamics and Thermodynamics of Compressible Fluid Flow*, Vol. 1, New York, Ronald Press, 1953.

PROBLEMS

1. Air flow enters a constant-area insulated duct with a Mach number of 0.20. For a duct diameter of 1 cm and friction coefficient of 0.02, determine the duct length required to reach Mach 0.60. Determine the length required to attain Mach 1. Now suppose an additional 75 cm of duct length is added. If the same initial stagnation conditions are maintained, determine the reduction in flow rate that would occur.

2. Air enters a constant-area insulated duct with a Mach number of 0.35, a stagnation pressure of 105 kPa, and stagnation temperature of 300 K. For a duct length of 50 cm, duct diameter of 1 cm, and friction coefficient of 0.022, determine the air force on the duct wall.

3. Hydrogen enters a constant-area insulated duct with a velocity of 2600 m/s, static temperature of 300 K, and stagnation pressure of 520 kPa. The duct is 2 cm in

diameter, and 10 cm long. For a friction coefficient of 0.02, determine the change of static pressure and temperature in the duct and the hydrogen exit velocity.

4. A constant-area duct, 25 cm in length by 1.3 cm in diameter, is connected to an air reservoir through a converging nozzle, as shown in Figure P9.4. For a constant reservoir pressure of 1 MPa and constant reservoir temperature of 600 K, determine the flow rate through the duct for a back pressure of 101 kPa.

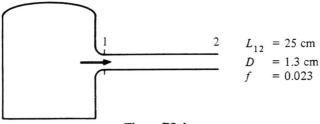

$$L_{12} = 25 \text{ cm}$$
$$D = 1.3 \text{ cm}$$
$$f = 0.023$$

Figure P9.4

5. Find the time required for the pressure in the tank of Figure P9.5 to drop from 1 MPa to 500 kPa. Assume adiabatic flow in the tube with $f = 0.018$ and back pressure = 101 kPa.

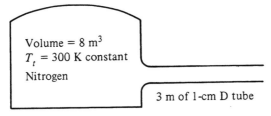

Volume = 8 m³
T_t = 300 K constant
Nitrogen

3 m of 1-cm D tube

Figure P9.5

6. A converging–diverging nozzle has an area ratio of 3.3 to 1. The nozzle is supplied from a tank containing air at 100 kPa at 270 K. For case A of Figure P9.6, find the maximum mass flow possible through the nozzle and the range of back pressures over which the mass flow can be attained. Repeat for case B, in which a constant-area insulated duct is added to the nozzle.

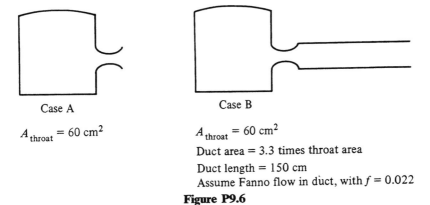

Case A

Case B

$A_{\text{throat}} = 60 \text{ cm}^2$

$A_{\text{throat}} = 60 \text{ cm}^2$

Duct area = 3.3 times throat area

Duct length = 150 cm

Assume Fanno flow in duct, with $f = 0.022$

Figure P9.6

7. A 3-m³-volume tank is to be filled to a pressure of 200 kPa (initial pressure 0 kPa). The tank is connected to a reservoir tank containing air at 3 MPa, whose volume is also 3 m³. A 30-m length of 2.5-cm-diameter tubing is available to connect the two vessels, as in Figure P9.7. Determine the time required to fill the tank to 200 kPa. Assume Fanno flow.

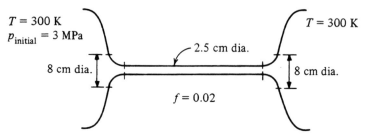

Figure P9.7

8. Find the mass flow rate through the system shown in Figure P9.8. Assume Fanno line flow in the duct and isentropic flow in the converging sections; $f = 0.01$.

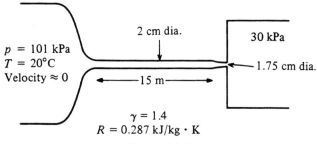

Figure P9.8

9. For the flow shown in Figure P9.9, assume isentropic flow in the convergent–divergent nozzle and Fanno flow in the constant-area duct.
   (a) Find the mass flow rate for a back pressure of 0 kPa.
   (b) For part (a), find the pressure at the exit plane of the duct.
   (c) Find the back pressure necessary for a normal shock to occur at the exit plane of the nozzle (2).
   (d) Find the back pressure necessary for a normal shock to appear just downstream of the nozzle throat (1).

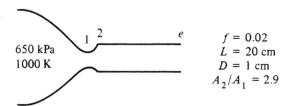

Figure P9.9

10. In which configuration of Figure P9.10, (a) or (b), will the high-pressure tank empty faster? Explain.

L = 75 cm
D_2 = 2.5 cm
f = 0.02
D_4 = 1.25 cm
D_1 = 5.0 cm

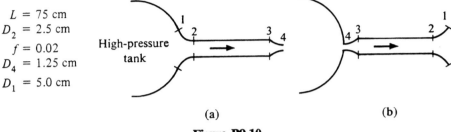

(a)   (b)

Figure P9.10

11. A converging–diverging nozzle with area ratio of 2.9 (Figure P9.11) exhausts into a constant-area insulated duct with a length of 50 cm and diameter of 1 cm. If the system back pressure is 50 kPa, determine the range of reservoir pressures over which a normal shock will appear in the duct.

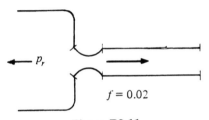

Figure P9.11

12. A converging–diverging nozzle with area ratio of 3.2 (Figure P9.12) exhausts into a constant-area insulated duct with a length of 50 cm and diameter of 1 cm. If the reservoir pressure is 500 kPa, determine the range of back pressures over which a normal shock will appear in the duct (f = 0.02).

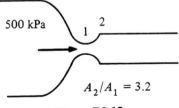

Figure P9.12

13. Oxygen is to be pumped through an uninsulated 2.5-cm pipe, 1000 m long (Figure P9.13). A compressor is available at the oxygen source capable of providing a pressure of 1 MPa. If the supply pressure is to be 101 kPa, determine the mass flow rate through the system and the compressor power required. Assume isothermal flow.

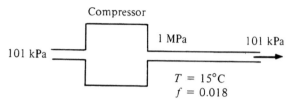

Figure P9.13

14. Natural gas is to be pumped over a long distance through a 7.5-cm-diameter pipe (Figure P9.14). Assume the gas flow to be isothermal, with $T = 15°C$. Compressor stations capable of delivering 20 kW to the flow are available, with each compressor capable of raising the gas pressure isothermally to 500 kPa (inlet compressor pressure is to be 120 kPa). How far apart should the compressor stations be located? Assume isothermal compression in each compressor, with $f = 0.017$.

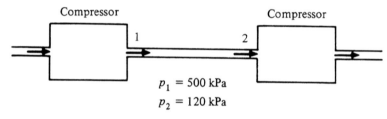

Figure P9.14

15. A rocket nozzle is operating with a stretched out throat, as shown in Figure P9.15. If the inlet stagnation conditions are $p_{t_1} = 1$ MPa and $T_{t_1} = 1500$ K, determine the nozzle exit velocity and mass flow for a back pressure of 30 kPa. Treat the exhaust gases as perfect, with $\gamma = 1.4$, and assume isentropic flow in variable-area sections and Fanno flow in constant-area sections. $R = 0.50$ kJ/kg · K.

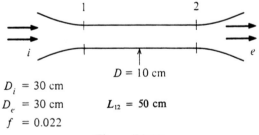

Figure P9.15

16. In a rocket nozzle of area ratio 8 to 1, combustion gases are expanded from a chamber pressure and temperature of 5 MPa and 2000 K. For a nozzle coefficient C_n equal to 0.96, determine the rocket exhaust velocity in space. Assume $\gamma = 1.2$ and $R = 0.50$ kJ/kg·K.

17. Air flows adiabatically in a tube of circular cross section with an initial Mach number of 0.5, initial $T_1 = 500$ K, and $p_1 = 600$ kPa. The tube is to be changed in cross-sectional area so that, taking friction into account, there is no change in the temperature of the stream. Assume the distance between inlet and exit, L, is equal to $100D_1$, with $D_1 = $ initial duct diameter; $f = 0.02$. Find the following:
(a) Mach number M_2
(b) D_2/D_1
(c) Static pressure p_2

10

FLOW WITH HEAT
ADDITION OR HEAT LOSS

10.1 Introduction

We have studied in previous chapters the effects on a gas flow of area change and friction. For these cases, flows were assumed to be adiabatic. In this chapter, the effect of heat addition or loss on a one-dimensional gas flow will be investigated. Flows with heat transfer occur in a wide variety of situations, for example, combustion chambers, in which the heat addition is supplied internally by a chemical reaction, or heat exchangers, in which heat flow occurs across the system boundaries.

The analysis followed will be similar to that of the previous chapter. The continuity, momentum, and energy equations will first be presented for the case of constant-area flow; later, combined effects of heat transfer and area change, as well as friction and heat exchange will be considered. However, in the application of the flow equations, whereas the frictional force term showed up in the momentum equation, the heat addition term will be present in the energy equation. Thus, in this chapter, the length of duct, or more properly the term fL/D, will not be the significant parameter, but rather an expression containing the amount of heat addition and the resultant change of enthalpy of the flow.

The same simplifying assumptions will be made as in previous chapters: one-dimensional, steady flow of a perfect gas with constant specific heats. In a flow with heat addition, where large temperature changes may result, the assumption of a perfect gas with constant specific heats may produce considerable error. Methods of treating nonperfect gases and perfect gases with variable specific heats will be presented in Chapter 12. Furthermore, in any chemical reaction, the composition of the products is different from that of the reactants; again, the assumption of a perfect gas with constant specific heats breaks down. However, in cases such as that of the jet-engine combustion chamber, the fuel–air ratio is small enough so that the chemical change involves only a small fraction of the overall flow; here the perfect gas assumption is reasonable. The student should keep in mind the assumptions made in the derivations of the flow equations. One of the most important tasks of the engineer in any field is to be able to analyze what may be an extremely complex physical situation, make suitable approximations so as to reduce the overall equations describing the situation to simplified form, and solve these equations, subject to pre-scribed boundary conditions.

10.2 Constant-Area Frictionless Flow with Heat Transfer—Rayleigh Line Flow

First we shall study constant-area steady flow with no friction. Refer to the control volume shown in Figure 10.1, with q equal to the heat addition in joules per kilogram (J/kg).
   The continuity equation is

$$\rho V = (\rho + d\rho)(V + dV)$$

so that

$$\frac{d\rho}{\rho} + \frac{dV}{V} = 0 \qquad\qquad (10.1)$$

or

$$\rho V = \text{constant}$$

   The momentum equation is

$$dp + \rho V \, dV = 0$$

since the only forces acting on the control volume are pressure forces. With

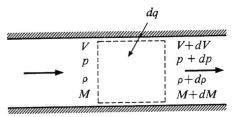

Figure 10.1

ρV = constant, we can integrate the preceding equation to obtain

$$p + \rho V^2 = \text{constant} \tag{10.2}$$

The energy equation is

$$dq = V\,dV + dh \tag{10.3}$$

For a perfect gas,

$$dh = c_p\,dT$$

so that

$$c_p\,dT + V\,dV = dq \tag{10.4}$$

From the definition of stagnation temperature (see Section 3.3),

$$c_p(T_t - T) = \frac{V^2}{2}$$

or

$$c_p\,dT_t = c_p\,dT + V\,dV$$

so that

$$dq = c_p\,dT_t \tag{10.5}$$

Equations (10.1), (10.2), and (10.5) represent the basic equations for constant-area flow with heat exchange. It is now necessary to express these equations in terms of Mach number and arrange the results in tabular form suitable for application to engineering problems.

Referring to Eq. (10.2),

$$p + \rho V^2 = \text{constant}$$

or

$$p\left(1 + \frac{V^2}{RT}\right) = \text{constant}$$

and, since $a^2 = \gamma RT$,

$$p(1 + \gamma M^2) = \text{constant} \tag{10.6}$$

From the perfect gas law,

$$T = \frac{p}{\rho R}$$

But

$$\rho = \frac{\dot{m}}{AV}, \quad \text{where} \quad V = M\sqrt{\gamma RT}$$

so that

$$\sqrt{T} = pM \times \text{constant}$$

or

$$T = \frac{M^2}{(1 + \gamma M^2)^2} \times \text{constant} \qquad (10.7)$$

Let us now apply Eqs. (10.6) and (10.7) to a control volume bounded by cross sections 1 and 2, as shown in Figure 10.2, between which there is a heat addition q. Now

$$p_1(1 + \gamma M_1^2) = p_2(1 + \gamma M_2^2) \qquad (10.8)$$

$$\frac{T_1(1 + \gamma M_1^2)^2}{M_1^2} = \frac{T_2(1 + \gamma M_2^2)^2}{M_2^2} \qquad (10.9)$$

Also, Eq. (10.5) can be integrated for a perfect gas with constant specific heats to yield

$$q = c_p(T_{t_2} - T_{t_1}) \qquad (10.10)$$

To facilitate the tabulation of these expressions, let state 2 be a reference state at which Mach number 1 occurs, as was done for Fanno flow and isentropic flow. More physical significance will be attached to the Mach 1 state later. Denoting the properties at Mach 1 by (*), we get

$$\frac{p}{p^*} = \frac{1 + \gamma}{1 + \gamma M^2} \qquad (10.11)$$

$$\frac{T}{T^*} = \frac{(1 + \gamma)^2 M^2}{(1 + \gamma M^2)^2} \qquad (10.12)$$

$$\frac{V}{V^*} = \frac{\rho^*}{\rho} = \frac{p^*}{p} \frac{T}{T^*}$$

So

$$\frac{V}{V^*} = \frac{(1 + \gamma)M^2}{1 + \gamma M^2} \qquad (10.13)$$

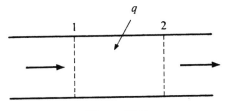

Figure 10.2

Also, since

$$T_t = T\left(1 + \frac{\gamma - 1}{2} M^2\right)$$

$$\frac{T_t}{T_t^*} = \frac{(1 + \gamma)^2 M^2}{(1 + \gamma M^2)^2} \frac{\left(1 + \frac{\gamma - 1}{2} M^2\right)}{\left(1 + \frac{\gamma - 1}{2}\right)}$$ **(10.14)**

Finally, with

$$p_t = p\left(1 + \frac{\gamma - 1}{2} M^2\right)^{\gamma/(\gamma-1)}$$

we obtain

$$\frac{p_t}{p_t^*} = \left[\frac{1 + \gamma}{1 + \gamma M^2}\right]\left[\frac{1 + \frac{\gamma - 1}{2}}{1 + \frac{\gamma - 1}{2} M^2}\right]^{\gamma/(\gamma-1)}$$ **(10.15)**

These equations are presented in Appendix F for $\gamma = 1.4$ from Mach 0 to Mach 5.0.

Example 10.1

Air is flowing in a constant-area duct at a Mach number of 0.20, a static temperature of 300 K, and a static pressure of 100 kPa. Heat is then added to the flow at a rate of 50 kJ/kg of air. Assume frictionless flow, and also that the air behaves as a perfect gas with constant specific heats; $c_p = 1.004$ kJ/kg·K. Determine the final state of the air (see Figure 10.3).

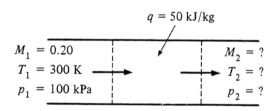

Figure 10.3

Solution

From Appendix A, at $M_1 = 0.20$, $T/T_{t_1} = 0.9921$, so that

$$T_{t_1} = \frac{300}{0.9921} = 302.4 \text{ K}$$

From Eq. (10.10),

$$T_{t_2} - T_{t_1} = \frac{q}{c_p} = \frac{50 \text{ kJ/kg}}{1.004 \text{ kJ/kg} \cdot \text{K}} = 49.8 \text{ K}$$

so that $T_{t_2} = 349.8$ K. From Appendix F, $T_{t_1}/T_t^* = 0.17355$. But

$$\frac{T_{t_2}}{T_t^*} = \frac{T_{t_2}}{T_{t_1}} \frac{T_{t_1}}{T_t^*} = \frac{349.8 \text{ K}}{302.4 \text{ K}} 0.17355 = 0.20008$$

Thus, from Appendix F, $M_2 = 0.217$. Other properties at state 2 can be found in similar fashion.

$$T_2 = \frac{T_2}{T^*} \frac{T^*}{T_1} T_1 = (0.2387) \frac{1}{0.2066} 300 = \underline{346.6 \text{ K}}$$

$$p_2 = \frac{p_2}{p^*} \frac{p^*}{p_1} p_1 = (2.252) \frac{1}{2.273} 100 = \underline{99.1 \text{ kPa}}$$

$$p_{t_2} = \frac{p_{t_2}}{p_2} p_2 = \frac{1}{0.9677} 99.1 = \underline{102.4 \text{ kPa}}$$

where p_{t_2}/p_2 was found in Appendix A. Since

$$p_{t_1} = p_1 \frac{p_{t_1}}{p_1} = \frac{100}{0.9725} = 102.8 \text{ kPa}$$

the difference is

$$p_{t_1} - p_{t_2} = \underline{0.4 \text{ kPa}}$$

To gain a better concept of the effect of heat addition on Mach number, let us determine the locus of states for a given mass flow on a Ts diagram; the resultant plot is termed the *Rayleigh line*. From Chapter 1, for a perfect gas,

$$ds = c_p \frac{dT}{T} - R \frac{dp}{p}$$

Integrating between an arbitrary point and the reference state at which $M = 1$, and assuming constant specific heats, we get

$$s - s^* = c_p \ln \frac{T}{T^*} - R \ln \frac{p}{p^*} \qquad \textbf{(10.16)}$$

To plot $s - s^*$ versus T/T^*, we must express p/p^* in terms of T/T^*. From Eq. (10.11),

$$M^2 = \frac{1 + \gamma}{\gamma} \frac{p^*}{p} - \frac{1}{\gamma}$$

Also, combining Eqs. (10.11) and (10.12) yields

$$\frac{p}{p^*} = \frac{1}{M} \sqrt{\frac{T}{T^*}}$$

or

$$\frac{p}{p^*}\sqrt{\frac{(1+\gamma)(p^*/p)-1}{\gamma}} = \sqrt{\frac{T}{T^*}}$$

Squaring yields

$$\frac{\gamma T}{T^*} = \left(\frac{p}{p^*}\right)^2\left[(1+\gamma)\left(\frac{p^*}{p}\right)-1\right]$$

Simplifying and arranging in quadratic form gives

$$\left(\frac{p}{p^*}\right)^2 - (1+\gamma)\left(\frac{p}{p^*}\right) + \frac{\gamma T}{T^*} = 0$$

Solving for p/p^* yields

$$\frac{p}{p^*} = \frac{1+\gamma}{2} \pm \frac{\sqrt{(1+\gamma)^2 - 4\gamma(T/T^*)}}{2}$$

Returning to Eq. (10.16),

$$\frac{s-s^*}{c_p} = \ln\frac{T}{T^*} - \frac{\gamma-1}{\gamma}\ln\left[\frac{(\gamma+1)\pm\sqrt{(\gamma+1)^2 - 4\gamma(T/T^*)}}{2}\right] \quad \textbf{(10.17)}$$

Equation (10.17) has been sketched in Figure 10.4. Several features of this plot are to be seen immediately. First, there is a point of tangency at which ds/dT is equal to zero. The Mach number at this point A can be obtained from the continuity and momentum equations. Thus

$$dp + \rho V\,dV = 0 \quad \textbf{(10.2)}$$

But

$$\frac{d\rho}{\rho} + \frac{dV}{V} = 0 \quad \textbf{(10.1)}$$

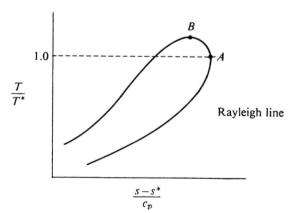

Figure 10.4

or

$$\rho\, dV = -V\, d\rho$$

Combining yields

$$dp = V^2\, d\rho$$

or

$$V^2 = \frac{dp}{d\rho} \quad \text{at } ds = 0$$

But

$$\left(\frac{\partial p}{\partial \rho}\right)_{ds=0} = a^2$$

so that at point A, the point of tangency $ds/dT = 0$, the Mach number is equal to 1.

Also, point B represents a point of tangency at which T/T^* is maximum. To determine this maximum, set

$$\frac{d(T/T^*)}{dM} = 0$$

Differentiating Eq. (10.12), we obtain

$$\frac{(1 + \gamma M^2)^2(1 + \gamma)^2 2M - (1 + \gamma)^2 M^2 2(1 + \gamma M^2)2\gamma M}{(1 + \gamma M^2)^2} = 0$$

Simplifying and setting the numerator equal to zero yields

$$1 + \gamma M^2 = 2\gamma M^2$$

or

$$M^2 = \frac{1}{\gamma}$$

$$M_B = \sqrt{\frac{1}{\gamma}} \qquad \qquad \textbf{(10.18)}$$

The maximum value of T/T^* is now obtained by substituting back into Eq. (10.12):

$$\left(\frac{T}{T^*}\right)_{max} = \frac{(1 + \gamma)^2 \dfrac{1}{\gamma}}{4} \qquad \qquad \textbf{(10.19)}$$

Having determined the significant points on the Rayleigh line, we are now in a position to discuss completely the effect of heat addition on flow properties. Starting at state point a in Figure 10.5, suppose we add heat reversibly

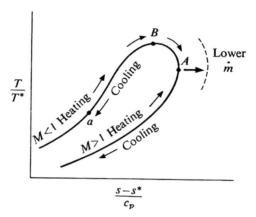

Figure 10.5

to flow in a constant-area duct. First, consider the velocity at a to be subsonic. Since $ds = (dq/T)_{rev}$, it follows that the addition of heat must result in an increase of entropy of the fluid, corresponding to a movement of the state point to the right on a Ts diagram. The top point B of the Rayleigh curve corresponds to subsonic flow [see Eq. (10.18)], so that as heat is added progressively to the flow, the state point moves from a to B and finally to A, at which $M = 1$. Point A represents a state of maximum entropy for the given mass flow rate; any further heat addition can only serve to reduce the mass flow rate in the duct or, in other words, to jump to another Rayleigh line of lower mass flow. There results a maximum amount of heat that can be added to flow in a duct; this maximum is determined by the attainment of Mach 1. Thus flow in a duct can be choked by heat addition, just as it was demonstrated in the last chapter that flow can be choked by friction.

Returning to Figure 10.5, we see that as heat is added to a subsonic flow the Mach number increases to 1. From the momentum equation, this yields a decrease in static pressure. Note, however, from Figure 10.5, that the static temperature first increases (up to T/T^*_{max} at B) and then decreases to T^* at A. For cooling, the reverse of the above variations occur, as shown on the Rayleigh line. For supersonic flow, refer to the lower part of the curve. As heat is added, the flow Mach number decreases to Mach 1. Thus, from momentum considerations, the static pressure increases up to p^*. From Figure 10.5, we see that the static temperature also increases up to $T = T^*$. Again, cooling has the reverse effect.

With Rayleigh line flow, changes are brought about by heat addition, which manifests itself in an increase in stagnation temperature. To use the tables presented in Appendix F to determine the changes in properties brought about by a certain heat addition to a flow in a duct, the initial step is to calculate the stagnation temperature change brought about by the heat addition, and hence the parameter T_t/T^*_t at the duct exit. Reference to the tables

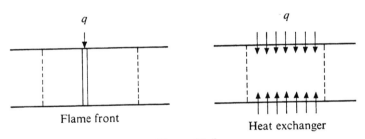

Figure 10.6

then reveals the desired properties for this value of T_t/T_t^*. This is analogous to the methods used for Fanno flow, where fL_{max}/D was the critical parameter.

For Rayleigh flow, it can be seen that the same overall changes in properties result whether the heat is added at one cross section (as a combustion flame front) or distributed over the length of the control volume, as a heat exchanger (see Figure 10.6). Also, no differentiation can be drawn between internal heat addition (a combustion reaction) or external heat addition (a heat exchanger). The following example problems illustrate these points.

Example 10.2

Air enters a ramjet combustion chamber with a velocity of 100 m/s at static temperature of 400 K. Determine the maximum amount of heat that can be added in the combustion chamber without reducing the mass flow rate. For this q_{max}, find the fuel–air ratio. If the fuel–air ratio were to be increased by 10 percent, determine the reduction in $\dot{m}$ for the same inlet stagnation pressure and temperature. Assume the heating value of the fuel to be 40 MJ/kg, neglect the fuel flow rate in comparison to the air flow rate, and assume the air to behave as a perfect gas with constant specific heats ($\gamma = 1.4$ and $c_p = 1.004$ kJ/kg·K). Neglect friction.

Solution

First, the inlet Mach number M_1 is equal to

$$\frac{V_1}{a_1} = \frac{100}{\sqrt{1.4(287)400}} = 0.249$$

From Appendix A,

$$T_{t_1} = T_1 \times \frac{T_{t_1}}{T_1} = \frac{400}{0.9878} = 404.9 \ K$$

Applying the energy equation to the flow in the combustion chamber, $\dot{q} = \dot{m}_{air}c_p(T_{t_2} - T_{t_1}) = \dot{m}_{fuel}(40,000)$, where $\dot{q} = $ rate of heat transfer in kilowatts. For $\dot{q}_{max}$, $T_{t_2} = T_t^*$. From Appendix F,

$$T_t^* = T_{t_1}\frac{T_t^*}{T_{t_1}} = \frac{404.9}{0.2551} = 1587.2 \ K$$

so that

$$\frac{\dot{m}_{fuel}}{\dot{m}_{air}} = \frac{1.004(1587.2 - 404.9)}{40,000} = \underline{0.0297} \quad \text{for } \dot{q}_{max}$$

and

$$q_{max} = 1.004(1587.2 - 404.9) = \underline{1187.0 \text{ kJ/kg}}$$

If the fuel–air ratio were to be increased by 10 percent, the new ratio would be $\dot{m}_{fuel}/\dot{m}_{air} = 0.0327$. From the energy equation,

$$T_{t_2} - T_{t_1} = \frac{0.0327(40,000)}{1.004} = 1302.8 \text{ K}$$

or, since $T_{t_2} = T_t^*$ for choked flow, $T_{t_1}^* = 1302.8 + 404.9 = 1707.7$ K.

$$\frac{T_{t_1}}{T_t^*} = \frac{404.9}{1707.7} = 0.237$$

From Appendix F, $M_1 = 0.239$. From Appendix A, at $M_1 = 0.239$, $T_1/T_{t_1} = 0.9887$ or $T_1 = 400.3$ K. Also from Appendix A, at $M_1 = 0.239$, $p_1/p_{t_1} = 0.9610$, whereas for $M_1 = 0.249$, $p_1/p_{t_1} = 0.9578$. Thus, the 10 percent increase in fuel–air ratio results in a slight increase of inlet static pressure. Since

$$\dot{m} = \rho A V \quad \text{or} \quad \dot{m} = \frac{p}{RT} A (M\sqrt{\gamma RT})$$

the resultant effect of increasing fuel flow is a decrease in $\dot{m}$ given by

$$\frac{\dot{m}_{10\% \text{ increase}}}{\dot{m}} = \frac{0.239}{0.249} \frac{0.9610}{0.9578} = 0.963$$

corresponding to a <u>3.7 percent decrease of mass flow</u>.

Example 10.3

Nitrogen enters an uninsulated duct at Mach 2.0, with a stagnation temperature of 1000 K and stagnation pressure of 1.4 MPa. Heat is lost from the nitrogen to the outside ambient air at 20°C, with the mean overall heat transfer coefficient $\bar{h}$ between fluid and air equal to 60 W/m² · K. [A differential heat loss from the

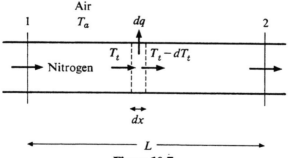

Figure 10.7

nitrogen can be expressed as $d\dot{q} = \bar{h}(dA_p)(T_t - T_a)$, with T_a the ambient air temperature and dA_p a differential area normal to the direction of heat flow.] The duct diameter is 5 cm and length is 2 m. Determine the outlet stagnation temperature, outlet Mach number, and percent of change of stagnation pressure (see Figure 10.7). Neglect friction.

Solution

Consider a differential control volume in the duct as shown in Figure 10.7, and write the energy equation

$$d\dot{q} = -\dot{m}c_p \, dT_t = \bar{h} \, dA_p (T_t - T_a)$$

where

$$dA_p = \pi D \, dx$$

To determine the overall change of stagnation temperature in the duct, integrate the energy equation between 1 and 2. Thus

$$-\int_1^2 \dot{m}c_p \, dT_t = \int_1^2 \bar{h}(T_t - T_a)\pi D \, dx$$

Assuming the nitrogen to have constant specific heat over the temperature range of interest, separate variables to obtain

$$-\int_1^2 \frac{dT_t}{T_t - T_a} = \int_1^2 \frac{\bar{h}\pi D}{\dot{m}c_p} \, dx = \frac{\bar{h}\pi D}{\dot{m}c_p} \int_1^2 dx$$

$$\ln \frac{T_{t_1} - T_a}{T_{t_2} - T_a} = \frac{\bar{h}\pi D L}{\dot{m}c_p}$$

At $M = 2$, from Appendix A, $T/T_t = 0.5556$ and $p/p_t = 0.1278$ so that $T_1 = 555.6$ K and $p_1 = 0.1789$ MPa.

$$V_1 = M_1\sqrt{\gamma R T_1}$$
$$= 2.0\sqrt{1.4(296.8)(555.6)}$$
$$= 961.0 \text{ m/s}$$

$$\dot{m} = \frac{p_1}{RT_1} A V_1$$

$$= \frac{(178.9 \text{ kN/m}^2)\left(\frac{\pi}{4}0.05^2 \text{ m}^2\right)(961 \text{ m/s})}{(0.2968 \text{ kJ/kg} \cdot \text{K})(555.6 \text{ K})}$$

$$= 2.047 \text{ kg/s}$$

$$\ln \frac{T_{t_1} - T_a}{T_{t_2} - T_a} = \frac{(60 \text{ W/m}^2 \cdot \text{K})\pi(0.05 \text{ m})2 \text{ m}}{(2.047 \text{ kg/s})(1.038 \text{ kJ/kg} \cdot \text{K})(1000 \text{ J/kJ})}$$

$$= 0.008871$$

$$\frac{T_{t_1} - T_a}{T_{t_2} - T_a} = 1.0089$$

or

$$T_{t_2} - T_a = \frac{1000 - 293}{1.0089} = 700.8 \text{ k}$$

$$T_{t_2} = 293 + 700.8 = 993.8 \text{ K}$$

To find M_2, use the Rayleigh tables in Appendix F.

$$\frac{T_{t_2}}{T_t^*} = \frac{T_{t_2}}{T_{t_1}} \frac{T_{t_1}}{T_t^*} = \frac{993.8}{1000} 0.7934$$

$$= 0.7885$$

$$\underline{M_2 = 2.025}$$

$$p_2 = \frac{p_2}{p^*} \frac{p^*}{p_1} p_1$$

$$= \frac{0.3560}{0.3636} 178.9 \text{ kPa}$$

$$= 175.2 \text{ kPa}$$

and

$$p_{t_2} = \frac{p_{t_2}}{p_2} p_2 = \frac{175.2}{0.1230}$$

$$= 1424 \text{ kPa}$$

or

$$\frac{p_{t_2} - p_{t_1}}{p_{t_i}} = \frac{1424 - 1400}{1400}$$

$$= 0.017$$

or 1.7 percent increase.

Example 10.4

A constant-area duct is connected to a high-pressure air reservoir through a converging nozzle (Figure 10.8). The walls of the duct are heated so as to supply 250 kJ/kg to the air passing through the duct. If the reservoir pressure and temperature are, respectively, 750 kPa and 300 K, and the system back pressure is 300 kPa, determine whether or not the duct is choked. Also find the mass flow rate of air passing through the duct. The duct diameter is 5 cm and duct length is 1.2 m. Assume isentropic flow in the nozzle and frictionless flow in the duct, with the air behaving as a perfect gas with constant specific heats.

Solution

First, consider a plot of p versus x for this system. For a back pressure of 750 kPa, of course, there is no flow; as the back pressure is lowered, subsonic flow occurs in

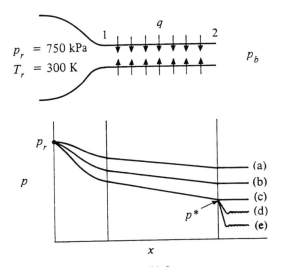

Figure 10.8

nozzle and duct, with the static pressure decreasing in both nozzle and duct, as shown by curves a and b in Figure 10.8. When the back pressure is decreased to that of curve c, sonic flow occurs at the duct exit. Any further decrease in back pressure cannot be sensed by the reservoir, so for all back pressures below that of curve c, the mass flow and pressure distribution in the duct remain the same as that for curve c. For back pressures such as d and e, adjustment to the back pressure takes place via expansion waves outside the duct. The duct is choked, then, when the back pressure is either equal to or below that of curve c. First, we shall calculate this back pressure necessary to choke the system. For $M_2 = 1$, $T_{t_2} = T_t^*$; from the energy equation, $q = (T_{t_2} - T_{t_1})c_p = (T_t^* - T_{t_1})c_p$, where $T_{t_1} = 300$ K. Thus $T_t^* - 300 = 250 \text{ kJ/kg}/1.004 \text{ kJ/kg} \cdot \text{K} = 249.0 \text{ K}$, or $T_t^* = 549.0$ K. From Appendix F, at $T_{t_1}/T_t^* = 300/549 = 0.5464$, $M_1 = 0.410$, and $p_1/p^* = 1.9428$. From Appendix A, at $M_1 = 0.410$, $p_1/p_{t_1} = 0.8907$ or $p_1 = 668.0$ kPa. Therefore, $p^* = 668.0/1.9428 = 343.8$ kPa; in other words, the system is choked for all back pressures below 343.8 kPa. Since the specified back pressure is 300 kPa, the duct is indeed choked and the pressure at the exit plane p_2 is greater than the back pressure. To calculate the mass flow rate, consider the flow conditions at state 1: Mach number 0.410, static pressure of 668.0 kPa, and, from Appendix A, $T_1/T_{t_1} = 0.9675$ so that $T_1 = 290.3$ K. Therefore,

$$\dot{m} = \frac{p_1}{RT_1} AM\sqrt{\gamma RT_1}$$

$$= \frac{(668 \text{ kN/m}^2)\left(\frac{\pi}{4}0.05^2 \text{ m}^2\right)(0.410)\sqrt{1.4(287.0 \text{ J/kg} \cdot \text{K})(290.3 \text{ K})}}{(0.2870 \text{ kJ/kg} \cdot \text{K})(290.3 \text{ K})}$$

$$= 2.204 \text{ kg/s}$$

10.3 Normal Shock on Rayleigh and Fanno Line *Ts* Diagrams

To develop the *Ts* diagram for Fanno flow, we used continuity and energy equations (a friction term does not appear in these equations). To develop the *Ts* diagram for Rayleigh flow, the continuity and momentum equations were employed (a term containing heat addition *q* does not appear in these equations). The Fanno continuity and energy equations and the Rayleigh continuity and momentum equations are exactly the same as the continuity, momentum, and energy equations developed for the normal shock in Chapter 4. Thus the locus of states before and after a normal shock appears on Rayleigh and Fanno lines. Figure 10.9 shows the Rayleigh and Fanno lines on a *Ts* diagram for the same mass flow rate. The intersections of the curves yield state points 1 and 2 for the flow ahead of and behind the shock, respectively. Whereas the increase of entropy for the shock is brought about by irreversibilities such as heat conduction and viscous dissipation occurring internal to the wave, the entropy rise for Fanno flow is brought about by friction, and for Rayleigh flow by heat addition. The line between 1 and 2 is dashed; since the fluid inside the shock is not at thermodynamic equilibrium, its thermodynamic state is not defined.

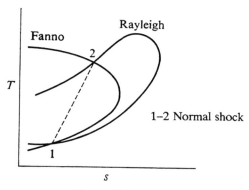

Figure 10.9

10.4 Flow with Heat Addition and Area Change

There are many cases in which a flow involves both area change and heat addition. A prime example is the exhaust of combustion gases through a rocket nozzle. As with flow with friction and area change, a combination of the two effects introduces considerable complexity in the resultant analysis.

Select a control volume as shown in Figure 10.10, with frictionless, reversible, one-dimensional flow. It is desired to derive an expression relating Mach number variation to changes in area and heat addition or loss. The

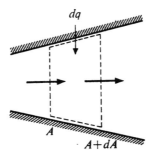

Figure 10.10

continuity, momentum, and energy equations yield

$$\frac{d\rho}{\rho} + \frac{dA}{A} + \frac{dV}{V} = 0 \qquad\qquad \textbf{(10.20)}$$

$$dp + \rho V\,dV = 0 \qquad\qquad \textbf{(10.21)}$$

$$dq = c_p\,dT_t \qquad\qquad \textbf{(10.22)}$$

The perfect gas law, in differential form, becomes

$$\frac{dp}{p} = \frac{d\rho}{\rho} + \frac{dT}{T} \qquad\qquad \textbf{(10.23)}$$

Divide Eq. (10.21) by p and substitute from Eq. (10.23):

$$\frac{d\rho}{\rho} + \frac{dT}{T} + \frac{\rho}{p} V\,dV = 0$$

or, using Eq. (10.20) and the definition of Mach number,

$$-\frac{dA}{A} - \frac{dV}{V} + \frac{dT}{T} + \gamma M^2 \frac{dV}{V} = 0$$

Since

$$\frac{dV}{V} = \frac{dM}{M} + \frac{1}{2}\frac{dT}{T}$$

then

$$-\frac{dA}{A} + (\gamma M^2 - 1)\frac{dM}{M} + \frac{\gamma M^2 + 1}{2}\frac{dT}{T} = 0 \qquad\qquad \textbf{(10.24)}$$

But $dq = c_p\,dT_t$, where, for a perfect gas with constant specific heats,

$$T_t = T\!\left(1 + \frac{\gamma - 1}{2} M^2\right)$$

so that

$$\frac{dT_t}{T} = \left(1 + \frac{\gamma - 1}{2} M^2\right) \frac{dT}{T} + (\gamma - 1)M^2 \frac{dM}{M}$$

or

$$\frac{dT}{T} = \frac{\dfrac{dq}{c_p T}}{1 + \dfrac{\gamma - 1}{2} M^2} - \frac{M^2(\gamma - 1)\dfrac{dM}{M}}{1 + \dfrac{\gamma - 1}{2} M^2}$$

Substituting into Eq. (10.24),

$$-\frac{dA}{A} + \frac{(\gamma M^2 + 1)\dfrac{dq}{c_p T}}{2\left(1 + \dfrac{\gamma - 1}{2} M^2\right)} + \frac{(M^2 - 1)\dfrac{dM}{M}}{1 + \dfrac{\gamma - 1}{2} M^2} = 0 \qquad \textbf{(10.25)}$$

The direct integration of this equation, except for special cases, is not possible. However, certain general conclusions can be drawn from it. For dq equal to zero, Eq. (10.25) reduces to the isentropic flow equation; for example, Mach 1 occurs at the minimum area or throat of a converging–diverging nozzle. However, with heat addition, it can be seen that the second term above is positive, so that the sonic point now occurs at a position where dA/A is positive, that is, the diverging portion of the nozzle. Conversely, if heat is rejected from the gas stream (as, for example, with hot gas flow through a rocket nozzle), the sonic point occurs in the converging part of the nozzle.

The question that the engineer must ask in trying to apply Eq. (10.25) concerns the relative importance of the term involving area change in comparison with that containing heat addition. It is generally the case that one effect far overrides the other. For example, with a rocket nozzle, sufficient accuracy can almost always be attained by neglecting heat losses from the gas stream; then the problem reduces to simple isentropic flow, as discussed in Chapter 3. If the effect of heat addition and area change are estimated to be of the same order of magnitude, numerical integration of Eq. (10.25) is required.

10.5 Flow with Friction and Heat Addition

When heat is added to a gas flowing in a duct, whether it be external heat transfer as in a heat exchanger, or internal heat addition, as via chemical reaction, the effect of wall friction is present. As has been discussed previously, it is often possible to neglect the lesser of the two effects. However, there is an important class of problems where the effects may be of the same order of magnitude; a prime example is the heat exchanger, where often the designer

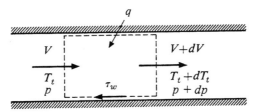

Figure 10.11

must compromise the heat transfer design to minimize pressure drops due to friction and hence minimize pumping power requirements.

For flow in a constant-area duct with heat transfer and friction, as shown in Figure 10.11, the equations of motion yield, for continuity,

$$\frac{d\rho}{\rho} + \frac{dV}{V} = 0 \tag{10.26}$$

For momentum,

$$dp + \frac{1}{2}\rho V^2 \frac{f\,dx}{D} + \rho V\,dV = 0$$

or, since

$$V = M\sqrt{\gamma RT} \quad \text{and} \quad \frac{dV}{V} = \frac{dM}{M} + \frac{1}{2}\frac{dT}{T}$$

then

$$\frac{dp}{p} + \frac{1}{2}\gamma M^2 \frac{f\,dx}{D} + \gamma M^2 \frac{dM}{M} + \frac{1}{2}\gamma M^2 \frac{dT}{T} = 0 \tag{10.27}$$

For energy,

$$dq = c_p\,dT_t$$

or, since

$$T_t = T\left(1 + \frac{\gamma - 1}{2}M^2\right)$$

then

$$\frac{dq}{c_p T} = \left(1 + \frac{\gamma - 1}{2}M^2\right)\frac{dT}{T} + (\gamma - 1)M\,dM \tag{10.28}$$

For a perfect gas,

$$\frac{dp}{p} = \frac{d\rho}{\rho} + \frac{dT}{T}$$

Combining with Eq. (10.26),

$$\frac{dp}{p} = -\frac{dV}{V} + \frac{dT}{T}$$

and with the definition of Mach number,

$$\frac{dp}{p} = -\frac{dM}{M} + \frac{1}{2}\frac{dT}{T}$$

Substitute this expression for dp/p into Eq. (10.27) along with dT/T from Eq. (10.28):

$$\frac{dq}{c_p T_t}\frac{1}{2}(1 + \gamma M^2) + \frac{1}{2}\gamma M^2 \frac{f\,dx}{D}$$

$$= \left[\frac{1 - \gamma M^2}{M} + \frac{(\gamma - 1)(1 + \gamma M^2)\frac{1}{2}M}{1 + \frac{\gamma - 1}{2}M^2}\right] dM \quad \textbf{(10.29)}$$

Just as Eq. (10.25) indicated the relative effects of area change and heat addition on Mach number, Eq. (10.29) shows the effects of heat addition and friction. To afford a better comparison of the friction and heat addition terms, express dq in terms of a convective heat transfer coefficient h (see Example 10.3). For compressible flow, the heat transfer rate between a differential wall area $dA_p = \pi D\, dx$ for a circular duct and the gas flowing in the duct is given by[1]

$$d\dot{q} = h(T_w - T_{aw})\, dA_p$$

with T_w the duct wall temperature and T_{aw} the adiabatic wall temperature. For a gas Prandtl number $(c_p\mu/k)$ close to 1, T_{aw} is approximately equal to the stagnation temperature. Therefore,

$$d\dot{q} = h(T_w - T_t)\pi D\, dx \quad \textbf{(10.30)}$$

where

$$dq = \frac{d\dot{q}}{\dot{m}} = \frac{d\dot{q}}{\rho A V}$$

A relation between the convective heat transfer coefficient h and friction factor f is given by Reynolds analogy[1]. For Prandtl numbers close to 1,

$$h = \frac{\rho V c_p f}{8} \quad \textbf{(10.31)}$$

Substituting Eqs. (10.30) and (10.31) into Eq. (10.29) yields

$$\frac{T_w - T_t}{T_t}\frac{1 + \gamma M^2}{4}\frac{f\,dx}{D} + \frac{1}{2}\gamma M^2 \frac{f\,dx}{D}$$

$$= \left[\frac{1 - \gamma M^2}{M} + \frac{(\gamma - 1)(1 + \gamma M^2)\frac{1}{2}M}{1 + \frac{\gamma - 1}{2}M^2}\right] dM \quad \textbf{(10.32)}$$

It can be seen that, if the duct wall temperature is maintained much higher than the fluid stagnation temperature, heat transfer effects dominate over friction effects. However, if there is only a small difference between wall temperature and stagnation temperature, friction predominates. If both effects are of equal magnitude, Eq. (10.32) must be numerically integrated (a very time consuming procedure). Similar integrations would have to be carried out to determine variations in pressure and other flow properties. Note that, for a given variation of wall temperature with x, Eq. (10.30) must be solved to obtain T_t versus x. Thus

$$d\dot{q} = \dot{m}c_p\,dT_t = h(T_w - T_t)\pi D\,dx$$

$$\frac{dT_t}{T_w - T_t} = \frac{h\pi D\,dx}{\dot{m}c_p} \tag{10.33}$$

Such a procedure was used in working Example 10.3. The variation of T_t with x as obtained from Eq. (10.33) must then be substituted into Eq. (10.32) and the resultant equation numerically integrated to give M versus x.

Often it is necessary to account for the combined effects of heat addition and friction, yet time is not available to carry out the lengthy numerical integrations involved in Eq. (10.32). One approximate method of handling this situation is to add the variation of a property due to heat addition with the variation of the same property due to friction. For example, pressure drop due to friction could be calculated from Fanno flow; then the pressure change due to heat addition or loss could be found from Rayleigh flow. The sum of these two pressure differences would then represent an approximation to the true overall pressure difference involved.

10.6 Summary

Chapter 10 has presented an analysis of compressible flow with heat addition. Using the continuity and momentum equations, the Ts diagram was plotted for constant-area frictionless flow. It was found that the point of maximum entropy on this diagram corresponds to the sonic point. Thus as heat is added to flow in a constant area duct with a resultant increase of fluid entropy, the Mach number of the flow changes toward Mach 1. This corresponds to an increase of Mach number for subsonic flow and a decrease for supersonic flow. If enough heat is added in the duct to reach Mach 1, the duct is choked. Any further heat addition can only serve to reduce the mass flow rate through the duct.

Due to the significance for Rayleigh flow of the sonic point, this was used as a reference state in the computation of tables expressing the variation of properties for this flow. Using the continuity, momentum, and energy equations, the ratios T/T^*, p/p^*, and T_t/T_t^* were found as functions of M; the resultant tables facilitate the solution of problems in this area.

Work in the subject of compressible flow with heat exchange was extended to cover the combined effects of heat addition and area change, as well as heat addition and friction. In the latter cases, it was impossible to integrate directly the flow equations to solve for property variations. Instead, approximate procedures were suggested. For the important case of combined friction and heat transfer, an approximate method would be to calculate separately the change of a property incurred by each effect and then add the two together.

In conclusion, the student should observe the similarity in the approach used for handling isentropic flow with area change, Fanno flow, and Rayleigh flow. In each case, continuity, momentum, and energy equations were written for a fixed differential control volume in the fluid, with the resultant equations integrated using the sonic point as a reference state. Tabular values of property variations were provided in each case, with area change manifesting itself in the term A/A^* for isentropic flow, friction in the term fL_{max}/D for Fanno flow, and heat addition in T_t/T_t^* for Rayleigh flow. The student should now be well versed in the use of tables for handling engineering problems in one-dimensional gas dynamics.

REFERENCES

Specific References

1. KREITH, F., and BLACK, W.Z., *Basic Heat Transfer*, New York, Harper & Row, Publishers, 1980, p. 217.

General References

2. SHAPIRO, A., *The Dynamics and Thermodynamics of Compressible Fluid Flow*, Vol. 1, New York, Ronald Press, 1953.

PROBLEMS

1. Air flows in a constant-area duct of 5-cm diameter at a rate of 2 kg/s. If the inlet stagnation pressure and temperature are, respectively, 700 kPa and 300 K, plot T versus s for Rayleigh line flow. For the same inlet conditions and mass flow rate, plot a Ts diagram for Fanno flow. From the points of intersection of Rayleigh and Fanno lines, show the states on either side of a normal shock. Assume the air to behave as a perfect gas with constant specific heats.

2. Air flows in a constant-area duct of diameter 1.5 cm with a velocity of 100 m/s, static temperature of 320 K, and static pressure of 200 kPa. Determine the rate of heat input to the flow necessary to choke the duct. Assume Rayleigh line flow; express your answer in kilowatts. Assume the air to behave as a perfect gas with constant specific heats.

3. Air flows in a constant-area duct of 10 cm diameter at a rate of 0.5 kg/s. The inlet stagnation pressure is 100 kPa; inlet stagnation temperature is 35°C. Find the following:
(a) Two possible values of inlet Mach number.

(b) For each inlet Mach number of part (a), determine the heat addition rate in kilowatts necessary to choke the duct.

4. A supersonic flow at $p_t = 1.0$ MPa and $T_t = 1000$ K enters a 5-cm-diameter duct at Mach 1.8. Heat is added to the flow via a chemical reaction taking place inside the duct. Determine the heat transfer rate in kilowatts necessary to choke the duct, Assume the air to behave as a perfect gas with constant specific heats; neglect changes in the composition of the gas stream due to the chemical reaction.

5. Heat is added to airflow in a constant-area duct at the rate of 30 kJ/m. If flow enters at Mach 0.20, $T_1 = 300$ K, and $p_1 = 100$ kPa, plot M versus x, p versus x, T versus x, and p_t versus x.

6. An airstream passing through a 5-cm-diameter, thin-walled tube is to be heated by high-pressure steam condensing on the outer surface of the tube at 160°. The overall heat transfer coefficient between steam and air can be assumed to be 140 W/m$^2 \cdot$ K, with the air entering at 30 m/s, 70 kPa, and 5°C. The air is to be heated to 65°C. Determine the tube length required. Assuming Rayleigh line flow, calculate the static pressure change due to heat addition. Also, for the same inlet conditions, calculate the pressure drop due to friction, assuming Fanno flow in the duct with $f = 0.018$. To obtain an approximation to the overall pressure drop in this heat exchanger, add the two results. Discuss the accuracy of this calculation.

7. Air enters a turbojet combustion chamber at 400 K and 200 kPa, with a temperature after combustion of 1000 K. If the heating value of the fuel is 48,000 kJ/kg, determine the required fuel–air ratio (on a mass basis). Assume Rayleigh line flow in the combustion chamber. What fuel–air ratio would be required to choke the combustion chamber? Inlet velocity is 35 m/s.

8. A constant-area duct is fed by a converging nozzle, as shown in Figure P10.8. Heat is lost at the rate of 250 kJ/kg. Discuss the variation of Mach number with x in the duct. Explain your answer.

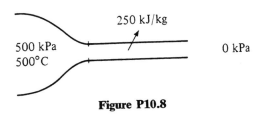

Figure P10.8

9. Consider flow in a constant-area duct with friction and heat transfer. To maintain a constant subsonic Mach number, should heat be added or removed? Repeat for supersonic flow.

10. For the system shown in Problem 8, determine the mass flow rate if 250 kJ/kg of heat energy is added to the flow in the duct. Assume the duct diameter to be 2 cm. Repeat for a back pressure of 100 kPa. Working fluid is air.

11. A detonation wave (Figure P10.11) represents a shock sustained by chemical reaction. Give the continuity, momentum, and energy equations for such a wave, assuming that a chemical reaction taking place in the wave liberates heat q. Denote properties of the unburned gas ahead of the wave by the subscript u and those of

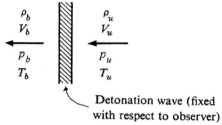

Detonation wave (fixed
with respect to observer)

Figure P10.11

the burned gases behind the wave by b. Write the equations for an observer
traveling with the wave.

12. Develop a computer program that will yield values of p/p^*, T/T^*, $T_t T_t^*$, and $p_t p_t^*$
for Rayleigh line flow with the working fluid consisting of a perfect gas with
constant $\gamma = 1.36$. Use Mach number increments of 0.10 over the range $M = 0$ to
$M = 2.5$.

11

FLOW WITH APPLIED ELECTRIC AND MAGNETIC FIELDS

11.1 Introduction

Previous chapters have dealt chiefly with the flow of perfect, nonconducting gases. Under certain conditions, however, a gas can be made into an electrically conducting fluid, possessing electrical properties similar to a solid conductor. If such a conducting gas stream is allowed to pass through a magnetic field aligned perpendicular to the flow, an electrical field is induced normal to the flow direction and the magnetic field. Since the gas is able to conduct electricity, the electric field can be used to generate a current between electrodes placed in the fluid, and this current is able to produce work through an external load.

Alternatively, if an external power source is used to generate an electric field greater than the induced field between the electrodes, the interaction between the electric current normal to the gas stream and an applied magnetic field yields an accelerating force in the direction of motion of the gas stream. Such a device can be used in a rocket propulsion system to attain high exhaust velocities. Thus an electrically conducting fluid interacting with applied magnetic and electric fields can be used both as a means to produce power and as a means to accelerate a gas flow.

One immediately asks how a gas can be made into an electrical conductor. One method is simply to raise the temperature of the gas. A gas such as oxygen will dissociate into its constituent atoms at sufficiently high temperature (several thousand Kelvin). At still higher temperatures, the atoms will further split into positively charged ions and negatively charged electrons. The resultant gas, termed a *plasma*, exhibits an electrical conductivity. To produce an electrically conducting gas at lower temperatures, the gas may be seeded with small percentages of materials of low ionization energy. In any case, using this and other methods, gases with appreciable electrical conductivity can be generated at temperatures of approximately 2000 K.

The study of the interaction of electric and magnetic fields with a conducting fluid flow is called *magnetohydrodynamics*, or more simply MHD. As can be inferred from the brief description given, this field encompasses many areas of both electrical and mechanical engineering. A detailed study is, therefore, beyond the scope of this text; whole textbooks have been written, for example, on plasma flows, electric and magnetic field theory, and gaseous conduction. Rather, it is the purpose of this chapter to introduce the student to a relatively new, yet rapidly developing field. Of key importance is the presentation of the equations of continuity, momentum, and energy, with terms to describe the effects of the electric and magnetic fields. The student should again appreciate the value of these three basic flow equations, now applied to a somewhat more complex physical situation. Several broad assumptions will have to be made concerning the nature of the flow in order to simplify the resultant equations of motion to the point where the effect of the prime variables on the flow can be ascertained. Such assumptions necessarily lead to error in the resultant analysis; the analysis can in no sense be considered quantitatively exact. However, the student should gain a knowledge of the general concepts and potential applications of MHD and learn how electromagnetic effects can be included in the equations of motion, just as were the effects of heat addition, friction, and area change in previous chapters.

11.2 Fundamentals

To understand better the effect of electric and magnetic fields on an electrically conducting gas flow, let us first review the appropriate electromagnetic equations. Suppose two like electric charges, q_1 and q_2, are separated by a distance d, as shown in Figure 11.1. The two charges repel each other with a force proportional to the product of the charge strengths and inversely proportional to the square of the distance separating the charges. The repulsive force $\mathbf{F}$ is directed along a line connecting the charges.

$$\mathbf{F} = K\frac{q_1q_2}{d^2}\mathbf{r}_{1\rightarrow2} \tag{11.1}$$

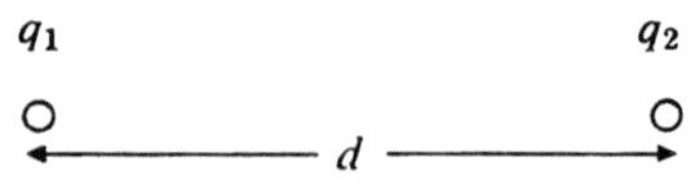

Figure 11.1

where K is a proportionality constant and $\mathbf{r}_{1\to2}$ is a unit vector along a line connecting the charges. An alternative way of expressing the force exerted by q_1 on q_2 is in terms of the electric field $\mathbf{E}$ generated by q_1. Thus

$$\mathbf{E} = \frac{\mathbf{F}}{q_2} \tag{11.2}$$

where $\mathbf{E}$ is expressed in newtons per coulomb (N/C). The electric field generated by a charge is a conservative field; that is, the work necessary to move a charge between any two points in the field is independent of the path, so that E can be expressed in terms of a scalar potential V where

$$\mathbf{E} = -\nabla V \tag{11.3}$$

or

$$V = \frac{kq}{d} \tag{11.4}$$

The units of electric potential are newton·meters/coulomb or, more familiarly, volts.

It is well known, from Ohm's law, that if a potential difference ΔV is applied across an electrical resistance $\mathcal{R}$, a current I flows through the resistance, where $I = \Delta V/\mathcal{R}$. Now suppose an electrically conducting gas is located between parallel plates across which is applied an electric potential (see Figure 11.2). The electrical resistance of the gas can be represented by

$$\mathcal{R} = \frac{1}{\sigma}\frac{d}{A} \tag{11.5}$$

where σ = gaseous electrical conductivity, A = gas cross-sectional area normal to current flow, and d = path length along the current flow direction. Again, $I = \Delta V/\mathcal{R}$. The application of a potential difference ΔV across the plates gives an electric field between the plates of magnitude $E = \Delta V/d$, so that

$$I = \frac{Ed}{d/(\sigma A)} = \sigma EA$$

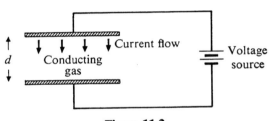

Figure 11.2

In its more general form, Ohm's law becomes

$$\mathbf{J} = \sigma\mathbf{E} \tag{11.6}$$

where $J =$ current density in amperes/square meter $= I/A$.

When an electric charge q moves with velocity $\mathbf{V}$ in an applied magnetic field B, there is a Lorentz force exerted on the charge, given by

$$\mathbf{F} = q\mathbf{V} \times \mathbf{B} \tag{11.7}$$

where $\mathbf{B}$ is expressed in newton·seconds/coulomb·meter or webers/square meter ($1 \text{ W/m}^2 = 1$ tesla).

If there exists a flow of charge, with charge density n per unit volume, moving with velocity $\mathbf{V}$, then the force per unit volume is given by

$$\mathbf{f} = nq\mathbf{V} \times \mathbf{B} \tag{11.8}$$

But $nq\mathbf{V}$ is just the charge flow per unit cross-section area, or $\mathbf{J}$, so that

$$\mathbf{f} = \mathbf{J} \times \mathbf{B} \tag{11.9}$$

Comparing Eqs. (11.2) and (11.8), it can be seen that the effect of the applied magnetic field on a flow of charged particles is equivalent to an induced electric field.

$$\mathbf{E}_{\text{ind}} = \mathbf{V} \times \mathbf{B} \tag{11.10}$$

Consider a situation in which a gas possessing a density of charged particles sufficient to provide an appreciable electrical conductivity flows in the positive x direction between two plates, or electrodes, as shown in Figure 11.3. An electrical field $\mathbf{E}_{\text{appl}}$ is applied between the plates, in the negative y direction, with a magnetic field $\mathbf{B}$ in the negative z direction (into the paper). The magnetic field induces an electric field $\mathbf{E}_{\text{ind}}$ in a direction opposite to $\mathbf{E}_{\text{appl}}$. From Ohm's law, a current is induced between the plates due to the induced field:

$$\mathbf{J}_{\text{ind}} = \sigma\mathbf{E}_{\text{ind}} \tag{11.11}$$

Denoting the current due to the applied field $\mathbf{E}_{\text{appl}}$ by $\mathbf{J}_{\text{appl}}$, the total current per unit area between the plates is

$$\mathbf{J} = \sigma(\mathbf{E}_{\text{appl}} + \mathbf{E}_{\text{ind}})$$
$$= \sigma(\mathbf{E}_{\text{appl}} + \mathbf{V} \times \mathbf{B}) \tag{11.12}$$

Figure 11.3

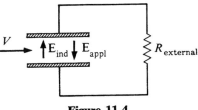

Figure 11.4

In scalar form,

$$J = \sigma(VB - E_{appl}) \qquad (11.13)$$

with J equal to the magnitude of the current density in the positive y direction and E, V, and B the scalar magnitudes of the corresponding vector quantities. (The vector $\mathbf{V} \times \mathbf{B}$ is in the direction of movement of a right-hand screw when turned from the direction of $\mathbf{V}$ to the direction of $\mathbf{B}$.)

Let us consider now the use of MHD flows to produce electrical power and to accelerate a gas stream. For the former case, let the electrodes be connected through an external load resistance (Figure 11.4). Again, a magnetic field $\mathbf{B}$ is applied into the paper with the resultant induced electric field in the positive y direction. In this case, with no external power source, the applied electric field is due to the voltage drop incurred by the passage of current through the load resistance, so the applied field is less than the induced field. The net current flow is thus in the positive y direction. The reaction between the current flow and the applied magnetic field yields a Lorentz force in the negative x direction, with the retarding force transmitted from the charged particles to the entire gas flow by molecular collisions. The gas must push against this retarding force, or do work, as it passes between the electrodes. The work done by the gas is then converted into useful electrical energy as the generated current passes through the external load.

Now let an external power source be used in place of the load resistance, so that $\mathbf{E}_{appl} > \mathbf{V} \times \mathbf{B}$. In this case, the net current is in the negative y direction, giving the flow an accelerating force. Thus this system acts to speed up the gas flow (see Figure 11.5).

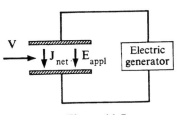

Figure 11.5

11.3 Gaseous Conduction

The operation of an MHD system, whether to produce power or to accelerate a gas stream, depends fundamentally on the ability to convert a neutral gas stream into one that possesses an appreciable electrical conductivity. To conduct a current, the gas must have an appreciable percentage of charge carriers, either electrons or positively charged ions. In other words, the gas must be ionized.

Perhaps the most straightforward method of ionizing a gas is heating, that is, supplying sufficient thermal energy to the gas molecules to remove electrons from their parent nuclei. For each gas molecule that is singly ionized, one electron and one positively charged ion are produced. The conductivity of such an ionized gas stream, or plasma, is primarily due to the electrons; the electron mass, for example, is only $\frac{1}{1800}$ of the mass of the hydrogen nucleus, which provides the electron with a proportionately increased mobility over the ion. The energy required to completely remove an electron from its parent nucleus is called the *ionization energy*, usually expressed in electron volts, where $1 \text{ eV} = 1.6 \times 10^{-19}$ J. The ionization energy is clearly dependent on the forces holding the electron to the nucleus of a given atom. Also, the energy required to remove an electron from the outmost energy level of the atom is less than that required to remove an electron from one of the inner energy shells. Table 11.1 presents the ionization energy required for removal of the outermost electron from the given atom.

Saha[2], using classical thermodynamics, developed an expression for the percent ionization of a gas at equilibrium as a function of gas temperature and

Table 11.1 Ionization energy (Ref. I)

Gas	Ionization Energy (Electron Volts)
Helium	24.46
Neon	21.47
Argon	15.68
Krypton	13.93
Atomic hydrogen (H)	13.53
Molecular hydrogen (H_2)	15.6
Atomic nitrogen (N)	14.48
Molecular nitrogen (N_2)	15.51
Atomic oxygen (O)	13.55
Molecular oxygen (O_2)	15.51
Sodium	5.12
Potassium	4.32
Cesium	3.87
Barium	5.19

pressure.

$$\frac{\alpha^2}{1 - \alpha^2} = C\frac{T^{5/2}}{p} \exp -\frac{(\Delta E)_i}{kT} \tag{11.14}$$

where

α = fraction of ionized atoms in gas
$(\Delta E)_i$ = ionization energy
k = Boltzmann constant = 1.38×10^{-23} J/K
p = gas pressure in atmospheres
C = constant dependent on gas

It can be seen that, for a given gas, the degree of ionization is only dependent on pressure and temperature. For argon, Eq. (11.14) yields[3]

$$\frac{\alpha^2}{1 - \alpha^2} p = (3.16 \times 10^{-7})T^{5/2} \exp \frac{-(\Delta E)_i}{kT} \tag{11.15}$$

An expression relating α with electrical conductivity σ for a slightly ionized gas ($\alpha < 10^{-4}$) is given by Cambel[3]:

$$\sigma = 3.34 \times 10^{-12} \frac{\alpha}{QT^{1/2}} \text{ S cm}^{-1} * \tag{11.16}$$

where Q = electron-atom cross section (an area about a particle such that a collision will occur as soon as another particle comes within that area).

Example 11.1

Determine the percent of ionization and electrical conductivity for argon at 5000 K and 7500 K and 1 atm. Assume $Q = 1 \times 10^{-16}$ cm^2.

Solution

From Eq. (11.15), at 1 atm and 5000 K,

$$\frac{\alpha^2}{1 - \alpha^2} = (3.16 \times 10^{-7})5000^{5/2}e^{-(15.68 \times 1.6 \times 10^{-19})/(1.38 \times 10^{-23} \times 5000)}$$

$$\underline{\alpha = 3.36 \times 10^{-7}}$$

and

$$\sigma = \frac{(3.34 \times 10^{-12})(3.36 \times 10^{-7})}{1 \times 10^{-16} \times \sqrt{5000}}$$

$$= \underline{1.59 \times 10^{-4} \, (\Omega \cdot \text{cm})^{-1}}$$

At 1 atm and 7500 K,

$$\frac{\alpha^2}{1 - \alpha^2} = (3.16 \times 10^{-7})(7500)^{5/2}e^{-(15.68 \times 1.6 \times 10^{-19})/(1.38 \times 10^{-23} \times 7500)}$$

$$\underline{\alpha = 2.23 \times 10^{-4}}$$

---

* Siemens (S) is a unit of electrical conductance, S = ampere/volt.

and

$$\sigma = \frac{(3.34 \times 10^{-12})(2.23 \times 10^{-4})}{(1 \times 10^{-16})\sqrt{7500}}$$

$$= 0.086(\Omega\cdot\text{cm})^{-1} = \underline{8.6\ (\Omega\cdot\text{m})^{-1}}$$

Successful operation of an MHD device may require conductivities of $10\ (\Omega\cdot\text{m})^{-1}$. It can be seen from the example problem that the temperatures required to produce such a gas conductivity are extremely high; materials are not available to withstand such temperatures. There exists a need to produce a conductive gas at much lower temperatures, say 2000 K.

One method that has met with some success is seeding of the gas with a material of low ionization energy. From Table 11.1, we can see such a material might be cesium or potassium. Conductivities of 10 to 100 S/m have been reported for argon seeded with 1 percent potassium. Note that the concentration of seed material need only be small (1 percent or less) to produce appreciable conductivity.

Other methods currently under investigation show promise of yielding a low-temperature electrical conductivity[4]. For example, several techniques of nonequilibrium ionization, such as photoionization and direct joule heating, produce a highly ionized gas at low temperature as a result of electron heating.

Gaseous conductivity has been shown to be dependent on pressure and temperature; it can also be dependent on the strength of the applied magnetic field. In this case, conductivity can no longer be spoken of as a scalar quantity, but must be considered a tensor. However, in the analysis that follows, the conductivity of the gas is assumed, at least as a first approximation, to be a constant scalar quantity.

11.4 Equations of Motion for MHD Flows

We shall now develop the equations of motion for flow of an electrically conducting gas in a duct with applied electric and magnetic fields. The electric conductivity is assumed to be a scalar constant. Furthermore, the flow itself is taken to be steady, one-dimensional, and frictionless with negligible heat transfer (adiabatic flow). Gas flow is in the positive x direction, with an applied magnetic field of magnitude B in the negative z direction, and an applied electric field of magnitude E in the negative y direction:

	x	y	z
B	0	0	B_z
E	0	E_y	0
V	V_x	0	0
J	0	J_y	0
J × B	$J_y B_z$	0	0

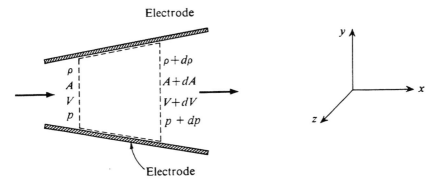

Figure 11.6

Select a control volume as shown in Figure 11.6. For steady flow, the continuity equation yields

$$\iint_s \rho \mathbf{V} \cdot d\mathbf{A} = 0$$

or

$$\frac{d\rho}{\rho} + \frac{dA}{A} + \frac{dV}{V} = 0 \qquad (11.17)$$

The momentum equation for steady flow in the x direction is

$$\Sigma F_x = \iint_s V_x (\rho \mathbf{V} \cdot d\mathbf{A})$$

where forces acting on the fluid in the control volume include pressure forces and electromagnetic forces. The pressure forces are

$$pA - (p + dp)(A + dA) + \left(p + \frac{dp}{2}\right) dA$$

where the last term accounts for the pressure force on the side walls of the control volume. The electromagnetic force per unit volume is $\mathbf{J} \times \mathbf{B}$. For our system, $B_z = -B$, whereas J_y can be either positive or negative. Thus the electromagnetic force acting on the control volume in the x direction is $-J_y BA\, dx$. Simplifying and dropping second-order terms, the momentum equation becomes

$$-A\, dp - J_y BA\, dx = \rho AV\, dV$$

or

$$\rho V \frac{dV}{dx} + \frac{dp}{dx} = -J_y B \qquad (11.18)$$

The energy equation for one-dimensional, adiabatic, steady flow, according to Eq. (1.19), yields

$$\iint_s \left(h + \frac{V^2}{2} \right) (\rho \mathbf{V} \cdot d\mathbf{A}) = -\frac{dW}{dt}$$

where W is the work done by the fluid in the control volume. The rate of doing work, or electrical power, can be expressed as

$$\frac{dW}{dt} = JEA\,dx$$

where dW/dt is expressed in watts. Substituting yields

$$\rho A V\,dh + \rho A V^2\,dV = -JEA\,dx$$

or

$$\rho V\,dh + \rho V^2\,dV = -JE\,dx \qquad \textbf{(11.19)}$$

In the following two sections, we shall consider the application of MHD flows to a power generator and to an accelerator. The specific examples to be worked out include a constant-velocity electric generator and a constant-area, constant-temperature gas accelerator. It is emphasized that the cases to be considered are not necessarily the most practical (from a manufacturing standpoint) or even the most efficient. However, they allow a simplification of the equations presented in this section so that closed-form solutions can be obtained in a relatively simple manner. The method of solution for these two examples follows essentially that presented by Angrist[5] for the generator and Hill and Peterson[6] for the accelerator.

11.5 MHD Power Generator

As a first application, we shall discuss the constant-velocity, MHD electric power generator. As shown in Figure 11.7, the generator consists of an ionization chamber, nozzle, and channel in which the power is generated. The top and bottom of the channel consist of the electrodes, made of electric conductors and connected through an external load resistance. The sides of the channel are constructed of electrical insulators. A magnetic field of magnitude B is applied in the negative z direction. The gas is assumed to be perfect with constant specific heats; that is, it is assumed that the small percentage of seeding compound required to produce an appreciable gas conductivity does not materially affect the gas properties. The electrical conductivity of the gas is assumed to be a scalar constant.

We are interested in the channel geometry required to produce constant velocity flow, pressure and temperature variations in the flow, and the power

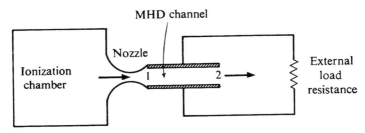

Figure 11.7

produced by such a generator. For the constant-velocity case, the equations of continuity, momentum, and energy reduce to

$$\frac{d\rho}{\rho} + \frac{dA}{A} = 0 \tag{11.20}$$

$$\frac{dp}{dx} = -J_y B \tag{11.21}$$

$$\rho V \frac{dh}{dx} = -JE \tag{11.22}$$

or, for a perfect gas with constant specific heats,

$$\rho V c_p \frac{dT}{dx} = -JE \tag{11.23}$$

To solve these equations, it is necessary to specify some characteristic of the external load. For example, define K equal to the ratio of the electric field across the load to the open circuit electric field, or $K = E/VB$. Note that K is also equal to

$$K = \frac{\text{useful electric power output}}{\text{total electric power generated}}$$

$$= \frac{EJ}{VBJ}$$

$$= \frac{E}{VB} \tag{11.24}$$

K must be less than 1, since some power must be used up in I^2R or joule heating of the plasma.

Combining the momentum and energy equations,

$$\rho V c_p \frac{dT}{dp} = \frac{E}{B}$$

or

$$\rho c_p \frac{dT}{dp} = K$$

From the perfect gas law, $p = \rho RT$, so

$$K \frac{dp}{p} = \frac{c_p}{R} \frac{dT}{T} = \frac{\gamma}{\gamma - 1} \frac{dT}{T}$$

Integrating between cross sections 1 and 2, for K independent of x,

$$K \ln \frac{p_2}{p_1} = \frac{\gamma}{\gamma - 1} \ln \frac{T_2}{T_1}$$

or

$$\frac{p_2}{p_1} = \left(\frac{T_2}{T_1}\right)^{\gamma/K(\gamma-1)} \tag{11.25}$$

For this nonisentropic flow, then, the actual temperature ratio is less than the isentropic temperature ratio for the same pressure ratio.

To produce a constant-velocity flow, the channel area must vary with x. Assume the distance between electrodes d is equal to a constant, with the distance s varying with x as shown in Figure 11.8. From Eqs. (11.20) and (11.25),

$$\frac{A_2}{A_1} = \frac{\rho_1}{\rho_2} = \frac{p_1}{p_2} \frac{T_2}{T_1} = \left(\frac{p_2}{p_1}\right)^{[K(\gamma-1)/\gamma]-1} \tag{11.26}$$

enabling a determination of channel exit area if the pressure ratio is specified. The current per unit electrode area, J, is given by

$$J_y = \sigma(VB - E)$$
$$= \sigma VB(1 - K) \tag{11.27}$$

or

$$\frac{I}{A} = \sigma VB(1 - K)$$

Thus

$$I = \sigma VB(1 - K)L \frac{s_1 + s_2}{2} \tag{11.28}$$

where I is the total current passing through the external load. The useful

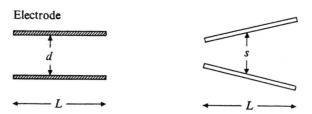

Electrode

d

s

L

L

Figure 11.8

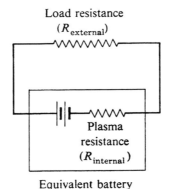

Load resistance
(R_{external})

Plasma
resistance
(R_{internal})

Equivalent battery

Figure 11.9

output power per unit volume JE can be written as

$$JE = \sigma V^2 B^2 (1 - K)K \qquad \text{(11.29)}$$

The total output power is then JE times the channel volume:

$$\mathscr{P} = \sigma V^2 B^2 (1 - K)KLd \frac{s_1 + s_2}{2} \qquad \text{(11.30)}$$

To determine the value of K required to produce maximum power output, differentiate $\mathscr{P}$ with respect to K and set the result equal to zero. For $\mathscr{P}_{\text{max}}$,

$$\frac{d\mathscr{P}}{dK} = 0$$

or

$$K = \frac{1}{2}$$

This corresponds to the case for which plasma and load resistance are equal. Note that this is exactly the same condition as for a battery circuit. This is not surprising, in that the circuit for an MHD generator can be reduced to an equivalent battery circuit, as shown in Figure 11.9.

Example 11.2

An electrically conducting gas flow enters a constant velocity MHD generator at Mach 0.6, stagnation temperature 2000 K, and static pressure 300 kPa. The duct exhaust pressure is 100 kPa. The distance between electrodes is a constant 0.5 m, with the inlet duct cross section square. The conductivity of the gas is assumed constant at 20 S/m, with a constant magnetic field applied of 2 tesla. Determine the maximum power output, the current and load voltage at maximum power, the duct geometry, and the exhaust temperature. Assume the gas to behave as a perfect gas with constant specific heats, with a molecular mass of 40. $c_p =$ 0.519 kJ/kg·K, $\gamma = 1.67$, and $R = 0.2081$ kJ/kg·K (properties of argon).

Solution

$$\frac{T_{t_1}}{T_1} = \left(1 + \frac{\gamma - 1}{2} M_1^2\right) \qquad \text{from Eq. (3.6)}$$

$$T_1 = \frac{2000}{1.1206} = 1784.8 \text{ K}$$

and

$$V_1 = M_1 a_1$$

$$= 0.6\sqrt{1.67(208.1)1784.8}$$

$$= 472.5 \text{ m/s}$$

For $\mathscr{P}_{\max}$, $K = \frac{1}{2}$. From Eq. (11.26),

$$\frac{A_2}{A_1} = \left(\frac{p_2}{p_1}\right)^{[K(\gamma-1)/\gamma]-1}$$

$$= \left(\frac{100}{300}\right)^{-0.7994} = 2.407$$

From Eq. (11.21),

$$\frac{dp}{dx} = -J_y B$$

$$= -\sigma V B(1 - K)B = -\sigma V B^2(1 - K)$$

$$= -\left(20\frac{1}{\Omega \cdot m}\right)(472.5 \text{ m/s})(4 \text{ Wb}^2/\text{m}^4)\frac{1}{2}$$

where $1 \text{ Wb} = 1 \text{ V}\cdot\text{s}$.

Integrate for dp/dx constant:

$$\frac{dp}{dx} = -18.9 \text{ kPa/m}$$

$$p_2 - p_1 = -(18.900 \text{ kN/m}^3)L$$

where

$$p_1 - p_2 = 200 \text{ kPa}$$

$$L = \frac{200 \text{ kPa}}{18.9 \text{ kPa/m}} = 10.58 \text{ m}$$

The channel volume is $[(A_1 + A_2)/2] \times L$, where

$$A_1 = 0.5^2 = 0.25 \text{ m}^2$$

$$A_2 = 2.407 \times 0.25 = 0.6018 \text{ m}^2$$

$$\text{volume} = 4.506 \text{ m}^3$$

$$\mathscr{P}_{\max} = \sigma V^2 B^2 (1 - K) K \left(Ld \frac{s_1 + s_2}{2} \right) \qquad (11.30)$$

$$= 20 \frac{1}{\Omega \cdot m} \times 472.5 \text{ m}^2/\text{s}^2 \times 4 \text{ V}^2 \cdot \text{s}^2/\text{m}^4 \times 0.25 \times 4.506 \text{ m}^3$$

$$= 20.12 \text{ MW}$$

For $\mathscr{P}_{\max}$,

$$J = \sigma V B \left(1 - \frac{1}{2} \right)$$

$$= \frac{1}{2} \left(20 \frac{1}{\Omega \cdot m} \right) (472.5 \text{ m/s})(2 \text{ V} \cdot \text{s/m}^2)$$

$$= 9450 \text{ A/m}^2$$

or

$$I = 9450 \times A_{\text{electrode}}$$

$$= 9450 \times L \times \frac{s_1 + s_2}{2}$$

$$s_1 = 0.5 \text{ m}$$

$$s_2 = 2.407 \times 0.5 = 1.204 \text{ m}$$

$$I = (9450 \text{ A/m}^2)(10.58 \text{ m}) \frac{1.704}{2} \text{ m}$$

$$= 85,180 \text{ A}$$

Load voltage is $(Ed) = BVKd = (2 \text{ V} \cdot \text{s/m}^2)(472.5 \text{ m/s}) \frac{1}{2}(0.5 \text{ m}) = 236.3 \text{ V}.$

$$\frac{T_2}{T_1} = \left(\frac{p_2}{p_1} \right)^{K(\gamma - 1)/\gamma} = \left(\frac{100}{300} \right)^{0.2006} = 0.8022$$

$$T_2 = 1784.8(0.8022) = 1431.8 \text{ K}$$

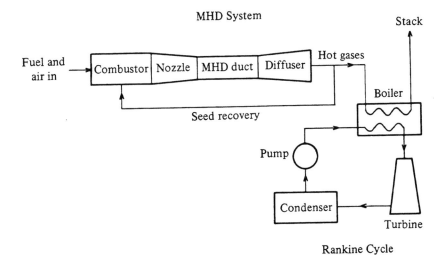

MHD System

Stack

Fuel and air in → Combustor | Nozzle | MHD duct | Diffuser → Hot gases

Seed recovery

Boiler

Pump

Condenser

Turbine

Rankine Cycle

Figure 11.10

The first use of MHD power generation will probably be as a topping device for a conventional Rankine cycle power plant. In this application, heat from the exhaust of the MHD duct will be used to provide input energy for a steam cycle, as shown in Figure 11.10. Basic power plant thermal efficiencies, currently a maximum of about 40 percent for a Rankine steam cycle, can be boosted to close to 50 percent with the combined cycle incorporating MHD.

11.6 MHD Accelerator

Now we shall consider a second application of flows with applied electric and magnetic fields. As has been shown previously, by connecting the electrodes to an external electric power source, it is possible to accelerate an electrically conducting gas flow (see Figure 11.11). The electric field produced by the external generator is in the negative y direction, with the applied magnetic field in the negative z direction. The magnitude of the current is given by

$$J = \sigma(E - VB) \tag{11.31}$$

with current flow now in the negative y direction for $E > VB$. With applied electric and magnetic fields as additional variables, it is necessary to specify two additional flow properties to obtain a solution. For simplicity, we shall assume constant area and constant temperature flow. Now we are interested in the acceleration that can be produced by given electric and magnetic fields.

For constant area and temperature, the flow equations yield

Continuity:

$$\frac{d\rho}{\rho} + \frac{dV}{V} = 0 \tag{11.32}$$

Momentum:

$$\rho V \frac{dV}{dx} + \frac{dp}{dx} = JB \tag{11.33}$$

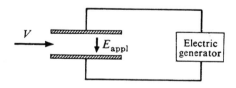

Figure 11.11

Energy:

$$\rho V^2 \frac{dV}{dx} = JE \qquad \text{(plus sign on RHS since work done on fluid)} \qquad \textbf{(11.34)}$$

Perfect gas law:

$$\frac{dp}{p} - \frac{d\rho}{\rho} = 0 \qquad\qquad \textbf{(11.35)}$$

Combining Eqs. (11.33) and (11.35) yields

$$V \frac{dV}{dx} + \frac{p}{\rho^2} \frac{d\rho}{dx} = \frac{JB}{\rho}$$

Substitute for $d\rho/dx$ from Eq. (11.32):

$$V \frac{dV}{dx} - \frac{RT}{V} \frac{dV}{dx} = \frac{JB}{\rho}$$

From the definition of Mach number,

$$M = \frac{V}{\sqrt{\gamma RT}}$$

for a perfect gas with c_p and c_v constant, so that

$$V \frac{dV}{dx} = \frac{JB}{\rho} \frac{\gamma M^2}{\gamma^2 - 1} \qquad\qquad \textbf{(11.36)}$$

Combining with Eq. (11.34), we obtain

$$\frac{E}{B} = \frac{V \gamma M^2}{\gamma M^2 - 1} \qquad\qquad \textbf{(11.37)}$$

for T and A constant.

In order to determine the variation of M with x, integrate Eq. (11.36):

$$V \frac{dV}{dx} = \frac{\dot{m}}{\rho A} \frac{dV}{dx}, \qquad \text{where } \frac{dV}{dx} = a \frac{dM}{dx} \text{ for } T = \text{constant}$$

$$V \frac{dV}{dx} = \frac{\dot{m}}{\rho A} \frac{a\, dM}{dx} = \frac{JB}{\rho} \frac{\gamma M^2}{\gamma M^2 - 1}$$

$$\frac{\dot{m}}{A} \frac{a\, dM}{dx} = [\sigma(E - VB)]B \frac{\gamma M^2}{\gamma M^2 - 1}$$

$$= \sigma V B^2 \left[\frac{\gamma M^2}{\gamma M^2 - 1} - 1 \right] \frac{\gamma M^2}{\gamma M^2 - 1}$$

$$\frac{\dot{m}}{A} \frac{dM}{dx} = \frac{\sigma B^2 \gamma M^3}{(\gamma M^2 - 1)^2}$$

Integrating yields

$$\int_1^2 \sigma B^2 \, dx = \frac{\dot{m}}{A} \int_1^2 \frac{(\gamma M^2 - 1)^2}{\gamma M^3} \, dM$$

For B = constant, independent of x,

$$\sigma B^2 (x_2 - x_1) = \frac{\dot{m}}{A} \left[\frac{\gamma}{2} (M_2^2 - M_1^2) - \frac{1}{2\gamma M_2^2} \left(1 - \frac{M_2^2}{M_1^2} \right) - 2 \ln \frac{M_2}{M_1} \right]$$

$$(11.38)$$

The variation of E that must be provided between the electrodes to produce constant temperature flow in a constant-area channel is given by Eq. (11.37).

Example 11.3

An MHD constant-area duct is to be installed at the exit of a rocket nozzle in order to accelerate the exhaust flow from Mach 2.0 to Mach 4.0. Determine the duct length required if the flow can be assumed isothermal. Assume a constant magnetic field of 1 tesla is applied (independent of x), with $\gamma = 1.4$, a gas conductivity of 10 S/m, and a mass flux of 10 kg/s/m^2.

Solution

From Eq. (11.38),

$$x_2 - x_1 = \frac{\dot{m}}{A} \frac{1}{\sigma B^2} \left[\frac{\gamma}{2} (M_2^2 - M_1^2) - \frac{1}{2\gamma M_2^2} \left(1 - \frac{M_2^2}{M_1^2} \right) - 2 \ln \frac{M_2}{M_1} \right]$$

$$= \left(10 \frac{kg}{s \cdot m^2} \right) \left(\frac{1}{10} \frac{m}{S} \right) \left(\frac{m^4}{1 \, V^2 \cdot s^2} \right) 7.081$$

$$= \underline{7.081 \text{ m}}$$

11.7 Summary

Chapter 1 has demonstrated the effects of applied electric and magnetic fields on the flow of an electrically conducting gas. The equations of fluid flow are complicated by the inclusion of a force term in the momentum equation and a work term in the energy equation. Also, an additional equation, Ohm's law, is necessary to describe the current flow through the plasma. However, the student should realize that, once the physics is understood and the appropriate expressions for forces and energies derived, the inclusion of these terms in the equations of motion presents no problem.

Solutions in closed form have only been derived for very simple cases. It must be emphasized that detailed quantitative solutions are beyond the scope of this text. Rather, the chapter is meant to serve as an introduction to a field in which the equations of fluid mechanics and electromagnetics are interwoven.

Applications are presented to demonstrate the potential of MHD in power generation and propulsion. Although many research experiments have been conducted at various installations throughout the country, the full potential of MHD systems has not been realized. One of the many problems that has arisen is that of finding materials to withstand the extremely high temperatures existent in the channel. High temperatures are necessary, of course, to generate a sufficient electrical conductivity in the gas stream. Such problems, however, do not seem insurmountable; the potential of the end product certainly warrants a continuation of the research efforts in this area. It seems quite probable that within the next few years operable MHD systems will be generating at least some of our electrical power and providing propulsive force for space rockets.

REFERENCES

Specific References

1. *Handbook of Chemistry and Physics*, 48th ed., Cleveland, Ohio, Chemical Rubber Co., 1967 p. E-69.
2. SAHA, M., "Ionization in the Solar Chromosphere," *Philosophical Magazine*, Vol. 40 (1920), pp. 472–488.
3. CAMBEL, A., *Plasma Physics and Magnetofluidmechanics*, New York, McGraw-Hill Book Company, 1963, Chapter 7.
4. KERREBROCK, J., "Magnetohydrodynamic Generators with Nonequilibrium Ionization," *AIAA Journal*, Vol. 3 (April 1965), p. 591.
5. ANGRIST, S., *Direct Energy Conversion*, 3rd ed., Boston, Allyn and Bacon, Inc., 1976, Chapter 7.
6. HILL, P., AND PETERSON, C., *Mechanics and Thermodynamics of Propulsion*, Reading, Mass., Addison-Wesley Publishing Co., Inc., 1965, Chapter 5.

General References

7. ANGRIST, S., *Direct Energy Conversion*, 3rd ed., Boston, Allyn and Bacon, Inc., 1976, Chapter 7.
8. HUGHES, W., AND YOUNG, F., *The Electromagnetodynamics of Fluids*, New York, John Wiley & Sons, Inc., 1966.
9. SUTTON, G., AND SHERMAN, A., *Engineering Magnetohydrodynamics*, New York, McGraw-Hill Book Company, 1965.
10. CULP, A., *Principles of Energy Conversion*, New York, McGraw-Hill Book Company, 1979.
11. CHAPMAN, J., STROM, S. AND WU, Y., "MHD Steam Power – Promises, Progress and Problems," *Mechanical Engineering* (September 1981), pp. 30–37.

PROBLEMS

1. Determine the electrical conductivity of argon at 0.1 atm, 0.01 atm, 10 atm, and 5000 K. Explain physically the resultant variation of σ with p.

2. Determine the electrical conductivity of helium at 5000 K, and 0.01 atm.

3. Determine the electrical conductivity of cesium at 2000 K and 0.1 atm, assuming Eq. (11.15) can be used for cesium vapor.

4. An MHD power generator is to produce a maximum power of 500 MW. A constant magnetic field of 1.5 tesla is applied to the flow, with a gas conductivity of 20 S/m. Assume a constant flow velocity of 500 m/s, with a square duct inlet 1 m on a side. For a duct pressure drop of 2 atm, determine the duct length and exit dimensions. Assume the distance between the electrodes is constant, with the gas possessing the properties of argon.

5. A constant velocity MHD generator has a velocity of 1000 m/s with $\sigma = 15$ S/m, and an applied magnetic field of 1.8 tesla. If the distance between the electrodes is 0.4 m and the electrode area is 0.3 m^2, determine the open-circuit voltage and the power output for $K = \frac{1}{2}$.

6. For Example 11.3, determine the strength of the applied electric field necessary to produce isothermal flow. Use Eq. 11.37.

7. If the electric field were to be kept constant with x, develop an expression for B versus x to produce a constant-area, constant-temperature accelerator.

8. A constant-area MHD duct is 1 m long, with inlet conditions of 1000 K, 300 kPa, and 1000 m/s. Assuming constant temperature flow with an applied magnetic field of 1.2 tesla, determine the duct exit velocity. The duct area is 0.1 m^2, with the gas having the properties of argon with a conductivity of 20 S/m. Also find the exit static pressure.

9. An MHD constant-area duct is to be installed at the exit of a rocket nozzle to accelerate exhaust flow from Mach 2.0. The duct is 5 m in length, duct diameter is 0.5 m, and inlet static pressure and temperature are 50 kPa and 2200 K. A constant field of 2 tesla is applied, with $\gamma = 1.67$; gas conductivity is 10 S/m and the molecular mass of gas = 20. Find the duct exit Mach number.

12

IMPERFECT GAS EFFECTS

12.1 Introduction

In Chapters 1 through 11, all analyses have involved the flow of a perfect gas $(p = \rho RT)$ with constant specific heats. However, under the pressure and temperature extremes that may occur in compressible flow, there can be significant departures of the gas from these assumptions. A gas for which the specific heats c_p and c_v can be assumed constant is termed a calorically perfect gas. A gas that obeys the equation of state $p = \rho RT$ is called a *thermally perfect gas*. We shall be interested in this chapter in a determination of the extent of caloric and thermal imperfections in real gases and their effect on the one-dimensional equations of motion.

For a great majority of cases encountered in gas flows, high pressures are associated with high temperatures (for example, combustors) and low temperatures with low pressures (for example, nozzle expansions), so that discrepancies from the perfect gas law, as expressed by a departure from unity of $Z = p/\rho RT$, are small in the real case. Even for a thermally perfect gas, however, specific heats are temperature dependent, so caloric imperfections must be considered. Large variations in the heat capacities of polyatomic molecules can occur, due to the excitation of the rotational and vibrational degrees of freedom of the molecule. At extremely high temperatures (for example, greater than 3000 K) electronic excitation may occur, or even dissociation of the gas molecules. In the latter case, the composition of the gas becomes a variable, dependent on pressure and temperature. Under these conditions, the perfect gas law is not a good approximation; it is difficult to find one equation of state that will prove satisfactory for describing a dissociated gas. Fortunately, Mollier charts have been derived for air in dissociated equilibrium, so the gas properties, for example, in the extreme high temperature region behind the attached shock on a reentry body, can be determined.

In this chapter, using theories from classical thermodynamics and quantum mechanics, we shall give expressions for the specific heats of monatomic and diatomic gases as a function of temperature. Using these expressions, we shall derive the appropriate gas dynamic formulas for a thermally perfect, calorically imperfect gas and compare these with the results already obtained for a calorically and thermally perfect gas.

Next, we shall discuss several different equations of state. Included will be a discussion of the perfect gas itself, and those assumptions that are consonant with this equation of state. Finally, the Mollier diagram for air will be used for the solution of several examples involving extremely high temperature flows.

12.2 Calorically Imperfect Gases

We have shown that for a thermally perfect gas, the internal energy is a function only of temperature. This implies that a gas which obeys the equation of state $p = \rho RT$ has specific heats c_p and c_v, which are likewise functions only of temperature, with $c_p - c_v = R$. We are interested in the nature of the variation of specific heats with temperature and the resultant effect of these caloric imperfections on the flow of a thermally perfect gas.

Consider first a monatomic gas, such as helium, argon, or neon. For each gas molecule there are three translational degrees of freedom (x,y,z directions). According to the classical theory of equipartition of energy, the total energy of a molecule with f degrees of freedom is equal to $(f/2)kT$ with k = Boltzmann's constant. With three degrees of freedom, this becomes $\frac{3}{2}kT$, so the internal energy of the gas can be written as

$$u = \left(\frac{3}{2}kT\right)\frac{N_A}{M}$$

with

$$u = \text{internal energy per unit mass}$$

$$N_A = \text{Avogardo's number} = \frac{\text{molecules}}{\text{mole}}$$

Since

$$N_A k = \bar{R},$$

we arrive at

$$u = \frac{3}{2}RT \quad \text{or} \quad c_v = \frac{3}{2}R$$

Table 12.1 (Ref. 1)

Gas	γ (Room T)
Helium	1.66
Neon	1.64
Argon	1.67
Krypton	1.68

But $c_p - c_v = R$; so for a monatomic gas,

$$\gamma = \frac{c_p}{c_v} = 1.67$$

This is in good agreement with experimental results, which show the specific heats of the monatomic gases to be practically independent of temperature. Caloric imperfections tend to be small, then, for monatomic gases. Experimental values of γ for several monatomic gases at room temperature are presented in Table 12.1.

A diatomic molecule, however, has additional degrees of freedom that lead to discrepancies with the theory of equipartition of energy. Consider a molecule having the dumbell structure shown in Figure 12.1. This molecule possesses the three translational degrees of freedom of the monatomic molecule, as well as two rotational degrees of freedom about the y and z axes. In addition, depending on the rigidity of the atomic bond between the atoms, the atoms may vibrate along the x axis. This introduces two vibrational degrees of freedom, since the energy is part kinetic and part potential. According to the theory of equipartition of energy, the specific heat c_v of a diatomic gas should be

$$c_v = \frac{f}{2}R = \frac{7}{2}R$$

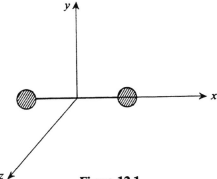

Figure 12.1

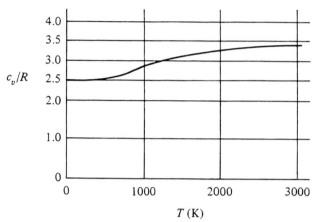

Figure 12.2 C_v vs. T for nitrogen

Experimental plots of c_v versus temperature for a diatomic molecule (nitrogen) appear as shown in Figure 12.2. Only at extremely high temperatures does c_v approach the value of $\frac{7}{2}R$ predicted from the equipartition of energy. It would appear, then, that either the rotational or vibrational modes are not fully excited until very high temperatures are reached. The equipartition principle is thus unable to determine the variation of specific heat with temperature when the rotational and vibrational degrees of freedom are present.

A reliable determination of the variation of the rotational and vibrational contributions to c_v with temperature for a diatomic gas can be made using quantum theory. According to quantum mechanics, the energy of a molecule is restricted to certain discrete values. By determining from spectroscopic data the energy levels and the way in which the molecules are distributed in these various possible energy states, it is possible to develop expressions for c_v versus T. From these expressions, it is customary to define characteristic temperatures θ_{rot} and θ_{vib}, expressing the degree of excitation of the rotational and vibrational degrees of freedom. Thus, for example, for $T \gg \theta_{rot}$ the rotational degrees of freedom are fully excited, and $c_{v\,rot}$, denoting the rotational contribution to c_v, is equal to R.

The characteristic temperatures for rotation of diatomic molecules are given in Table 12.2. It can be seen that at room temperature and above, the

Table 12.2   (Ref. 2)

Gas	θ_{rot} (K)
H_2	87.2
N_2	2.86
O_2	2.08
CO	2.76

rotational degrees of freedom of the common diatomic gases are fully excited, and any variation in specific heat with temperature at higher temperatures must be due to the vibrational modes. In other words,

$$c_v = \frac{5}{2}R + c_{v\,vib}$$

Assuming the diatomic molecule to behave as a simple harmonic oscillator with one natural frequency of vibration v, it can be shown[1] that

$$c_{v\,vib} = R\left(\frac{\theta_{vib}}{T}\right)^2 \frac{e^{\theta_{vib}/T}}{(e^{\theta_{vib}/T} - 1)^2} \qquad (12.1)$$

where

$$\theta_{vib} = \frac{hv}{k}$$

and

$$h = \text{Planck's constant}$$

Values of v have been obtained from spectroscopic measurements, so it is possible to evaluate θ_{vib} for different diatomic gases (see Table 12.3). It can be seen that only at extremely high temperatures will the vibrational modes be fully excited. Note that for $T \gg \theta_{vib}$, the term $(e^{\theta_{vib}/T} - 1)^2$ can be approximated by θ_{vib}^2/T^2, and Eq. (12.1) reduces to $c_{v\,vib} = R$, as predicted by the equipartition principle. Also, for $T \ll \theta_{vib}$, $c_{v\,vib} \to 0$.

Therefore, over the range from room temperature up to 3000 K, above which molecular dissociation becomes important,

$$c_v = \frac{5}{2}R + R\left(\frac{\theta_{vib}}{T}\right)^2 \frac{e^{\theta_{vib}/T}}{(e^{\theta_{vib}/T} - 1)^2} \qquad (12.2)$$

These results help to explain the curve of c_v/R versus T shown in Figure 12.2. At room temperature and above, both the rotational and translational degrees are fully excited, adding to give a contribution to c_v of $\frac{5}{2}R$, while the vibrational modes are partially excited, the degree of excitation and hence the contribution to specific heat increasing with temperature.

For a thermally perfect gas, $c_p - c_v = R$, and theory predicts a value for

Table 12.3 (Ref. 2)

Gas	θ_{vib} (K)
H_2	6340
O_2	2270
N_2	3390
CO_4	3120

the ratio of specific heats of a diatomic molecule at room temperature (vibrational modes not excited) of

$$\gamma = \frac{c_p}{c_v} = \frac{\frac{7}{2}R}{\frac{5}{2}R} = 1.4$$

This is well borne out by experimental data (see Appendix G).

For triatomic and other polyatomic molecules, the evaluation of the vibrational and rotational modes becomes considerably more complex. Certainly, more degrees of freedom are involved, so that c_v for a triatomic molecule, for example, should be higher than c_v for a diatomic molecule, whereas γ for the triatomic molecule should be less than 1.4 at room temperature. These conclusions are again supported by experimental evidence. However, due to the added complexity of the calculations for polyatomic molecules, we shall restrict our analysis to gases composed of diatomic molecules.

We are primarily interested in the effect of caloric imperfections on the one-dimensional flow equations already derived for a calorically and thermally perfect gas. An expression for stagnation temperature can be derived from the energy equation. For one-dimensional adiabatic flow of a thermally perfect gas, with no external work,

$$c_p \, dT + d\frac{V^2}{2} = 0 \qquad (12.3)$$

To integrate this equation for a calorically imperfect diatomic gas, use Eq. (12.2):

$$\frac{V^2}{2} = \int_T^{T_t} R\left[\frac{7}{2} + \left(\frac{\theta_{vib}}{T}\right)^2 \frac{e^{\theta_{vib}/T}}{(e^{\theta_{vib}/T} - 1)^2}\right] dT$$

$$= R\Big|_T^{T_t}\left[\frac{7}{2}T + \theta_{vib}\left(\frac{1}{e^{\theta_{vib}/T} - 1}\right)\right]$$

$$= R\left[\frac{7}{2}(T_t - T) + \theta_{vib}\left(\frac{1}{e^{\theta_{vib}/T_t} - 1} - \frac{1}{e^{\theta_{vib}/T} - 1}\right)\right] \qquad (12.4)$$

To express this in terms of Mach number, we need an expression for a^2. For a thermally perfect gas,

$$a^2 = \gamma RT \qquad (12.5)$$

For a calorically imperfect gas,

$$\gamma = \frac{\dfrac{7}{2} + \left(\dfrac{\theta_{vib}}{T}\right)^2 \dfrac{e^{\theta_{vib}/T}}{(e^{\theta_{vib}/T} - 1)^2}}{\dfrac{5}{2} + \left(\dfrac{\theta_{vib}}{T}\right)^2 \dfrac{e^{\theta_{vib}/T}}{(e^{\theta_{vib}/T} - 1)^2}} \qquad (12.6)$$

Combining Eqs. (12.4) and (12.5), we obtain

$$M^2 = \frac{V^2}{a^2} = \frac{2}{\gamma}\left[\frac{7}{2}\left(\frac{T_t}{T} - 1\right) + \frac{\theta_{vib}}{T}\left(\frac{1}{e^{\theta_{vib}/T_t} - 1} - \frac{1}{e^{\theta_{vib}/T} - 1}\right)\right] \quad (12.7)$$

where γ can be found from Eq. (12.6). Thus, for a given T_t, we can relate M to static temperature.

It is of interest to derive an expression for total pressure. For an isentropic process, thermally perfect gas,

$$ds = c_p\frac{dT}{T} - R\frac{dp}{p} = 0$$

or

$$c_p\frac{dT}{T} = R\frac{dp}{p}$$

Integrating between a point in the flow and the stagnation point conditions yields

$$\int_T^{T_t} c_p\frac{dT}{T} = R\ln\frac{p_t}{p} \quad (12.8)$$

For a diatomic gas with caloric imperfections

$$\int_T^{T_t}\left[\frac{7}{2} + \left(\frac{\theta_{vib}}{T}\right)^2 \frac{e^{\theta_{vib}/T}}{(e^{\theta_{vib}/T} - 1)^2}\right]\frac{dT}{T} = \ln\frac{p_t}{p}$$

$$\left|_T^{T_t}\left[\frac{7}{2}\ln T + \frac{\theta_{vib}}{T}\frac{e^{\theta_{vib}/T}}{e^{\theta_{vib}/T} - 1} - \ln(e^{\theta_{vib}/T} - 1)\right]\right. = \ln\frac{p_t}{p}$$

$$\frac{p_t}{p} = \left(\frac{T_t}{T}\right)^{7/2}\left(\frac{e^{\theta_{vib}/T} - 1}{e^{\theta_{vib}/T_t} - 1}\right)e^{\left\{\left[\frac{\theta_{vib}}{T_t}\frac{(e^{\theta_{vib}/T_t})}{e^{\theta_{vib}/T_t} - 1}\right] - \left[\frac{\theta_{vib}}{T}\frac{e^{\theta_{vib}/T}}{(e^{\theta_{vib}/T} - 1)}\right]\right\}} \quad (12.9)$$

Example 12.1

Nitrogen is expanded isentropically in a nozzle from 1.0 MPa at 1500 K to 100 kPa. Calculate the nozzle exit velocity, assuming nitrogen to behave as a calorically imperfect, thermally perfect gas. Compare with the results obtained under the assumption that nitrogen behaves as a thermally and calorically perfect gas, with $\gamma = 1.4$. Repeat for $T_t = 1000$ K and $T_t = 500$ K.

Solution

With $p_t/p = 10$ and $\theta_{vib} = 3390$ K, Eq. (12.9) can be solved for static temperature, and then Eq. (12.4) for velocity. The results are as follows:

$T_t = 1500$ K:	$T = 842$ K,	$V = 1251.0$ m/s
$T_t = 1000$ K:	$T = 539$ K,	$V = 1012.0$ m/s
$T_t = 500$ K:	$T = 260$ K,	$V = 707.9$ m/s

For a calorically and thermally perfect gas, for isentropic flow, use Appendix A. Thus, for $p/p_t = 0.1$, $M = 2.157$ and $T/T_t = 0.5180$. With $V = M\sqrt{\gamma RT}$, the results are as follows:

$$T_t = 1500 \text{ K:} \qquad T = 727 \text{ K,} \qquad V = 1225.6 \text{ m/s}$$
$$T_t = 1000 \text{ K:} \qquad T = 518 \text{ K,} \qquad V = 1000.7 \text{ m/s}$$
$$T_t = 500 \text{ K:} \qquad T = 209 \text{ K,} \qquad V = 707.6 \text{ m/s}$$

12.3 Thermally Imperfect Gases

One measure of the departure of a real gas from a thermally perfect gas is the compressibility Z, defined by $Z = p/\rho RT$. For a thermally perfect gas, $Z = 1.0$. Curves of Z versus reduced pressure p_r and reduced temperature T_r, as provided by reference (3), are given in Figure 12.3 ($p_r = p/p_{crit}$, $T_r = T/T_{crit}$). Such a plot is quite accurate for almost all the gases of interest. It can be seen that deviations from thermally perfect gases are greatest near the critical temperature and pressure. Fortunately, in the great majority of engineering problems in gas dynamics, temperatures are much higher than the critical temperature (note that, for room temperature, $T_r \approx 2.0$). It is of interest, however, to study the assumptions made in the derivation of the perfect gas

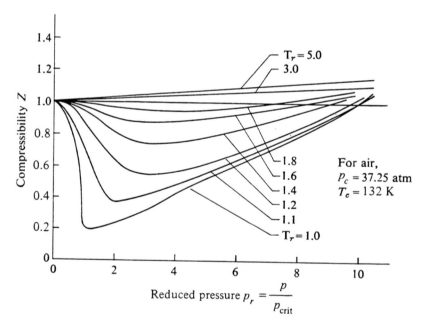

Figure 12.3 (Adapted from Dodge, *Chemical Engineering Thermodynamics*, McGraw-Hill Book Company, Inc., New York, 1944, and is reprinted by permission of the publisher.)

law, to consider some of the other equations of state that have been proposed, and to look at the complexities involved in the inclusion of these other equations of state in the flow equations.

In the derivation of the perfect gas law from kinetic theory, the volume of the molecules and the intermolecular forces are neglected. In the strictest sense, these assumptions are only valid at zero pressure. Thus the *Clausius equation of state* takes into account that the volume available to a molecule in a total volume must be reduced by the volume of the other molecules present.

$$p(v - b) = RT, \quad \text{where } v = \text{specific volume} \qquad \textbf{(12.10)}$$

The *van der Waals equation of state* takes into account the intermolecular forces as well as molecular volumes. Due to the forces of attraction between molecules, the pressure exerted by a group of molecules is reduced due to the net inward force on the outer layer of molecules. Therefore,

$$\left(p + \frac{a}{v^2}\right)(v - b) = RT \qquad \textbf{(12.11)}$$

where a and b are constants for a particular gas. Rewritten in terms of density,

$$p = \frac{\rho RT}{1 - b\rho} - a\rho^2$$

This equation of state has been found to describe accurately the properties of a gas near the liquefaction point.

For gas dynamic calculations, better agreement with experimental data has been provided by the *Berthelot equation of state*, in which a temperature dependence is added to the pressure term of the van der Waals equation. Thus

$$p = \frac{\rho RT}{1 - b\rho} - \frac{a\rho^2}{T} \qquad \textbf{(12.12)}$$

Another equation of state that seems to fit experimental data over a wide range of pressure and temperature is the *Beattie–Bridgman equation*,

$$p = \frac{RT(1 - \varepsilon)}{v^2}(v + b) - \frac{A}{v^2} \qquad \textbf{(12.13)}$$

where

$$A = A_0\left(1 - \frac{a}{v}\right) \qquad B = B_0\left(1 - \frac{b}{v}\right) \qquad \varepsilon = \frac{c}{vT^3}$$

where A_0, B_0, a, b, and c are constants dependent on the particular gas being considered.

Now that several equations of state have been introduced, we are interested in the problems involved in applying these expressions to the equations of motion. For a gas with thermal imperfections, internal energy is no longer a function of temperature alone, but must be expressed in terms of two independent thermodynamic coordinates. Furthermore, the difference $c_p - c_v$, which arises many times in gas dynamic calculations, is no longer simply equal to R. Thus solutions involving thermally and calorically imperfect gases require long, laborious numerical calculations.

Real gas thermodynamic properties have been calculated and tabulated by Hilsenrath et al.[4], with some results given in Appendix G. These results were obtained using the virial equation of state, which expresses the compressibility factor Z as an infinite series in terms of density or pressure. The values presented in Appendix G are therefore valid for a gas that is imperfect both calorically and thermally.

Example 12.2

Using real gas properties for nitrogen, repeat Example 12.1.

Solution

For nitrogen at 10 atm, from Appendix G: at 1500 K, $(s - s_{ref})/R = 26.7721$; at 1000 K, $(s - s_{ref})/R = 25.1223$; at 500 K, $(s - s_{ref})/R = 22.5390$. For isentropic flow, $s = $ constant. Therefore, at 1 atm, $T = 845$ K, 540 K, and 259 K for the three cases being considered.

From the first law, $V^2/2 = \Delta h$. Using results from the tables in Appendix G, for $T_t = 1500$ K, $\Delta h = 1681.3 - 903.3 = 788$ kJ/kg, or $V = 1255.4$ m/s. For $T_t = 1000$ K, $\Delta h = 1076.2 - 562.8 = 513.4$ kJ/kg, or $V = 1013.3$ m/s. For $T_t = 500$ K, $\Delta h = 520.0 - 268.5 = 251.5$ kJ/kg, or $V = 709.2$ m/s.

Comparing the results of this example with those of Example 12.1, it can be seen that, over the temperature and pressure range of these examples, caloric imperfections outweigh the thermal imperfections of the gas. In fact, even with $T_t = 1500$ K, the error in the calculation of velocity is less than one-half of 1 percent if we use Eqs (12.4) and (12.9) for a thermally perfect, calorically imperfect gas. In fact, for $T_t = 1500$ K, the error in velocity incurred by assuming a perfect gas with constant specific heats is only 2 percent. It can be seen that, in the majority of cases over the range from room temperature up to the point of dissociation, caloric imperfections are more significant than thermal imperfections.

12.4 Air in Dissociated Equilibrium

As the temperature of a diatomic or polyatomic gas such as oxygen or nitrogen is raised above 1500 K, the bonds holding the molecule together loosen and, at higher temperatures, break. The molecule is then split, or dissociated, into its constituent atoms. At still higher temperatures, the atoms themselves break down into positively charged ions and negatively charged electrons.

Let us suppose that air, composed primarily of oxygen and nitrogen, is raised to 5000 K or above. At these temperatures, with dissociation and ionization, the following components will be present: N_2, O_2, N, O, N^+, O^+, e^-, and even NO and NO^+, the latter two formed by chemical reaction between oxygen and nitrogen. At equilibrium, the amount of each constituent depends on the gas pressure and temperature. For a group of gaseous species undergoing chemical reaction or dissociation, the partial pressure of each constituent at equilibrium can be determined from the appropriate equilibrium constant K_p. To illustrate the definition and use of K_p, consider the chemical reaction between four gaseous compounds A, B, C, and D proceeding according to $\nu_A A + \nu_B B \rightleftharpoons \nu_C C + \nu_D D$, where the ν's are the respective stoichiometric number of mols of each gas. (For example, for the reaction $H_2O \rightleftharpoons H_2 + \frac{1}{2}O_2$, $\nu_A = 1$, $\nu_C = 1$, $\nu_D = \frac{1}{2}$). For this general reaction, the equilibrium constant K_p is defined as $K_p = (p_C^{\nu_C} p_D^{\nu_D})/(p_A^{\nu_A} p_B^{\nu_B})$, where p_c, for example, is the partial pressure of component C, and pressures are expressed in atmospheres. This expression for K_p is derived by taking, as a criterion of chemical equilibrium, that the change of Gibbs function is equal to zero. [The derivation can be found in most standard texts on classical thermodynamics[2,5].]

At equilibrium, then, each of the four components A, B, C, and D is present; the partial pressures and mol fraction of each constituent can be determined from K_p. Consider, for example, the familiar reaction

$$H_2O \rightleftharpoons H_2 + \tfrac{1}{2}O_2$$

where $K_p = (p_{H_2})(p_{O_2})^{1/2}/p_{H_2O}$. At room temperature, H_2O is quite stable, very little dissociated into H_2 and O_2. From reference (5), $\ln K_p$ for this reaction is only -16.754 at 1000°C, indicating, as expected, that the partial pressures of H_2 and O_2 are negligible. However, at 5000°C, $\ln K_p$ for this reaction is 1.312; now the partial pressures of H_2 and O_2 are comparable to that of H_2O, indicating appreciable dissociation of H_2O into H_2 and O_2. It should be noted that since $p_A = (n_A/n_T)p_T$, where n_A = number of mols of A, n_T = total number of mols present, and p_T = total pressure, K_p can be written as

$$K_p = \frac{n_C^{\nu_C} n_D^{\nu_D} p_T^{\nu_C + \nu_D - \nu_A - \nu_B}}{n_A^{\nu_A} n_B^{\nu_B} n_T^{\nu_C + \nu_D - \nu_A - \nu_B}} \tag{12.14}$$

Table 12.4 Equilibrium constant ($\ln K_p$) for various reactions (Ref. 5)

T (°C)	$H_2O \rightleftharpoons H_2 + \frac{1}{2}O_2$	$H_2O \rightleftharpoons \frac{1}{2}H_2 + OH$	$O_2 \rightleftharpoons 2O$	$N_2 \rightleftharpoons 2N$	$\frac{1}{2}N_2 + \frac{1}{2}O_2 \rightleftharpoons NO$
1000	-16.754	-18.635	-32.103	-74.534	-7.021
2000	-6.332	-6.647	-10.942	-34.717	-3.259
3000	-2.245	-1.976	-2.655	-19.138	-1.802
4000	-0.161	0.506	1.770	-10.809	-1.038
5000	1.312	2.041	4.522	-5.599	-0.574

so that the number of moles of each constituent present at equilibrium is dependent on the total pressure (as long as $v_c + v_D$ is not equal to $v_A + v_B$).

The equilibrium constant itself is only a function of temperature, so tabular values for the common equilibrium reactions can be found in the literature[5]. K_p values for some typical reactions are provided in Table 12.4. As an example of the use of K_p, consider the following example problem.

Example 12.3

Oxygen at 0.1 atm is heated to 4000°C in a constant-pressure process. Determine the amounts of O_2 and O present at this temperature. (At 4000°C, the equilibrium constant for $O \rightleftharpoons O^+ + e^-$ is orders of magnitude less than that of $O_2 \rightleftharpoons 2O$.)

Solution

Considering the dissociation of molecular oxygen into atomic oxygen, for each mol of molecular oxygen initially present, let there be a mols of molecular oxygen and b mols of atomic oxygen at 4000°C.

$$O_2 \rightarrow aO_2 + bO$$

Taking a mass balance yields

$$2 = 2a + b$$

Also,

$$K_p = \frac{(p_O)^2}{p_{O_2}}$$

$$= \frac{\left(\dfrac{b}{a + b}\right)^2 (0.1)^2}{\left(\dfrac{a}{a + b}\right) 0.1} = 5.87$$

Substituting for b from the mass balance equation, and solving the resultant quadratic gives

$$a = 0.03, \qquad b = 1.94$$

To determine the composition of dissociated air at a given pressure and temperature, calculations such as have been made in Example 12.3 must be performed for each of the constituent gases. Once the composition has been determined, the thermodynamic properties of the mixture can be found from a knowledge of the properties of each of the constituents. For example,

$$u = m_{f_A} u_A + m_{f_B} u_B + \cdots \tag{12.15}$$

where u is the internal energy per unit mass of the dissociated air, u_A is the internal energy per unit mass of constituent A, and m_{f_A}, for example, is the mass fraction of constituent A in the mixture.

No single equation of state is adequate to describe the properties of dissociated air. Thermodynamic properties such as internal energy and enthalpy are dependent on gas composition, which is here a function of pressure as well as temperature. Fortunately, numerical computations have been made to determine the composition of dissociated and ionized air as a function of pressure and temperature, and also the thermodynamic properties of the high-temperature air. These calculations have been arranged in the form of Mollier diagrams for air in dissociated equilibrium[6], presenting $(h - h_{ref})/RT_0$ versus s/R with lines of constant temperature, pressure, and density, over the the temperature range from 5000 up to 10,000 K (see Figures 12.4 and 12.5).

It must be emphasized here that the determination of the thermodynamic properties of air at such extremely high temperatures is more than just an academic curiosity. The air surrounding a reentry body or a long-range ballistic missile reaches temperatures at which dissociation and ionization of the air take place. For accurate gas dynamic calculations in such situations, a Mollier diagram for air must be used. Two examples will be presented to illustrate the use of these diagrams, the first involving an isentropic expansion, the second the calculation of the conditions behind a normal shock on a high-speed body. The use of a Mollier diagram is based on the assumption that the dissociated air is in a state of thermodynamic equilibrium. Unfortunately, there are many cases in high-speed flow in which the air does not have a chance to reach

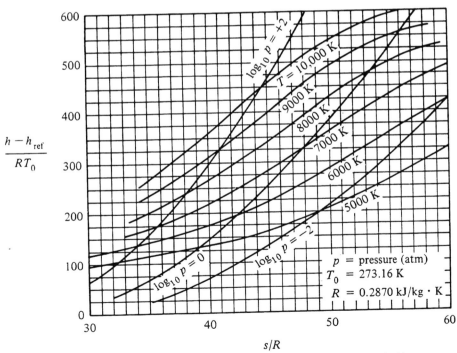

Figure 12.4 Mollier diagram for air. (Adapted from Ref. 6.)

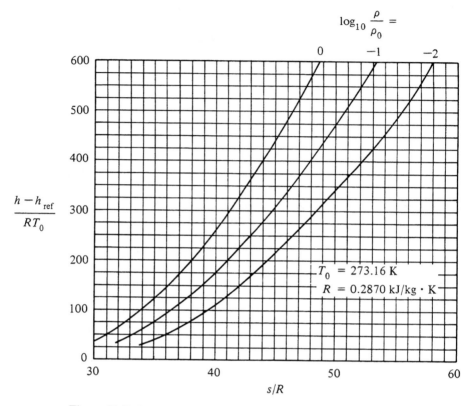

Figure 12.5   Mollier diagram for air. (Adapted from Ref. 6.)

equilibrium; for example, the time of passage of air through a high-strength shock may be short enough so that the air just downstream of the shock is not in a state of thermodynamic equilibrium. For cases such as this, use of the Mollier diagram can only yield a solution that is a good first approximation to the exact solution.

Example 12.4

Air in a stagnation region is at a temperature of 8000 K and a pressure of 100 atm. The air undergoes an isentropic expansion in a nozzle to a pressure of 1 atm. Determine the velocity at the end of the expansion.

Solution

From the Mollier diagram for air in dissociated equilibrium presented in Figure 12.4, at 8000 K and 100 atm $(h_1 - h_{ref})/RT_0 = 240$ and $s/R = 38$. Follow down the isentrope, as shown in Figure 12.6, until the pressure of 1 atm is reached. At this point, $(h_2 - h_{ref})/RT_0 = 110$. Now

$$V_2 = \sqrt{2(h_1 - h_2)} \quad \text{for} \quad V_1 = 0$$

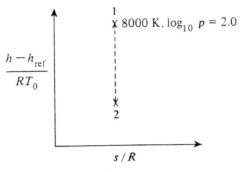

Figure 12.6

so that

$$V_2 = \sqrt{2(240 - 110)RT_0}$$
$$= \sqrt{2(130)(287 \text{ J/kg·K})(273.16 \text{ K})} = \underline{4515 \text{ m/s}}$$

Let us now compute the exit velocity based on the assumption that air behaves as a thermally and calorically perfect gas with $\gamma = 1.4$ and $R = 0.287$ kJ/kg·K. From Appendix A, for $p/p_t = 0.01$, $M_2 = 3.69$ and $T/T_t = 0.2686$. Thus $T_2 = 2150$ K and $V_2 = M_2\sqrt{\gamma RT_2} = 3.69\sqrt{1.4(287)2150} = 3430$ m/s. It is evident that, at the high temperatures and pressures of this example problem, a Mollier diagram must be used. Large errors result from the assumption that air behaves as a perfect gas with constant specific heats.

Example 12.5

A blunt-nosed body is traveling through the atmosphere with a velocity of 7500 m/s at 30 km. A shock is in front of the body, with this shock normal to the flow at the nose. Determine the conditions immediately behind this normal shock, assuming the air to be in a state of dissociated equilibrium (see Figure 12.7).

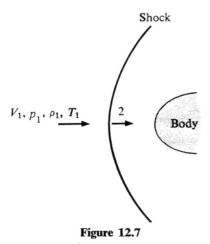

Figure 12.7

Solution

From Chapter 4, the equations for a normal shock are

$$\rho_1 V_1 = \rho_2 V_2$$

$$p_1 + \rho_1 V_1^2 = p_2 + \rho_2 V_2^2$$

$$h_1 + \frac{V_1^2}{2} = h_2 + \frac{V_2^2}{2}$$

Combining, we obtain

$$h_1 + \frac{V_1^2}{2} = h_2 + \frac{p_1 - p_2 + \rho_1 V_1^2}{2\rho_2} \qquad \text{(a)}$$

$$p_1 - p_2 = \rho_1 V_1^2 \left(\frac{\rho_1}{\rho_2} - 1 \right) \qquad \text{(b)}$$

Equations (a) and (b), in conjunction with the Mollier diagram, must be solved for the unknowns p_2, ρ_2, V_2, and T_2. Values will be assumed for h_2 and p_2 until the above equations are satisfied. Ambient conditions at 30 km are $T = 226.5$ K, $p = 1.20$ kPa, and $\rho = 0.0184$ kg/m³ (see Appendix H). First,

$$h_1 + \frac{V_1^2}{2} = 227 \text{ kJ/kg} + \frac{7500^2}{2} \text{ m}^2/\text{s}^2 = 227 + 28{,}125 = 28{,}352 \text{ kJ/kg}$$

Since V_2 should be quite small, it is logical to assume that h_2 will be close to 28,352 kJ/kg. Assume $h_2 = 28{,}200$ kJ/kg. To select a reasonable first value for p_2, use Eq. (4.5). For $M_1 \gg M_2$,

$$\frac{p_2}{p_1} = 1 + \gamma M_1^2, \qquad \text{where } M_1 = \frac{7500}{\sqrt{1.4(287)(226.5)}} = 24.9$$

Thus $p_2/p_1 = 869$. This value is only a starting point, since Eq. (4.5) was derived on the assumption that air is a thermally and calorically perfect gas. Thus, for our first trial,

$$p_2 = 1043 \text{ kPa} = 10.3 \text{ atm} \quad \text{or} \quad \log_{10} p_2 = 1.01$$

From Figures 12.6 and 12.7, for $\log_{10} p_2 = 1.01$ and $h_2/RT_0 = 360$, $s/R = 45.8$ and $\log_{10} \rho/\rho_0 = -0.74$. Thus $\rho_2 = 0.18(1.29)$ kg/m³ $= 0.23$ kg/m³. Substitute into (a) and (b):

$$\text{(a): } 28{,}352 \overset{?}{=} 28{,}200 + \frac{(101 - 1043) + 0.0184\left(\frac{7500^2}{1000}\right)}{2(0.23)} = 28{,}402$$

$$\text{(b): } 1.20 - 1043 \overset{?}{=} (0.0184)\left(\frac{7500^2}{1000}\right)\left(\frac{0.0184}{0.23} - 1\right) = -952$$

Since these equations do not quite balance, second trials must be made with new values of h_2 and p_2. The correct solution is $p_2 = 953$ kPa, $h_2 = 28{,}000$ kJ/kg, $\rho_2 = 0.23$ kg/m³, and $T_2 = 8100$ K.

If air were assumed to behave as a thermally and calorically perfect gas with

$\gamma = 1.4,$

$$M_1 = 24.9, \qquad \frac{T_2}{T_1} = 121.5 \quad \text{or} \quad T_2 = 27{,}500 \text{ K}$$

Again, a large error is incurred by these assumptions. Note that the temperature is much greater than that calculated by the more exact method, due to the large amount of energy absorbed by the gas during dissociation.

12.5 Summary

A gas can be assumed to behave as a perfect gas with constant specific heats only over a limited range of pressure and temperature. For an accurate calculation of the behavior of a real gas, account must be taken of gas imperfections. Fortunately, in a great many cases of interest in gas dynamics, high pressures occur in conjunction with high temperatures and low pressures with low temperatures, so deviations from the perfect gas equation of state do not contribute large errors. Thus, up to 1500 K, where dissociation must be considered, only caloric imperfections are significant. As the gas temperature is raised above room temperature, the vibrational degrees of freedom of diatomic and other polyatomic molecules are excited, yielding a contribution to specific heat. The quantum mechanical expression for specific heat versus temperature can then be included in the energy integral to obtain an expression for total temperature versus Mach number. Comparison of the resultant expression for T/T_t with that of a calorically perfect gas indicates that large errors can result from neglecting caloric imperfections at high temperatures.

Above 1500 K, the bonds holding molecules together loosen and, at higher temperatures, break. At still higher temperatures, the constituent atoms break down into electrons and ions. With dissociation, the gas composition and specific heats become functions of pressure as well as temperature. To calculate gas properties with dissociation, it is necessary to use a Mollier or hs diagram.

The determination of the properties of air and other gases at extreme temperatures can be seen to have great importance at the present; in the future as materials are developed to enable the utilization of the high temperatures and as vehicles are designed to travel at supersonic and even hypersonic velocities, this will assume even greater importance. It is hoped that the material presented in Chapter 11 on gaseous conduction and the material in this chapter will serve as a stimulus to the student to look further into this field.

REFERENCES

Specific References

1. SEARS, F.W., *An Introduction to Thermodynamics, the Kinetic Theory of Gases, and Statistical Mechanics*, Reading Mass., Addison-Wesley Publishing Company, Inc., 1953.

2. WARK, K., *Thermodynamics*, 3rd ed., New York, McGraw-Hill Book Company, 1977, pp. 446–450.

3. DODGE, B.F., *Chemical Engineering Thermodynamics*, New York, McGraw-Hill Book Company, 1944.

4. HILSENRATH, J., et al., *Tables of Thermodynamic and Transport Properties*, Pergamon Press, Elmsford, N.Y., 1960. Originally published as NBS Circular 564.

5. HABERMAN, W.L., and JOHN, J.E.A., *Engineering Thermodynamics*, Boston, Allyn and Bacon, Inc., 1980, p. 498.

6. "Mollier Chart for Air in Dissociated Equilibrium at Temperatures of 2000 K to 15000 K," NAVORD Report 4446, U.S. Naval Ordnance Lab, White Oak, Md., May 1957.

General References

7. ANDERSON, J.D., *Modern Compressible Flow*, McGraw-Hill Book Company, New York, 1982.

PROBLEMS

1. Equation (12.9) relates p_t/p to T_t/T for a diatomic gas. Show that, for $\theta_{vib} \gg T$, this reduces to the expression previously derived for a calorically perfect gas, $p_t/p = (T_t/T)^{\gamma/\gamma-1}$.

2. Air at a static temperature of 1500 K and static pressure of 20 kPa is flowing at Mach 2.0. Determine the stagnation pressure and temperature, assuming the air to behave as a diatomic gas with $\theta_{vib} = 3050$ K. Compare with results assuming the air to behave as a calorically and thermally perfect gas, with $\gamma = 1.4$.

3. Repeat Problem 2, using the real properties of air in Appendix G.

4. In Example 12.4, determine the stagnation pressure and temperature behind the shock wave.

5. A shock wave travels outward from an explosion at a velocity of 5000 m/s, and moves into ambient air at 300 K and 100 kPa. Using the Mollier diagram, determine the static and stagnation properties and the air velocity behind the wave. Treat the spherical wave using equations for a normal shock.

6. Air is expanded isentropically in a nozzle from 1800 K and 10 atm to 1 atm. Determine the nozzle exit velocity, using the real properties of air given in Appendix G. Repeat assuming air to be a calorically and thermally perfect gas with $\gamma = 1.4$.

7. Repeat Problem 6, with nitrogen as the working fluid.

8. Air is expanded isentropically in a nozzle from 800 K and 10 atm to 1 atm. Determine the nozzle exit air temperature and velocity, using, first, the real properties of air in Appendix G, and then assuming air to behave as a calorically and thermally perfect gas, with $\gamma = 1.4$.

9. Air at 1 atm and 20°C is raised to 4000 K in a constant-pressure process. Determine the composition of the air at this elevated temperature. Assume the air to consist initially of 3.76 mols of nitrogen per mol of oxygen.

10. Repeat Problem 5 for the case in which the air is confined in a constant volume. For this case, also determine the final pressure in the closed volume.

11. Plot c_v and γ versus T according to Eq. (12.1).

12. Water vapor at 1 atm is heated in a steady flow process to 4000°C. Determine the final composition of the products.

13. One mole of nitrogen and one mol of oxygen are contained in a rigid vessel at 1 atm and 300 K. The mixture is heated to 2000°C. Determine the final pressure in the vessel and the mixture composition.

14. A blunt-nosed body is moving at 5000 m/s at an altitude of 50 km. Determine the stagnation conditions behind the attached shock at the nose of the body. (Consider the point where the shock is normal to the flow direction.)

15. Air is expanded isentropically in a nozzle from 10,000 K and 100 atm to a pressure of 1 atm. Determine the nozzle exit velocity. For an exit area of 5.0 cm², find the air flow rate through the nozzle.

13

EQUATIONS OF MOTION
FOR MULTIDIMENSIONAL FLOW

13.1 Introduction

Previous chapters have dealt almost exclusively with one-dimensional flow. Exceptions were the treatment of oblique shocks in Chapter 6 and Prandtl Meyer flow in Chapter 7, yet even these cases were handled componentwise as equivalent one-dimensional flows. One-dimensional analysis has been shown to be useful for obtaining good engineering approximations to a wide variety of flow problems. However, such an analysis is necessarily an approximation; no real flow exists that is truly one dimensional. Furthermore, many problems cannot even be approached with a one-dimensional analysis. For example, whereas the procedures used in Chapter 3 enabled the prediction of supersonic nozzle area ratios for a required isentropic Mach number, the one-dimensional equations are inadequate for the design of the contour of such a nozzle (A versus x). Likewise, the flow over a cambered supersonic wing cannot be predicted on the basis of a simple one-dimensional theory. For these, and a great many other practical cases, we must develop the equations of motion for multidimensional gas dynamics and find means for solving these equations subject to prescribed boundary conditions.

In this chapter, we shall derive the equations of continuity and momentum for two- and three-dimensional compressible flows. Due to the complexity of these equations, we shall restrict ourselves to flows in which viscous forces, electromagnetic forces, heat transfer, and external work can be neglected. Later chapters will include methods for solving these partial differential equations.

13.2 Continuity Equation

Consider a differential control volume in a compressible flow, as shown in Figure 13.1. Let the velocity components in the x, y, and z directions at the center of the control volume be, respectively, u, v, and w. From the continuity equation for a control volume, given by Eq. (1.8),

$$\frac{\partial}{\partial t} \iiint_{\text{c.v.}} \rho \, d\Psi + \iint_{s} \rho \mathbf{V} \cdot d\mathbf{A} = 0$$

The mass flow crossing the left face is equal to

$$\left(\rho u - \frac{\partial \rho u}{\partial x} \frac{dx}{2} \right) dy \, dz$$

The mass flow leaving the control volume through the right-hand face is

$$\left(\rho u + \frac{\partial \rho u}{\partial x} \frac{dx}{2} \right) dy \, dz$$

The mass flow entering through the bottom face is equal to

$$\left(\rho v - \frac{\partial \rho v}{\partial y} \frac{dy}{2} \right) dx \, dz$$

The mass flow leaving the control volume through the top face is

$$\left(\rho v + \frac{\partial \rho v}{\partial y} \frac{dy}{2} \right) dx \, dz$$

The mass flow entering through the back face is

$$\left(\rho w - \frac{\partial \rho w}{\partial z} \frac{dz}{2} \right) dx \, dy$$

The mass flow leaving through the front is

$$\left(\rho w + \frac{\partial \rho w}{\partial z} \frac{dz}{2} \right) dx \, dy$$

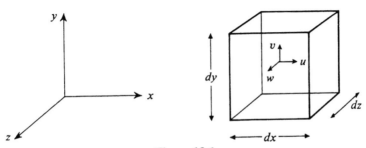

Figure 13.1

According to the continuity equation, the net rate of mass flow into the control volume is equal to the rate of mass buildup inside the control volume, $(\partial/\partial t)\rho\, dx\, dy\, dz$. Combining yields

$$\frac{\partial \rho u}{\partial x} + \frac{\partial \rho v}{\partial y} + \frac{\partial \rho w}{\partial z} + \frac{\partial \rho}{\partial t} = 0 \tag{13.1}$$

This represents the general form of the continuity equation. Rewriting in vector notation gives

$$\frac{\partial \rho}{\partial t} + \nabla \cdot \rho \mathbf{V} = 0 \tag{13.2}$$

For steady flow, the preceding simplifies to

$$\frac{\partial \rho u}{\partial x} + \frac{\partial \rho v}{\partial y} + \frac{\partial \rho w}{\partial z} = 0 \tag{13.3}$$

and

$$\nabla \cdot \rho \mathbf{V} = 0 \tag{13.4}$$

For steady, incompressible flow,

$$\frac{\partial u}{\partial x} + \frac{\partial v}{\partial y} + \frac{\partial w}{\partial z} = 0 \tag{13.5}$$

or

$$\nabla \cdot \mathbf{V} = 0 \tag{13.6}$$

13.3 Momentum Equation

The momentum equation for a control volume is given by Eq. (1.13):

$$\sum \mathbf{F} = \frac{\partial}{\partial t} \iiint_{\text{c.v.}} \rho \mathbf{V}\, d\mathcal{V} + \iint_{s} \mathbf{V}(\rho \mathbf{V} \cdot d\mathbf{A})$$

We shall neglect electromagnetic, gravity, and all other forces acting on the control volume with the exception of the pressure forces. Select a differential control volume as shown in Figure 13.1, with p the pressure at the center of the control volume. Consider the momentum equation in the x direction, with pressure forces acting as shown in Figure 13.2. The net external force acting on the control volume in the x direction is $-(\partial p/\partial x)\, dx\, dy\, dz$.

The x-momentum flux across the right-hand face out of the control volume of Figure 13.1 is

$$\left(u + \frac{\partial u}{\partial x}\frac{dx}{2}\right)\left(\rho u + \frac{\partial \rho u}{\partial x}\frac{dx}{2}\right) dy\, dz$$

$$\left(p - \frac{\partial p}{\partial x}\frac{dx}{2}\right) dy\, dz$$

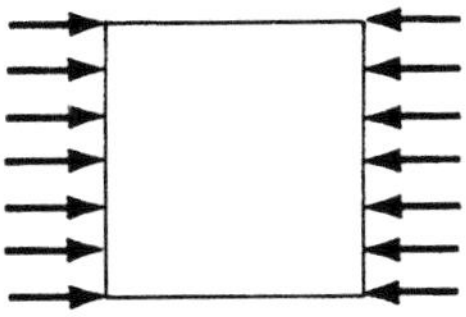

$$\left(p + \frac{\partial p}{\partial x}\frac{dx}{2}\right) dy\, dz$$

Figure 13.2

The x-momentum flux crossing the left-hand face into the control volume is

$$\left(u - \frac{\partial u}{\partial x}\frac{dx}{2}\right)\left(\rho u - \frac{\partial \rho u}{\partial x}\frac{dx}{2}\right) dy\, dz$$

The mass flow crossing the top face takes with it an x-momentum flux:

$$\left(\rho v + \frac{\partial \rho v}{\partial y}\frac{dy}{2}\right)\left(u + \frac{\partial u}{\partial y}\frac{dy}{2}\right) dx\, dz$$

The mass flow crossing the lower face brings with it an x-momentum flux:

$$\left(\rho v - \frac{\partial \rho v}{\partial y}\frac{dy}{2}\right)\left(u - \frac{\partial u}{\partial y}\frac{dy}{2}\right) dx\, dz$$

Similarly, the x-momentum flux leaving the control volume across the front face is

$$\left(\rho w + \frac{\partial \rho w}{\partial z}\frac{dz}{2}\right)\left(u + \frac{\partial u}{\partial z}\frac{dz}{2}\right) dx\, dy$$

The x-momentum flux crossing the back face into the control volume is

$$\left(\rho w - \frac{\partial \rho w}{\partial z}\frac{dz}{2}\right)\left(u - \frac{\partial u}{\partial z}\frac{dz}{2}\right) dx\, dy$$

Combining, substituting into the x-momentum equation, and dropping second-order terms yields

$$-\frac{\partial p}{\partial x} dx\, dy\, dz = \frac{\partial}{\partial t}\rho u\, dx\, dy\, dz$$
$$+\left(u\frac{\partial \rho w}{\partial z} + \rho w\frac{\partial u}{\partial z} + \rho v\frac{\partial u}{\partial y} + u\frac{\partial \rho v}{\partial y} + \rho u\frac{\partial u}{dx} + u\frac{\partial \rho u}{\partial x}\right) dx\, dy\, dz$$

Canceling out the product $dx\, dy\, dz$ and using the continuity equation gives

$$\frac{\partial p}{\partial x} + \frac{\partial \rho u}{\partial t} - u\frac{\partial \rho}{\partial t} + \rho\left(u\frac{\partial u}{\partial x} + v\frac{\partial u}{\partial y} + w\frac{\partial u}{\partial z}\right) = 0$$

But

$$\frac{\partial \rho u}{\partial t} = \rho\frac{\partial u}{\partial t} + u\frac{\partial \rho}{\partial t}$$

so that

$$\frac{\partial p}{\partial x} + \rho \frac{\partial u}{\partial t} + \rho \left(u \frac{\partial u}{\partial x} + v \frac{\partial u}{\partial y} + w \frac{\partial u}{\partial z} \right) = 0$$

or

$$-\frac{1}{\rho}\frac{\partial p}{\partial x} = \frac{\partial u}{\partial t} + u \frac{\partial u}{\partial x} + v \frac{\partial u}{\partial y} + w \frac{\partial u}{\partial z} \qquad (13.7)$$

The right-hand side of Eq. (13.7) can be recognized as the substantial derivative of u, or Du/Dt. Similar equations in the y and z directions yield

$$-\frac{1}{\rho}\frac{\partial p}{\partial y} = \frac{\partial v}{\partial t} + u \frac{\partial v}{\partial x} + v \frac{\partial v}{\partial y} + w \frac{\partial v}{\partial z} \qquad (13.8)$$

and

$$-\frac{1}{\rho}\frac{\partial p}{\partial z} = \frac{\partial w}{\partial t} + u \frac{\partial w}{\partial x} + v \frac{\partial w}{\partial y} + w \frac{\partial w}{\partial z} \qquad (13.9)$$

In vector notation, Eqs. (13.7), (13.8), and (13.9) can be written as

$$-\frac{1}{\rho}\nabla p = (\mathbf{V} \cdot \nabla)\mathbf{V} + \frac{\partial \mathbf{V}}{\partial t} \qquad (13.10)$$

13.4 Irrotational Flow

Direct integration of the equations of motion over the entire flow field is difficult unless further assumptions are made about the nature of the flow. An important type of flow, and one that leads to simplification in the equations of motion, is irrotational flow. Fluid rotation is defined as the average angular velocity of two mutually perpendicular differential elements in a fluid. For example, consider the rotation in the xy plane about P of two elements PR and PS of lengths dx and dy, as shown in Figure 13.3. Since the velocity

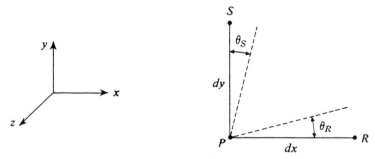

Figure 13.3

components u and v are different at P, R, and S, it follows that the elements will move with respect to each other. Let the velocity components at P be u and v. Then the velocity component in the x direction at S is $u + (\partial u/\partial y)\, dy$, and the y component at R is $v + (\partial v/\partial x)\, dx$. After a time dt, the point S will have moved in the x direction a distance $(\partial u/\partial y)\, dy\, dt$ with respect to P, so that the angle θ_s is equal to $(\partial u/\partial y)\, dt$. Likewise the point R will have moved in the y direction a distance $(\partial v/\partial x)\, dx\, dt$, so that $\theta_R = (\partial v/\partial x)\, dt$. Using the right-hand rule to find the net rotation about the z axis, it follows that counterclockwise angles in Figure 13.3 are positive. Therefore, the rotation about the z axis is

$$\omega_z = \frac{\dfrac{\partial v}{\partial x}\, dt - \dfrac{\partial u}{\partial y}\, dt}{2dt} = \frac{1}{2}\left(\frac{\partial v}{\partial x} - \frac{\partial u}{\partial y}\right) \tag{13.11}$$

Similarly, it can be shown that

$$\omega_x = \frac{1}{2}\left(\frac{\partial w}{\partial y} - \frac{\partial v}{\partial z}\right) \tag{13.12}$$

$$\omega_y = \frac{1}{2}\left(\frac{\partial u}{\partial z} - \frac{\partial w}{\partial x}\right) \tag{13.13}$$

Rotation can be defined in vector form very simply in terms of the curl of the velocity vector:

$$\boldsymbol{\omega} = \mathbf{i}\omega_x + \mathbf{j}\omega_y + \mathbf{k}\omega_z = \tfrac{1}{2}(\nabla \times \mathbf{V}) \tag{13.14}$$

where $\mathbf{i}$, $\mathbf{j}$, and $\mathbf{k}$ are unit vectors in the x, y, z directions. Two times the rotation vector is called the vorticity vector $(2\boldsymbol{\omega})$. For irrotational flow, it follows that

$$\frac{\partial u}{\partial z} = \frac{\partial w}{\partial x}, \quad \frac{\partial w}{\partial y} = \frac{\partial v}{\partial z}, \quad \frac{\partial u}{\partial y} = \frac{\partial v}{\partial x} \tag{13.15}$$

The simplification afforded by the assumption of irrotationality can be seen by substituting Eq. (13.15) into Eqs. (13.8), (13.9), and (13.10):

$$-\frac{1}{\rho}\frac{\partial p}{\partial x} = \frac{\partial u}{\partial t} + u\frac{\partial u}{\partial x} + v\frac{\partial v}{\partial x} + w\frac{\partial w}{\partial x}$$

$$-\frac{1}{\rho}\frac{\partial p}{\partial y} = \frac{\partial v}{\partial t} + u\frac{\partial u}{\partial y} + v\frac{\partial v}{\partial y} + w\frac{\partial w}{\partial y} \tag{13.16}$$

and

$$-\frac{1}{\rho}\frac{\partial p}{\partial z} = \frac{\partial w}{\partial t} + u\frac{\partial u}{\partial z} + v\frac{\partial v}{\partial z} + w\frac{\partial w}{\partial z}$$

For steady flow,

$$\frac{\partial u}{\partial t} = \frac{\partial v}{\partial t} = \frac{\partial w}{\partial t} = 0$$

Also,

$$dp = \frac{\partial \rho}{\partial x}\, dx + \frac{\partial p}{\partial y}\, dy + \frac{\partial p}{\partial z}\, dz$$

Furthermore,

$$u\frac{\partial u}{\partial x} + v\frac{\partial v}{\partial x} + w\frac{\partial w}{\partial x} = \frac{\partial}{\partial x}\frac{u^2 + v^2 + w^2}{2} = \frac{\partial}{\partial x}\frac{V^2}{2}$$

Substituting the preceding relations into Eq. (13.16) and adding yields

$$-\frac{dp}{\rho} = d\frac{V^2}{2}$$

Integrating yields

$$\int \frac{dp}{\rho} + d\frac{V}{2} = \text{constant} \tag{13.17}$$

Thus the equations of motion can be integrated for the case of irrotational flow. Assumptions made in the derivation are that the flow is irrotational, steady, frictionless, with no external forces except pressure acting on the control volume.

Equation (13.17) can also be derived using vector notation. From Eq. (13.10),

$$-\frac{\nabla p}{\rho} = (\mathbf{V}\cdot\nabla)\mathbf{V} + \frac{\partial \mathbf{V}}{\partial t}$$

It is possible to establish the following vector identity[1]; for vectors $\mathbf{A}$, $\mathbf{B}$,

$$\nabla(\mathbf{A}\cdot\mathbf{B}) = \mathbf{A}\times(\nabla\times\mathbf{B}) + (\mathbf{A}\cdot\nabla)\mathbf{B} + \mathbf{B}\times(\nabla\times\mathbf{A}) + (\mathbf{B}\cdot\nabla)\mathbf{A}$$

For our case, let $\mathbf{A} = \mathbf{B} = \mathbf{V}$, so that

$$\nabla(V^2) = 2[\mathbf{V}\times(\nabla\times\mathbf{V})] + 2[(\mathbf{V}\cdot\nabla)\mathbf{V}]$$

or

$$(\mathbf{V}\cdot\nabla)\mathbf{V} = \nabla\frac{V^2}{2} - \mathbf{V}\times(\nabla\times\mathbf{V}) \tag{13.18}$$

Substituting Eq. (13.18) into Eq. (13.10) yields

$$-\frac{\nabla p}{\rho} = -2(\mathbf{V}\times\boldsymbol{\omega}) + \nabla\frac{V^2}{2} + \frac{\partial \mathbf{V}}{\partial t} \tag{13.19}$$

For irrotational flow, $\boldsymbol{\omega} = 0$. For steady flow, $\partial \mathbf{V}/\partial t = 0$. With the assumptions of steady, irrotational flow, the resultant equation is

$$-\frac{\nabla p}{\rho} = \nabla\frac{V^2}{2}$$

agreeing with Eq. (13.17).

It is interesting to observe that Eq. (13.17) is the same equation already derived in Chapter 3, Eq. (3.2), from energy considerations, for isentropic flow. There seems to be a connection between irrotational flow and isentropic flow. To demonstrate this connection, let us write the second law, Eq. (1.22), in general form:

$$TVs = \nabla h - \frac{\nabla p}{\rho} \qquad\qquad \textbf{(13.20)}$$

Also, since

$$h_t = h + \frac{V^2}{2}$$

we have

$$\nabla h_t = \nabla h + \frac{\nabla(V \cdot V)}{2} \qquad\qquad \textbf{(13.21)}$$

Now, using Eq. (13.19), we obtain

$$-\frac{\nabla p}{\rho} = -2(V \times \omega) + \frac{\nabla(V \cdot V)}{2} + \frac{\partial V}{\partial t}$$

Substituting Eqs. (13.19) and (13.21) into Eq. (13.20) gives

$$TVs - \nabla h_t + \frac{\nabla(V \cdot V)}{2} = -2(V \times \omega) + \frac{\nabla(V \cdot V)}{2} + \frac{\partial V}{\partial t}$$

Simplifying yields

$$TVs - \nabla h_t = 2(\omega \times V) + \frac{\partial V}{\partial t} \qquad\qquad \textbf{(13.22)}$$

For adiabatic steady flow with no energy exchange ($\nabla h_t = 0$), Eq. (13.22) reduces to

$$TVs = 2(\omega \times V) \qquad\qquad \textbf{(13.23)}$$

Irrotational, steady flow, therefore, with $\omega = 0$, corresponds to isentropic flow. This explains the equivalence of Eqs. (3.2) and (13.10), the former derived for isentropic flow, the latter for irrotational flow. As an example, flow in a boundary layer, with viscous dissipation, is irreversible and nonisentropic. Hence, from Eq. (13.23), boundary layer flow is rotational.

13.5 Velocity Potential

The velocity potential ϕ is defined as a function of the space coordinates and time such that, when differentiated with respect to a space coordinate, there results the velocity component in the coordinate direction. In component form,

$$u = -\frac{\partial \phi}{\partial x}$$

$$v = -\frac{\partial \phi}{\partial y}$$ (13.24)

$$w = -\frac{\partial \phi}{\partial z}$$

In vector form,

$$\mathbf{V} = -\nabla \phi$$ (13.25)

If such a function can be defined at each point in the flow and has continuous derivatives, the flow is termed potential flow. The velocity potential is analogous to electric potential in an electric field, in which current flow can be related to the potential gradient, or to temperature in a heat-conducting material, where heat flow is proportional to temperature gradient. The negative sign in Eq. (13.24) conforms with convention, in that flow is from a higher to a lower potential. The advantage of introducing velocity potential into multidimensional fluid flow problems is simply that of being able to express the flow in terms of one scalar potential instead of the three velocity components u, v, and w. For incompressible flow, the continuity equation has been shown to yield

$$\frac{\partial u}{\partial x} + \frac{\partial v}{\partial y} + \frac{\partial w}{\partial z} = 0$$

In terms of ϕ, this reduces to the simple form

$$\frac{\partial^2 \phi}{\partial x^2} + \frac{\partial^2 \phi}{\partial y^2} + \frac{\partial^2 \phi}{\partial z^2} = 0$$ (13.26)

This equation can be recognized as *Laplace's equation*, a linear partial differential equation, for which the solutions are harmonic functions.

Let us now discuss the connection between irrotational and potential flow. From calculus, if a function is continuous and has continuous derivatives, the order of partial differentiation has no effect on the result. That is,

$$\frac{\partial^2 \phi}{\partial x \, \partial y} = \frac{\partial^2 \phi}{\partial y \, \partial x}$$

$$\frac{\partial^2 \phi}{\partial x \, \partial z} = \frac{\partial^2 \phi}{\partial z \, \partial x}$$

$$\frac{\partial^2 \phi}{\partial y \, \partial z} = \frac{\partial^2 \phi}{\partial z \, \partial y}$$

Substituting Eq. (13.24) yields

$$\frac{\partial u}{\partial y} = \frac{\partial v}{\partial x}$$

$$\frac{\partial u}{\partial z} = \frac{\partial w}{\partial x}$$

$$\frac{\partial v}{\partial z} = \frac{\partial w}{\partial y}$$

But, according to Eq. (13.15) these are the conditions for the flow to be irrotational. In other words, a potential flow is also irrotational. Let us now demonstrate the converse. We shall assume that the conditions for irrotational flow, as given above, are satisfied. From calculus[2], if functions P, Q, and R, each functions of x, y, and z, satisfy the conditions

$$\frac{\partial P}{\partial y} = \frac{\partial Q}{\partial x}, \qquad \frac{\partial Q}{\partial z} = \frac{\partial R}{\partial y}, \qquad \frac{\partial R}{\partial x} = \frac{\partial P}{\partial z}$$

and are continuous with continuous partial derivatives, then $P\,dx + Q\,dy + R\,dz$ is an exact differential of some function ϕ of x, y, and z, where

$$d\phi = P\,dx + Q\,dy + R\,dz$$

For our case, with $P = -u$, $Q = -v$, and $R = -w$,

$$+d\phi = -u\,dx - v\,dy - w\,dz$$

Also, since ϕ is a function of three space coordinates,

$$d\phi = \frac{\partial \phi}{\partial x}\,dx + \frac{\partial \phi}{\partial y}\,dy + \frac{\partial \phi}{\partial z}\,dz$$

Comparing, we have

$$u = -\frac{\partial \phi}{\partial x}, \qquad v = -\frac{\partial \phi}{\partial y}, \qquad w = -\frac{\partial \phi}{\partial z}$$

which is the definition of velocity potential. Thus we have proved the corollary, that for irrotational flow, a potential can be defined. In other words, potential flow and irrotational flow are equivalent and can be used interchangeably.

13.6 Equations of Motion in Terms of Velocity Potential

We shall now take the equations of motion, as derived previously in this chapter, and reduce them to one partial differential equation for the velocity potential ϕ. It will be assumed in this derivation that the flow is steady, irrotational, and isentropic, and that the only forces acting on the control volume are pressure forces (viscous forces, magnetic forces, and so on, are assumed zero).

The continuity equation for steady flow, Eq. (13.1), reduces to

$$\frac{\partial}{\partial x}\left(\rho\frac{\partial \phi}{\partial x}\right) + \frac{\partial}{\partial y}\left(\rho\frac{\partial \phi}{\partial y}\right) + \frac{\partial}{\partial z}\left(\rho\frac{\partial \phi}{\partial z}\right) = 0$$

or, expanding,

$$\frac{\partial \rho}{\partial x}\frac{\partial \phi}{\partial x} + \frac{\partial \rho}{\partial y}\frac{\partial \phi}{\partial y} + \frac{\partial \rho}{\partial z}\frac{\partial \phi}{\partial z} + \rho\left(\frac{\partial^2 \phi}{\partial x^2} + \frac{\partial^2 \phi}{\partial y^2} + \frac{\partial^2 \phi}{\partial z^2}\right) = 0 \qquad \textbf{(13.27)}$$

To obtain a partial differential equation for ϕ alone, we shall have to eliminate ρ using the momentum equation. Under the assumptions made, the momentum equation is given by Eq. (13.17):

$$\frac{dp}{\rho} + \frac{d(u^2 + v^2 + w^2)}{2} = 0$$

or

$$dp = -\frac{\rho}{2}d\left[\left(\frac{\partial \phi}{\partial x}\right)^2 + \left(\frac{\partial \phi}{\partial y}\right)^2 + \left(\frac{\partial \phi}{\partial z}\right)^2\right] \qquad \textbf{(13.28)}$$

For isentropic flow,

$$\frac{dp}{d\rho} = a^2$$

so that

$$a^2\, d\rho = -\frac{\rho}{2}d\left[\left(\frac{\partial \phi}{\partial x}\right)^2 + \left(\frac{\partial \phi}{\partial y}\right)^2 + \left(\frac{\partial \phi}{\partial z}\right)^2\right]$$

Now,

$$\frac{\partial \rho}{\partial x} = -\frac{\rho}{a^2}\left(\frac{\partial \phi}{\partial x}\frac{\partial^2 \phi}{\partial x^2} + \frac{\partial \phi}{\partial y}\frac{\partial^2 \phi}{\partial y\, \partial x} + \frac{\partial \phi}{\partial z}\frac{\partial^2 \phi}{\partial z\, \partial x}\right)$$

$$\frac{\partial \rho}{\partial y} = -\frac{\rho}{a^2}\left(\frac{\partial \phi}{\partial x}\frac{\partial^2 \phi}{\partial x\, \partial y} + \frac{\partial \phi}{\partial y}\frac{\partial^2 \phi}{\partial y^2} + \frac{\partial \phi}{\partial z}\frac{\partial^2 \phi}{\partial z\, \partial y}\right) \qquad \textbf{(13.29)}$$

$$\frac{\partial \rho}{\partial z} = -\frac{\rho}{a^2}\left(\frac{\partial \phi}{\partial x}\frac{\partial^2 \phi}{\partial x\, \partial z} + \frac{\partial \phi}{\partial y}\frac{\partial^2 \phi}{\partial y\, \partial z} + \frac{\partial \phi}{\partial z}\frac{\partial^2 \phi}{\partial z^2}\right)$$

Substituting into Eq. (13.27) yields

$$\frac{\partial^2 \phi}{\partial x^2} + \frac{\partial^2 \phi}{\partial y^2} + \frac{\partial^2 \phi}{\partial z^2} - \frac{1}{a^2}\left[\left(\frac{\partial \phi}{\partial x}\right)^2\frac{\partial^2 \phi}{\partial x^2} + \left(\frac{\partial \phi}{\partial y}\right)^2\frac{\partial^2 \phi}{\partial y^2} + \left(\frac{\partial \phi}{\partial z}\right)^2\frac{\partial^2 \phi}{\partial z^2}\right]$$

$$-\frac{2}{a^2}\left[\frac{\partial \phi}{\partial x}\frac{\partial \phi}{\partial y}\frac{\partial^2 \phi}{\partial x\, \partial y} + \frac{\partial \phi}{\partial y}\frac{\partial \phi}{\partial z}\frac{\partial^2 \phi}{\partial y\, \partial z} + \frac{\partial \phi}{\partial z}\frac{\partial \phi}{\partial x}\frac{\partial^2 \phi}{\partial z\, \partial x}\right] = 0 \qquad \textbf{(13.30)}$$

It can be seen that, for incompressible flow, with $a \to \infty$, Eq. (13.30) reduces to Laplace's equation, $(\partial^2 \phi / \partial x^2) + (\partial^2 \phi / \partial y^2) + (\partial^2 \phi / \partial z^2) = 0$, as mentioned previously.

To complete the derivation, it is necessary to obtain an expression for the velocity of sound in terms of the velocity potential and space coordinates. Such an expression can be obtained from the energy equation (see Sections 3.2 and 3.4). For a perfect gas with constant specific heats,

$$c_p T_t = c_p T + \frac{V^2}{2}, \qquad \text{where } c_p = \frac{R\gamma}{\gamma - 1} \quad \text{and} \quad a^2 = \gamma RT$$

Substituting yields

$$a^2 = \gamma RT_t - \frac{\gamma - 1}{2} V^2$$

$$a^2 = a_t^2 - \frac{\gamma - 1}{2} \left[\left(\frac{\partial \phi}{\partial x}\right)^2 + \left(\frac{\partial \phi}{\partial y}\right)^2 + \left(\frac{\partial \phi}{\partial z}\right)^2 \right] \qquad \textbf{(13.31)}$$

where a_t is the velocity of sound at the stagnation temperature.

Substituting Eq. (13.31) into Eq. (13.30), we obtain a partial differential equation for ϕ in terms of x, y, and z for steady, irrotational, isentropic flow of a perfect gas with constant specific heats. Solution of this equation for ϕ will yield the entire flow for prescribed boundary conditions.

The equation can be recognized as nonlinear, since terms appear such as $[(\partial \phi / \partial x)(\partial \phi / \partial y)(\partial^2 \phi / \partial x\, \partial y)]$ containing the products of the partial derivatives of the dependent variable. Furthermore, the partial differential equation for ϕ is second order, since the highest-order derivative is of the form $(\partial^2 \phi / \partial x^2)$. The nonlinearity of the equation presents extreme mathematical difficulties, which render a general solution impossible. The advantage of a linear equation, such as Laplace's equation, is that solutions may be added; complex solutions can be built up from the superposition of simple solutions. Thus one method of approach to Eq. (13.30) is to try to linearize the general partial differential equation by making suitable assumptions. This necessarily restricts the validity of the solution to a special class of problems, yet it does afford good engineering approximations to cases that might otherwise prove insoluble. In the following chapters, methods of linearizing the potential equation will be discussed, as well as approximate methods for solving the general potential equation.

13.7 Summary

Chapter 13 has presented the generalized equations of motion for three-dimensional compressible flow. Once again, the student must keep in mind the assumptions made in the derivation, that is, isentropic, steady flow, with pressure forces the only external forces acting on the control volume. When the complexity of the resultant nonlinear partial differential equations is seen, one can appreciate the great advantages inherent in the assumption of one-dimensional flow made in the first 12 chapters.

REFERENCES

Specific References

1. SPIEGEL, M.R., *Advanced Mathematics for Engineers and Scientists*, Schaums Outline Series, New York, McGraw-Hill Book Company, 1971, p. 127.
2. SPIEGEL, M.R., *Advanced Mathematics for Engineers and Scientists*, Schaums Outline Series, New York, McGraw-Hill Book Company, 1971, p. 152.

General References

3. SHAPIRO, A.H., *The Dynamics and Thermodynamics of Compressible Fluid Flow*, Vol. I, New York, Ronald Press, 1953.
4. ZUCROW, M.J., AND HOFFMAN, J.D., *Gas Dynamics*, Vol. 1, New York, John Wiley & Sons, Inc., 1976.

PROBLEMS

1. According to the generalized continuity equation given by Eq. (13.1), for steady, incompressible, one-dimensional flow, $\partial u/\partial x = 0$, or, in other words, u is equal to a constant. Previously, for incompressible flow, however, it has been customary to assume that, for steady, one-dimensional flow, the product of velocity and cross-sectional area (AV) is a constant. Explain this seeming contradiction.

2. Expand Eq. (13.10) into the three component equations, and show that Eqs. (13.7), (13.8), and (13.9) result.

3. Under what conditions can it be assumed that $p + \frac{1}{2}\rho V^2$ is equal to a constant?

4. The velocity components for a possible flow field are given by $u = -3x^2 + 2y$ and $v = 2x + 2y$. Is the flow irrotational? If so, determine the velocity potential.

5. Consider a steady, uniform flow with velocity components $u = 120$ m/s and $w = 0$. Determine the velocity potential, substitute into Eq. (13.31), and find the resultant difference between static and stagnation temperature. Use the properties of air in your solution ($\gamma = 1.4$, $R = 0.2870$ kJ/kg·K).

14

LINEARIZED FLOWS

14.1 Introduction

The equations of motion for two-dimensional compressible steady flow, as derived in Chapter 13, turn out to be nonlinear partial differential equations. The mathematics associated with a system of nonlinear partial differential equations is extremely complex; in many cases, solutions are not possible. To reduce the equations to a more workable form, we shall, in this chapter, place certain restrictions on the flow and thereby linearize the partial differential equations.

If a body when placed in a uniform flow causes only small disturbances to that flow, then it is possible to linearize the equations of fluid motion and thereby effect a solution. According to this method of small disturbances or small perturbations, the flow is considered to be made up of a uniform flow on which is superposed perturbation velocities. We shall first discuss the assumptions inherent in small perturbation theory, both as regards the flow and the body shape. Then we shall go over solutions, both for subsonic and supersonic flows, of the linearized equations. Of interest here is the contrast between the nature of subsonic and supersonic flow. Although the method of small perturbations is approximate, it does succeed in pointing up the very basic differences between the two types of flow. Furthermore, it will be shown possible, at least for subsonic flow, to derive similarity laws relating a compressible flow to the corresponding incompressible flow. This is quite important, in that incompressible solutions are already available for a wide variety of physical situations.

14.2 Linearization of the Potential Equation

One case for which the potential equation, Eq. (13.30), for isentropic steady flow can be linearized is that of flow over a body that presents only a small disturbance to the flow. Let us consider uniform parallel flow in the positive x direction over a thin body aligned with the flow (see Figure 14.1). The potential of the total velocity field can be considered to be made up of two parts, one part due to the uniform flow and the other part due to the perturbation introduced into the flow by the thin body. We shall assume that the flow is steady and two dimensional, so flow properties do not vary in the z direction. The presence of the body gives rise to small perturbation velocities, u_p and v_p in the x and y directions, respectively. From the definition of ϕ, we have

$$\frac{\partial \phi}{\partial x} = -U_\infty - u_p$$

$$\frac{\partial \phi}{\partial y} = -v_p$$

(14.1)

Substituting into Eq. (13.30) for two-dimensional flow,

$$\frac{\partial^2 \phi}{\partial x^2} + \frac{\partial^2 \phi}{\partial y^2} - \frac{1}{a^2}\left[(-U_\infty - u_p)^2 \frac{\partial^2 \phi}{\partial x^2} + v_p^2 \frac{\partial^2 \phi}{\partial y^2}\right]$$

$$-\frac{2}{a^2}(-U_\infty - u_p)(-v_p)\frac{\partial^2 \phi}{\partial x\, \partial y} = 0 \quad (14.2)$$

It is necessary also to get an expression for the sound velocity. From Eq. (13.31),

$$a^2 = a_t^2 - \frac{\gamma - 1}{2}[(-U_\infty - u_p)^2 + (-v_p)^2] \quad (14.3)$$

where

$$a_t^2 = \gamma R T_t$$

From the first law, for adiabatic flow, we have demonstrated that

$$c_p T_t = c_p T_\infty + \frac{U_\infty^2}{2} = \text{constant}$$

Figure 14.1

so that

$$a_t^2 = \gamma R \left(T_\infty + \frac{U_\infty^2}{2c_p} \right)$$

$$= a_\infty^2 + \frac{\gamma - 1}{2} U_\infty^2 \qquad (14.4)$$

We can now combine Eqs. (14.4) and (14.3) to yield

$$a^2 = a_\infty^2 - \frac{\gamma - 1}{2} [u_p^2 + 2u_p U_\infty + v_p^2] \qquad (14.5)$$

Substituting Eq. (14.5) into Eq. (14.2), we obtain

$$\left(\frac{\partial^2 \phi}{\partial x^2} + \frac{\partial^2 \phi}{\partial y^2} \right) a_\infty^2 + \left(\frac{\partial^2 \phi}{\partial x^2} + \frac{\partial^2 \phi}{\partial y^2} \right) \left(-\frac{\gamma - 1}{2} \right) (u_p^2 + 2u_p U_\infty + v_p^2)$$

$$= (U_\infty + u_p)^2 \frac{\partial^2 \phi}{\partial x^2} + v_p^2 \frac{\partial^2 \phi}{\partial y^2} + 2(U_\infty + u_p) v_p \frac{\partial^2 \phi}{\partial x\, \partial y} \qquad (14.6)$$

Since $M_\infty = U_\infty / a_\infty$, divide Eq. (14.6) by a_∞^2 and simplify to obtain

$$\frac{\partial^2 \phi}{\partial x^2} (1 - M_\infty^2) + \frac{\partial^2 \phi}{\partial y^2} = \frac{\gamma - 1}{2} M_\infty^2 \left(\frac{\partial^2 \phi}{\partial x^2} + \frac{\partial^2 \phi}{\partial y^2} \right) \left(\frac{u_p^2}{U_\infty^2} + \frac{2u_p}{U_\infty} + \frac{v_p^2}{U_\infty^2} \right)$$

$$+ \left(\frac{u_p^2}{U_\infty^2} + \frac{2u_p}{U_\infty} \right) M_\infty^2 \frac{\partial^2 \phi}{\partial x^2} + \frac{v_p^2}{U_\infty^2} M_\infty^2 \frac{\partial^2 \phi}{\partial y^2}$$

$$+ 2 \left(1 + \frac{u_p}{U_\infty} \right) \frac{v_p}{U_\infty} M_\infty^2 \frac{\partial^2 \phi}{\partial x\, \partial y} \qquad (14.7)$$

Since the perturbation velocities, u_p and v_p, are assumed small compared with U_∞,

$$M_\infty^2 \left(\frac{u_p}{U_\infty} \right)^2 \ll 1, \qquad M_\infty^2 \frac{v_p^2}{U_\infty^2} \ll 1, \qquad M_\infty^2 \frac{u_p v_p}{U_\infty^2} \ll 1 \qquad (14.8)$$

and Eq. (14.7) reduces to

$$\frac{\partial^2 \phi}{\partial x^2} (1 - M_\infty^2) + \frac{\partial^2 \phi}{\partial y^2} = \frac{\gamma - 1}{2} M_\infty^2 \left(\frac{\partial^2 \phi}{\partial x^2} + \frac{\partial^2 \phi}{\partial y^2} \right) \frac{2u_p}{U_\infty}$$

$$+ \frac{2u_p}{U_\infty} M_\infty^2 \frac{\partial^2 \phi}{\partial x^2} + \frac{2v_p}{U_\infty} M_\infty^2 \frac{\partial^2 \phi}{\partial x\, \partial y} = \left(M_\infty^2 (\gamma + 1) \frac{u_p}{U_\infty} \right) \frac{\partial^2 \phi}{\partial x^2}$$

$$+ \left(M_\infty^2 (\gamma - 1) \frac{u_p}{U_\infty} \right) \frac{\partial^2 \phi}{\partial y^2} + 2 M_\infty^2 \frac{v_p}{U_\infty} \frac{\partial^2 \phi}{\partial x\, \partial y} \qquad (14.9)$$

Unfortunately, this equation is still nonlinear, containing terms of the form $u_p(\partial^2 \phi / \partial x^2)$. To linearize the equation, it is therefore necessary to place additional restrictions on the flow. If we assume

$$\frac{M_\infty^2 (u_p / U_\infty)}{1 - M_\infty^2} \ll 1 \qquad (14.10)$$

then terms containing the nonlinear $(u_p/U_\infty)(\partial^2\phi/\partial x^2)$ are eliminated. If we further assume that $M_\infty^2(u_p/U_\infty)$ and $M_\infty^2(v_p/U_\infty) \ll 1$, then terms containing $u_p(\partial^2\phi/\partial y^2)$ and $v_p(\partial^2\phi/\partial x\,\partial y)$ are likewise eliminated. The resultant equation is

$$(1 - M_\infty^2)\frac{\partial^2\phi}{\partial x^2} + \frac{\partial^2\phi}{\partial y^2} = 0 \tag{14.11}$$

This equation is valid for steady, two-dimensional, irrotational and isentropic flow of a perfect gas with constant specific heats around a thin body. From the assumptions made in Eq. (14.10), it is further necessary to state that for Eq. (14.11) to hold M_∞ cannot be 1 or close to 1, so Eq. (14.11) is not valid for transonic flow. For transonic flow, the term containing $(u_p/U_\infty)(\partial^2\phi/\partial x^2)$ would have to be retained, so that for transonic flow

$$(1 - M_\infty^2)\frac{\partial^2\phi}{\partial x^2} + \frac{\partial^2\phi}{\partial y^2} = M_\infty^2(\gamma + 1)\frac{u_p}{U_\infty}\frac{\partial^2\phi}{\partial x^2} \tag{14.12}$$

Furthermore, for $M_\infty^2(u_p/U_\infty) \ll 1$, M_∞ cannot be too great, so Eq. (14.11) does not hold for hypersonic flow ($M_\infty \gg 3$ to 5).

Under these assumptions, then, we are left with Eq. (14.11), a linear partial differential equation. For $M_\infty < 1$, the equation is of the elliptic type, with solutions possessing similarity to the solutions of Laplace's equation (incompressible flow). For supersonic flow, however, with $M_\infty^2 - 1 > 0$, the equation is of the hyperbolic type, which can be transformed into the wave equation. Thus supersonic solutions are similar to those for a vibrating string or other wave phenomena. Needless to say, subsonic solutions have entirely different features than supersonic solutions.

It is advantageous to split the potential ϕ into two parts, one due to the uniform flow and the other due to the perturbation. For the linearized equation, for which solutions can be added,

$$\phi = \phi_p - U_\infty x \quad \text{with} \quad u_p = -\frac{\partial\phi_p}{\partial x}, \quad v_p = -\frac{\partial\phi_p}{\partial y}$$

Note that ϕ_p satisfies Eq. (14.11):

$$(1 - M_\infty^2)\frac{\partial^2\phi_p}{\partial x^2} + \frac{\partial^2\phi_p}{\partial y^2} = 0 \tag{14.13}$$

A solution of Eq. (14.13) for ϕ_p requires a specification of the boundary conditions both on the body and at infinity. For this irrotational flow, in which we neglect the effects of viscosity and hence the boundary layer on the body, the boundary conditions on the body surface require that the flow velocity at the surface be tangent to the body. Place the body on the x axis, as shown in Figure 14.2. The slope of the body surface $(dy/dx)_b$ must be equal to the slope of the velocity vector at the surface.

$$\left(\frac{dy}{dx}\right)_b = \frac{v_p}{u_p + U_\infty} \tag{14.14}$$

Figure 14.2

Since

$$\frac{u_p}{U_\infty} \ll 1$$

$$\left(\frac{dy}{dx}\right)_b = \frac{v_p}{U_\infty} \qquad (14.15)$$

According to the assumptions necessary for small perturbation flow, the y coordinate of the body, y_b, must be small. It is possible, then, to write a Taylor's series expansion for the perturbation velocities at a point (x_b, y_b) on the body surface:

$$v_p(x_b, y_b) = v_p(x_b, 0) + \left(\frac{\partial v_p}{\partial y}\right)_{x_b,0} y_b + \cdots$$

Neglecting all terms but the first yields

$$v_p(x_b, y_b) = v_p(x_b, 0)$$

The linearized boundary condition, Eq. (14.15), becomes

$$\left(\frac{dy}{dx}\right)_b = \frac{v_p(x_b, 0)}{U_\infty} \qquad (14.16)$$

The boundary condition at infinity for linearized flow is that the perturbation velocities must be finite or zero; they cannot go to infinity in magnitude.

Before proceeding to an example, we shall first derive an expression for pressure in a linearized two-dimensional flow. It is customary to express pressure in terms of a nondimensional pressure coefficient C_p where

$$C_p = \frac{p - p_\infty}{\frac{1}{2}\rho_\infty U_\infty^2} = \frac{p/p_\infty - 1}{\frac{1}{2}\gamma M_\infty^2} \qquad (14.17)$$

The energy equation for isentropic flow yields

$$c_p T_\infty + \frac{U_\infty^2}{2} = c_p T + \frac{(U_\infty + u_p)^2 + v_p^2}{2}$$

Simplifying gives

$$T_\infty - T = \frac{2U_\infty u_p + u_p^2 + v_p^2}{2c_p}$$

where

$$c_p = \frac{R\gamma}{\gamma - 1}$$

Dividing by T_∞,

$$\frac{T}{T_\infty} = 1 - \frac{\gamma - 1}{2}\frac{2U_\infty u_p + u_p^2 + v_p^2}{R\gamma T_\infty}$$

$$= 1 - \frac{\gamma - 1}{2}M_\infty^2\left(2\frac{u_p}{U_\infty} + \frac{u_p^2 + v_p^2}{U_\infty^2}\right)$$

For an isentropic process and a perfect gas with constant specific heats,

$$\frac{p}{p_\infty} = \left(\frac{T}{T_\infty}\right)^{\gamma/(\gamma-1)}.$$

so that

$$C_p = \frac{\left[1 - \dfrac{\gamma - 1}{2}M_\infty^2\left(2\dfrac{u_p}{U_\infty} + \dfrac{u_p^2 + v_p^2}{U_\infty^2}\right)\right]^{\gamma/(\gamma-1)} - 1}{\frac{1}{2}\gamma M_\infty^2}$$

The term in brackets can be expanded using the binomial theorem (for $x < 1, (1 + x)^m = 1 + mx + (m(m - 1)x^2/2!) + \cdots)$. We obtain

$$\left[1 - \frac{\gamma - 1}{2}M_\infty^2\left(2\frac{u_p}{U_\infty} + \frac{u_p^2 + v_p^2}{U_\infty^2}\right)\right]^{\gamma/(\gamma-1)}$$

$$= 1 - \frac{\gamma M_\infty^2}{2}\left(2\frac{u_p}{U_\infty} + \frac{u_p^2 + v_p^2}{U_\infty^2}\right) + \cdots$$

Under the assumptions of the small perturbation theory, the rest of the terms in the series can be dropped, so that

$$C_p = -2\frac{u_p}{U_\infty} - \frac{u_p^2 + v_p^2}{U_\infty^2} \tag{14.18}$$

For

$$\frac{u_p^2}{U_\infty^2} \quad \text{and} \quad \frac{v_p^2}{U_\infty^2} \ll 1 \qquad \text{[see Eq. (14.8)]}$$

$$C_p = -\frac{2u_p}{U_\infty} \tag{14.19}$$

Having established the linearized partial differential equation for small perturbations and the appropriate boundary conditions, we are now in a position to solve a simple boundary-value problem. In the next section, we shall discuss such a simple problem for subsonic flow and then return, in a later section, to the same problem for supersonic flow.

14.3 Subsonic Flow Over a Wavy Wall

Consider a uniform subsonic flow in the positive x direction of magnitude U_∞, as shown in Figure 14.3. The flow passes over a two-dimensional wave-shaped wall, with the equation of the wall being

$$y_b = A \sin \frac{2\pi x}{\lambda}$$

We shall treat this case using small perturbation theory, so we must assume $A \ll \lambda$. We wish to solve the partial differential equation

$$(1 - M_\infty^2) \frac{\partial^2 \phi_p}{\partial x^2} + \frac{\partial^2 \phi_p}{\partial y^2} = 0$$

subject to the boundary conditions,

$$v_p(x_b, 0) = U_\infty \left(\frac{dy}{dx}\right)_b = U_\infty A \frac{2\pi}{\lambda} \cos \frac{2\pi x}{\lambda}$$

and $u_p(x, \infty)$ and $v_p(x, \infty)$ are finite.

One of the methods of solving a linear partial differential equation is to assume that the potential function can be written as the product of functions X and Y, where X is a function of x alone, and Y is a function of y alone. This method will be tried here. The assumption of the product form can only be justified if it is possible to find a solution of this form that satisfies the boundary conditions. If this is not possible, then clearly another method of solution must be employed.

Assume $\phi_p = X(x)Y(y)$ and substitute into

$$\frac{\partial^2 \phi_p}{\partial x^2} + \frac{1}{1 - M_\infty^2} \frac{\partial^2 \phi_p}{\partial y^2} = 0$$

Then

$$\frac{1}{X} \frac{d^2 X}{dx^2} = -\frac{1}{1 - M_\infty^2} \frac{1}{Y} \frac{d^2 Y}{dy^2}$$

In this expression, partial derivatives are not used since X and Y are each functions of one variable. Since the left-hand side of the preceding equation is

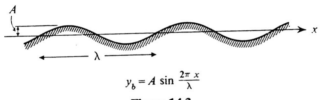

$$y_b = A \sin \frac{2\pi x}{\lambda}$$

Figure 14.3

dependent on x alone, and the right-hand side is dependent on y, with x and y independent variables, it follows that the left-hand side and the right-hand side are each equal to a constant. Letting this constant be $-k^2$, we obtain

$$\frac{d^2X}{dx^2} + k^2 X = 0 \tag{14.20}$$

$$\frac{d^2Y}{dy^2} - (1 - M_\infty^2)k^2 Y = 0 \tag{14.21}$$

Now we have to solve these two ordinary differential equations. The solution to Eq. (14.20) is

$$X = c_1 \cos kx + c_2 \sin kx \tag{14.22}$$

The solution to Eq. (14.21) is

$$Y = c_3 e^{\sqrt{1-M_\infty^2}\,ky} + c_4 e^{-\sqrt{1-M_\infty^2}\,ky} \tag{14.23}$$

Combining yields

$$\phi_p = (c_1 \cos kx + c_2 \sin kx)(c_3 e^{\sqrt{1-M_\infty^2}\,ky} + c_4 e^{-\sqrt{1-M_\infty^2}\,ky}) \tag{14.24}$$

The constants $c_1, c_2, c_3,$ and c_4 can be determined from the boundary conditions. At the wall,

$$v_p = -\frac{\partial \phi_p}{\partial y} = U_\infty A \frac{2\pi}{\lambda} \cos \frac{2\pi x}{\lambda} \quad \text{for} \quad y = 0$$

At infinity, $\partial \phi_p/\partial x$ and $\partial \phi_p/\partial y$ must be finite or zero.

From the second condition, we find that c_3 is equal to zero; from the first condition,

$$U_\infty A \frac{2\pi}{\lambda} \cos \frac{2\pi x}{\lambda} = -[(c_1 \cos kx + c_2 \sin kx)][-c_4 \sqrt{1 - M_\infty^2}\,k]$$

Matching terms, we obtain

$$c_2 = 0$$

$$c_1 c_4 \sqrt{1 - M_\infty^2} \frac{2\pi}{\lambda} = U_\infty A \frac{2\pi}{\lambda} \quad \text{with} \quad k = \frac{2\pi}{\lambda}$$

so that

$$c_1 c_4 = \frac{U_\infty A}{\sqrt{1 - M_\infty^2}}$$

Therefore,

$$\phi_p = \frac{U_\infty A}{\sqrt{1 - M_\infty^2}} e^{-(1-M_\infty^2)(2\pi/\lambda)y} \cos \frac{2\pi x}{\lambda} \tag{14.25}$$

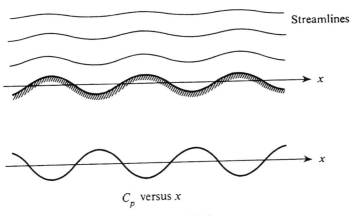

Figure 14.4

For the pressure distribution on the wall, use Eq. (14.19):

$$C_p = -\frac{2u_p}{U_\infty}$$

$$= \frac{2(\partial \phi_p / \partial x)}{U_\infty}$$

$$= -\frac{4\pi A}{\lambda \sqrt{1 - M_\infty^2}} \sin \frac{2\pi x_b}{\lambda} \qquad (14.26)$$

The flow streamlines and the pressure variation along the wall have been sketched in Figure 14.4. It can be seen that, for subsonic flow, the disturbance created by the wall dies out as y goes to infinity. Also, the pressure peaks occur in the troughs of the wall, and vice versa, so the pressure is in phase with the wall. Thus there is no drag force (force in the x direction) on the wall.

14.4 Similarity Laws for Subsonic Flow

For small-perturbation, linearized compressible flow, it has been shown that

$$(1 - M_\infty^2)\frac{\partial^2 \phi}{\partial x^2} + \frac{\partial^2 \phi}{\partial y^2} = 0 \qquad (14.11)$$

For incompressible, two-dimensional, steady potential flow, the potential equation reduces to Laplace's equation:

$$\frac{\partial^2 \phi_i}{\partial x_i^2} + \frac{\partial^2 \phi_i}{\partial y_i^2} = 0 \qquad (14.27)$$

where ϕ_i is the velocity potential and x_i and y_i the space coordinates for an incompressible flow field. Solutions to Laplace's equation are available for a wide variety of boundary conditions. Thus it would seem logical to try to transform the linearized, compressible potential equation into the incompressible potential equation, so as to utilize available incompressible flow solutions for problems in compressible flow. Consider a thin body in the x, y plane immersed in a uniform compressible flow U_∞ of Mach number M_∞. We shall transform this flow to the incompressible plane x_i, y_i. Let

$$x_i = k_1 x$$

$$y_i = k_2 y$$

$$\phi_i = k_3 \phi \tag{14.28}$$

$$U_{\infty_i} = k_4 U_\infty$$

and substitute into Eq. (14.11).

$$k_1^2 \frac{(1 - M_\infty^2) \partial^2 \phi_i}{k_3 \quad \partial x_i^2} + \frac{k_2^2}{k_3} \frac{\partial^2 \phi_i}{\partial y_i^2} = 0 \tag{14.29}$$

In order to transform Eq. (14.11) into Laplace's equation, using the relationships given by Eq. (14.28), it follows that

$$\frac{k_1}{k_2} = \frac{1}{\sqrt{1 - M_\infty^2}} \tag{14.30}$$

Notice that the term $\sqrt{1 - M_\infty^2}$ becomes imaginary for supersonic flow; the transformation can only be applied to subsonic flow. Physically, we have shown that, if a thin two-dimensional body is present in a subsonic compressible flow of Mach number M_∞, the transformed body in an incompressible flow field will have its dimensions altered according to Eq. (14.30). For example, consider compressible flow over a thin airfoil, of chord c and thickness t. The transformed airfoil in incompressible flow will have its thickness-to-chord ratio t/c altered, so that

$$\frac{t_i}{c_i} = \frac{t}{c} \sqrt{1 - M_\infty^2} \tag{14.31}$$

As shown in Figure 14.5, if the chord is maintained the same, then the incompressible thickness will be less, $t_i = \sqrt{1 - M_\infty^2}\, t$. However, if the thickness is maintained the same, the chord will be stretched $c_i = c/\sqrt{1 - M_\infty^2}$.

It is necessary for the compressible flow in the (x, y) plane and the transformed flow in the (x_i, y_i) plane to satisfy the boundary conditions on the body. We have established that, for small perturbation flows,

$$\left(\frac{dy}{dx}\right)_b = \frac{v_p}{U_\infty} \tag{14.15}$$

where

$$v_p = -\frac{\partial \phi_p}{\partial y}$$

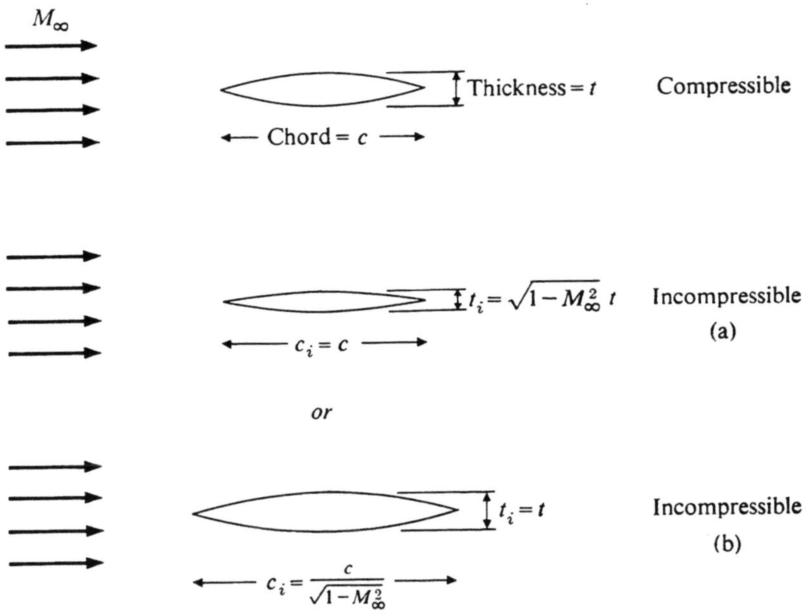

Figure 14.5

Transforming to the incompressible plane, we obtain

$$\frac{k_1}{k_2}\left(\frac{dy_i}{dx_i}\right)_b = -\frac{1}{U_{\infty_i}}\frac{\partial \phi_{ip}}{\partial y_i}\frac{k_4 k_2}{k_3}$$

To satisfy the boundary conditions at the body surface in the incompressible plane, it is necessary that

$$\frac{k_1}{k_2} = -\frac{k_4 k_2}{k_3}$$

or, since

$$\frac{k_1}{k_2} = \frac{1}{\sqrt{1 - M_\infty^2}}$$

it follows that

$$\frac{k_2 k_4}{k_3} = -\frac{1}{\sqrt{1 - M_\infty^2}} \tag{14.32}$$

It is important also to compare the pressure coefficients in the two planes. For compressible flow, it has been shown that $C_p = -2u_p/U_\infty$. For incompressible irrotational flow, Bernoulli's equation in the form $p + \frac{1}{2}\rho V^2 = $ constant can be applied. By definition,

$$C_{p_i} = \frac{p - p_\infty}{\frac{1}{2}\rho U_{\infty_i}^2}$$

where $p - p_\infty$, for incompressible flow, is equal to

$$\tfrac{1}{2}\rho U_{\infty_i}^2 - \tfrac{1}{2}[(u_{p_i} + U_{\infty_i})^2 + v_{p_i}^2]$$

or

$$p - p_\infty = \frac{1}{2}\rho U_{\infty_i}^2 \left[-2u_{p_i} \cdot U_{\infty_i} - \left(\frac{u_{p_i}}{U_{\infty_i}}\right)^2 - \left(\frac{v_{p_i}}{U_{\infty_i}}\right)^2 \right]$$

Dropping smaller terms, we receive

$$C_{p_i} = -\frac{2u_{p_i}}{U_{\infty_i}} \tag{14.33}$$

For the compressible plane,

$$C_p = -\frac{2u_p}{U_\infty} = \frac{2\dfrac{\partial \phi_p}{\partial x}}{U_\infty} = \frac{2\dfrac{k_1 k_4}{k_3}\dfrac{\partial \phi_{p_i}}{\partial x_i}}{U_{\infty_i}}$$

Since

$$\frac{k_2 k_4}{k_3} = -\frac{1}{\sqrt{1 - M_\infty^2}} \quad \text{and} \quad \frac{k_1}{k_2} = \frac{1}{\sqrt{1 - M_\infty^2}}$$

we have the result that

$$C_p = \frac{C_{p_i}}{1 - M_\infty^2} \tag{14.34}$$

From the transformation relationships given in Eq. (14.28), it is possible to determine the geometrical properties of an airfoil in the (x_i, y_i) plane from those in the compressible plane. If we assume the chord length to be the same in the two planes, then the thickness ratio t_i/c_i for the incompressible airfoil has been shown to be $\sqrt{1 - M_\infty^2}$ times the thickness ratio for the compressible foil. The same relationship holds for other foil characteristics involving a ratio of y dimension to x dimension. For example, the angles of attack of the airfoils (Figure 14.6) are related by

$$\alpha_i = \alpha\sqrt{1 - M_\infty^2} \tag{14.35}$$

Similarly, the camber ratios (Figure 14.7) are in proportion:

$$\frac{(\text{camber}/c)_i}{(\text{camber}/c)} = \sqrt{1 - M_\infty^2} \tag{14.36}$$

Figure 14.6

----- Mean line (equidistant between upper and lower surfaces)

Figure 14.7

The lift L of an airfoil, the upward force normal to the uniform flow direction, is obtained by integrating the pressure over the airfoil surface. Thus

$$L = \int_0^c p_L \, dx - \int_0^c p_u \, dx$$

where p_u and p_L are the pressures on the upper and lower airfoil surfaces. Define a lift coefficient C_L, where

$$C_L = \frac{\text{lift}}{\frac{1}{2}\rho U_\infty^2 c}$$

$$= \int_0^c C_{p_L} \frac{dx}{c} - \int_0^c C_{p_u} \frac{dx}{C} \qquad (14.37)$$

Thus the lift coefficients for the airfoils in the compressible and incompressible planes are in the same ratio as the pressure coefficient:

$$C_{L_i} = C_L(1 - M_\infty^2) \qquad (14.38)$$

The similarity laws as stated here are called the *Goethert rules*. It is emphasized that these rules are only valid for thin airfoils with small angles of attack, small camber ratios, and so on. Also, the Goethert rules only apply to subsonic flow, for which $\sqrt{1 - M_\infty^2}$ is real. The extension of these rules to three-dimensional flow will be left to the student as an exercise. The following example illustrates the application of the Goethert rules to a two-dimensional airfoil.

Example 14.1

A two-dimensional airfoil has a thickness ratio (maximum thickness to chord) of 0.04 and a camber ratio of 0.015. When tested in a low-speed wind tunnel (incompressible flow, $M_\infty = 0$) at an angle of attack of 3°, the lift coefficient C_L is measured to be 0.6. It is desired to determine the performance of a similar airfoil at $M_\infty = 0.5$. Using the Goethert rules, determine the geometrical characteristics of the related airfoil in compressible flow at $M_\infty = 0.5$; determine, also, the lift coefficient.

Solution

$$\sqrt{1 - M_\infty^2} = \sqrt{1 - (0.5)^2} = 0.866$$

From Eq. (14.31),

$$\left(\frac{t}{c}\right)_{M_\infty=0.5} = \left(\frac{t}{c}\right)_i \frac{1}{0.866} = (1.15)\left(\frac{t}{c}\right)_i = \underline{0.046}$$

From Eq. (14.36),

$$\left(\frac{\text{camber}}{c}\right)_{M_\infty=0.5} = \left(\frac{\text{camber}}{c}\right)_i 1.15 = \underline{0.0173}$$

From Eq. (14.35),

$$\alpha_{M_\infty=0.5} = \alpha_i \, 1.15 = \underline{3.45°}$$

From Eq. (14.38),

$$C_{L_{M_\infty=0.5}} = \frac{C_{L_i}}{1 - M_\infty^2} = \frac{0.6}{0.75} = \underline{0.80}$$

Whereas the Goethert rules have been shown to possess some application, it would seem far more useful to have a comparison between performance of the same airfoil in compressible and incompressible flows. So far, the foils have been distorted according to Eq. (14.28), as shown in Figure 14.5.

To this end, let us consider two airfoils in the incompressible plane, related to a third foil in the compressible plane. The first incompressible airfoil, as shown in Figure 14.8, is distorted according to Eq. (14.28) with its thickness-to-chord ratio, for example, given by

$$\left(\frac{t}{c}\right)_{i_1} = \frac{t}{c}\sqrt{1 - M_\infty^2} \tag{14.31}$$

The second incompressible foil is assumed to possess exactly the same dimensions as the compressible foil, so that, for example

$$\left(\frac{t}{c}\right)_{i_2} = \frac{t}{c}$$

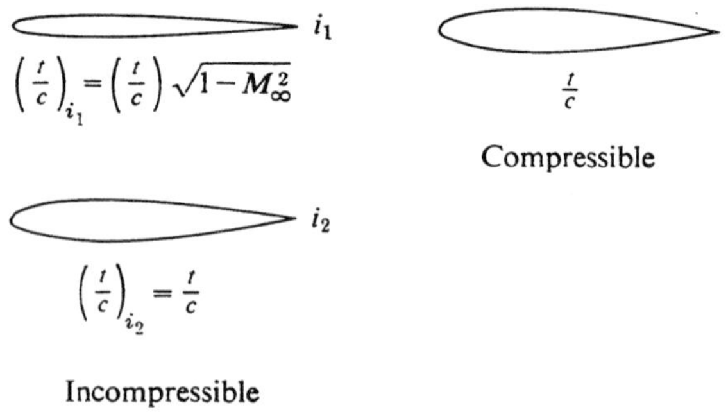

Compressible

Incompressible

Figure 14.8

or

$$\left(\frac{t}{c}\right)_{i_1} = \sqrt{1 - M_\infty^2} \left(\frac{t}{c}\right)_{i_2}$$

Since we have a relationship between the pressure coefficients for the compressible airfoil and incompressible airfoil i_1, we need a relation between the pressure coefficients for the incompressible airfoils. It can be shown that for thin bodies in incompressible, two-dimensional potential flow, related by $(y/c)_i = Kf(x/c)_i$ with f the same function for all bodies, C_{p_i} at corresponding points x/c is proportional to K. For our case,

$$\frac{\left(\dfrac{y}{c}\right)_{i_2}}{\left(\dfrac{y}{c}\right)_{i_1}} = \frac{K_{i_2}}{K_{i_1}} = \frac{1}{\sqrt{1 - M_\infty^2}}$$

which means that

$$\frac{C_{p_{i_1}}}{C_{p_{i_2}}} = \frac{K_{i_1}}{K_{i_2}} = \sqrt{1 - M_\infty^2} \tag{14.39}$$

Since

$$C_{p_{i_1}} = C_p(1 - M_\infty^2) \tag{14.34}$$

then

$$C_{p_{i_2}} = C_p\sqrt{1 - M_\infty^2} \tag{14.40}$$

This expression relates the pressure coefficient on a body immersed in two-dimensional compressible flow of Mach number M_∞ to the pressure coefficient on the same body immersed in incompressible flow.

Let us now attempt to determine the shape of a third incompressible airfoil i_3, which will have the same pressure coefficient at corresponding points as the compressible foil.

$$C_{p_{i_3}} = C_p$$

From Eq. (14.40),

$$\frac{C_{p_{i_3}}}{C_{p_{i_2}}} = \frac{1}{\sqrt{1 - M_\infty^2}}$$

But, from Eq. (14.39),

$$\frac{C_{p_{i_3}}}{C_{p_{i_2}}} = \frac{\left(\dfrac{y}{c}\right)_{i_3}}{\left(\dfrac{y}{c}\right)_{i_2}} = \frac{\left(\dfrac{t}{c}\right)_{i_3}}{\left(\dfrac{t}{c}\right)_{i_2}}$$

Therefore, for $C_{p_{i_3}} = C_p$, it follows that

$$\left(\frac{t}{c}\right)_{i_3} = \frac{1}{\sqrt{1 - M_\infty^2}}\left(\frac{t}{c}\right)_{i_2}$$

$$= \frac{1}{\sqrt{1 - M_\infty^2}}\frac{t}{c} \tag{14.41}$$

Likewise,

$$\alpha_{i_3} = \frac{1}{\sqrt{1 - M_\infty^2}}\alpha$$

$$\left(\frac{camber}{c}\right)_{i_3} = \frac{1}{\sqrt{1 - M_\infty^2}}\frac{camber}{c}$$

The similarity laws for subsonic compressible flow, as given by Eqs. (14.40) and (14.41), are called the *Prandtl Glauert rules*. Their validity is restricted to two-dimensional flows, since Eq. (14.39) is not valid for three-dimensional flows.

Example 14.2

The two-dimensional airfoil of Example 14.1, when tested in a low-speed wind tunnel at an angle of attack of 3°, is found to have a lift coefficient C_L of 0.6. Determine the lift coefficient of the same airfoil at $M_\infty = 0.50$.

Solution

From the Prandtl Glauert similarity rule, Eq. (14.40), we can write

$$CL_{M_\infty=0} = C_L\sqrt{1 - M_\infty^2}$$

Therefore,

$$C_L = \frac{0.6}{\sqrt{1 - (0.5)^2}}$$

$$= \frac{0.6}{0.866} = \underline{0.692}$$

Note that in Example 14.1 we compared the lift coefficients of two different airfoils, one in incompressible flow and the other, of different dimensions, in compressible flow. In Example 14.2, we are comparing the lift coefficients of the same airfoil in incompressible and compressible flow.

The upper limit of applicability of the similarity laws for subsonic flow is reached when the Mach number at some point on the airfoil is 1. The Mach number increases over the airfoil surface and reaches a maximum at the point of minimum pressure, so that Mach 1 will occur at a point on the surface when M_∞ is somewhat less than 1. Once Mach 1 or greater is reached at a point on

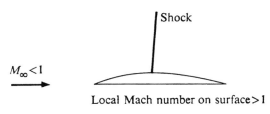

Local Mach number on surface>1

Figure 14.9

the airfoil (presumably on the upper surface), a shock will form (Figure 14.9), and the assumption of irrotational, isentropic flow will be invalid. With the pressure increase on the upper surface due to the shock (an adverse pressure gradient), the boundary layer will separate and cause a resultant decrease in airfoil lift–drag ratio. It becomes important to be able to calculate the critical free stream Mach number, $M_{\infty_{crit}}$, at which the Mach number will be equal to 1 at a point on the airfoil surface.

Assuming isentropic flow from M_∞ up to the point at which sonic flow is first reached on the airfoil (at the point of minimum pressure), we obtain, for p_t equal to a constant,

$$p_\infty\left(1 + \frac{\gamma - 1}{2} M_{\infty_{crit}}^2\right)^{\gamma/(\gamma-1)} = p_{M=1}\left(1 + \frac{\gamma - 1}{2}\right)^{\gamma/(\gamma-1)} \qquad (14.42)$$

We have defined pressure coefficient as

$$C_p = \frac{p - p_\infty}{\frac{1}{2}\rho U_\infty^2} = \frac{2(p - p_\infty)}{\gamma p_\infty M_\infty^2}$$

or

$$C_{p_{M-1}} = \frac{2}{\gamma M_{\infty_{crit}}^2}\left(\frac{p_{M=1}}{p_\infty} - 1\right)$$

Substituting from Eq. (14.42), we obtain

$$C_{p_{M-1}} = \frac{2}{\gamma M_{\infty_{crit}}^2}\left[\frac{\left(1 + \frac{\gamma - 1}{2} M_{\infty_{crit}}^2\right)^{\gamma/(\gamma-1)}}{\left(1 + \frac{\gamma - 1}{2}\right)^{\gamma/(\gamma-1)}} - 1\right] \qquad (14.43)$$

The minimum pressure coefficient for incompressible flow can be determined from low-speed wind-tunnel tests; then the minimum pressure coefficient for the same airfoil in compressible flow can be calculated from the Prandtl Glauert rule.

$$C_{p_{M_\infty}} = \frac{C_{p_{M_\infty=0}}}{\sqrt{1 - M_\infty^2}} \qquad (14.44)$$

Example 14.3

The minimum pressure coefficient on a two-dimensional airfoil, as measured in a low-speed wind tunnel ($M_\infty = 0$), is -0.8. Determine the critical Mach number for the foil.

Solution

From Eq. (14.44), $C_{P_{M_\infty}} = (-0.8/\sqrt{1 - M_\infty^2})$. This expression is plotted in Figure 14.10 and gives the minimum pressure coefficient as a function of free-stream Mach number. From Eq. (14.43), we have the pressure coefficient at which Mach 1 occurs versus free-stream Mach number. This relationship is also plotted in Figure 14.10; the intersection of the two curves is the solution to the problem. It can be seen that $\underline{M_{\infty_{crit}} = 0.64.}$

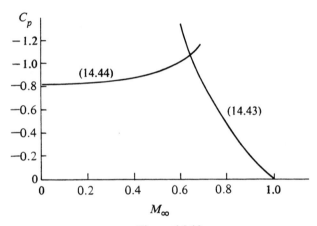

Figure 14.10

14.5 Supersonic Flow Over a Wavy Wall

The equation for the small-perturbation velocity potential for isentropic steady flow was found to be

$$(1 - M_\infty^2)\frac{\partial^2 \phi_p}{\partial x^2} + \frac{\partial^2 \phi_p}{\partial y^2} = 0 \qquad \textbf{(14.13)}$$

This equation is valid for supersonic flow as well as subsonic, with the aforementioned restriction that it becomes invalid for transonic and hypersonic flows. Furthermore, shocks that might occur in this supersonic flow must be assumed to be extremely weak, so as not to cause a significant departure from isentropic flow. For flows in which $M_\infty > 1$, the coefficient of the first term in Eq. (14.13) becomes negative. This changes the entire nature of the partial differential equation. For $M_\infty < 1$, Eq. (14.13) is an elliptic type of partial differential equation and, as we have shown, may be reduced to Laplace's equation by suitable transformation. However, for $M_\infty > 1$, Eq. (14.13) is of the hyperbolic type, with the form being similar to that of the wave equation. The general solution to this equation can be written as the sum of two arbitrary functions f and g such that

$$\phi_p = f(x + \sqrt{M_\infty^2 - 1}\,y) + g(x - \sqrt{M_\infty^2 - 1}\,y) \qquad (14.45)$$

To verify this, substitute Eq. (14.45) into Eq. (14.13):

$$\frac{\partial \phi_p}{\partial x} = \frac{\partial f}{\partial(x + \sqrt{M_\infty^2 - 1}\,y)} + \frac{\partial g}{\partial(x - \sqrt{M_\infty^2 - 1}\,y)}$$

$$\frac{\partial^2 \phi_p}{\partial x^2} = \frac{\partial^2 f}{\partial(x + \sqrt{M_\infty^2 - 1}\,y)^2} + \frac{\partial^2 g}{\partial(x - \sqrt{M_\infty^2 - 1}\,y)^2}$$

$$\frac{\partial \phi_p}{\partial y} = \left[\frac{\partial f}{\partial(x + \sqrt{M_\infty^2 - 1}\,y)}\right]\sqrt{M_\infty^2 - 1} - \left[\frac{\partial g}{\partial(x - \sqrt{M_\infty^2 - 1}\,y)}\right]\sqrt{M_\infty^2 - 1}$$

$$\frac{\partial^2 \phi_p}{\partial y^2} = \left[\frac{\partial^2 f}{\partial(x + \sqrt{M_\infty^2 - 1}\,y)^2}\right](M_\infty^2 - 1) + \left[\frac{\partial^2 g}{\partial(x - \sqrt{M_\infty^2 - 1}\,y)^2}\right](M_\infty^2 - 1)$$

It can be seen that

$$(1 - M_\infty^2)\frac{\partial^2 \phi_p}{\partial x^2} = -\frac{\partial^2 \phi_p}{\partial y^2}$$

satisfying Eq. (14.13).

For this linearized flow, lines of constant f, or lines for which $x + \sqrt{M_\infty^2 - 1}\,y = $ constant, are straight with slope given by

$$\frac{dy}{dx} = -\frac{1}{\sqrt{M_\infty^2 - 1}} \qquad (14.46)$$

Lines of constant g are also straight, inclined to the flow direction by

$$\frac{dy}{dx} = +\frac{1}{\sqrt{M_\infty^2 - 1}} \qquad (14.47)$$

From the expressions for the slopes, there appears to be a connection between lines of constant f and g and the Mach waves or lines discussed in Chapter 2. In Chapter 2, we found that if a point disturbance were placed in a uniform flow, the disturbance would be felt in the zone of action, defined by the Mach waves. This is depicted for two-dimensional flow in Figure 14.11. From Eq. (2.8), the Mach angle μ is given by

$$\sin \mu = \frac{1}{M_\infty} \qquad (2.8)$$

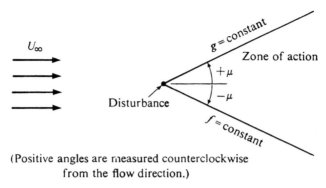

(Positive angles are measured counterclockwise
from the flow direction.)

Figure 14.11

Using trigonometry, we obtain

$$\tan \mu = \frac{1}{\sqrt{M_\infty^2 - 1}} = \frac{dy}{dx}$$

or

$$\tan(-\mu) = -\frac{1}{\sqrt{M_\infty^2 - 1}}$$

We see, therefore, that a line of constant g is the Mach line or wave in Figure 14.11 that makes a positive angle with the flow direction; a line of constant f is the Mach wave that makes a negative angle with the flow direction.

For a linearized supersonic flow over a thin body, lines of constant f and g form a network as shown in Figure 14.12. These lines extend to infinity in both directions. However, from Figure 14.11, it can be seen that an f = constant line is not a Mach wave above and to the left of the disturbance, since this would imply that the effect of the disturbance could be felt outside the zone of action. Likewise, a line of g = constant is not a Mach wave below and to the left of the disturbance.

For flow over a thin body, as shown in Figure 14.13, we need only consider the lines of g = constant above the body, since these are the only

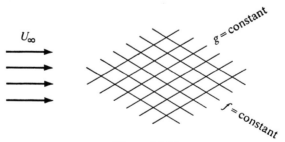

Figure 14.12

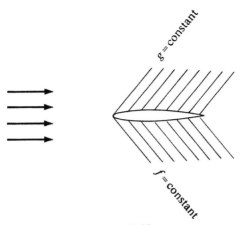

Figure 14.13

lines physically present in the flow as Mach waves. Likewise, below the body, we need only consider lines of f = constant.

The nature of the functions f and g is arbitrary; their particular form is dependent on the boundary conditions. To illustrate, we shall return to the problem of flow over a wavy wall. Consider uniform, irrotational supersonic flow over a wavy wall, as shown in Figure 14.3. The boundary condition at the wall, just as with the subsonic case, is

$$-\left(\frac{\partial \phi_p}{\partial y}\right)_{y=0} = U_\infty A \frac{2\pi}{\lambda} \cos \frac{2\pi x}{\lambda} \qquad (14.48)$$

As shown in Figure 14.14, and discussed previously, only the g function is applicable for this case, so

$$\phi_p = g(x - \sqrt{M_\infty^2 - 1}\, y) \qquad (14.49)$$

Substituting Eq. (14.49) into Eq. (14.48), we obtain

$$\sqrt{M_\infty^2 - 1}\left[\frac{\partial g}{\partial(x - \sqrt{M_\infty^2 - 1}\, y)}\right]_{y=0} = U_\infty A \frac{2\pi}{\lambda} \cos \frac{2\pi x}{\lambda}$$

$M_\infty > 1$

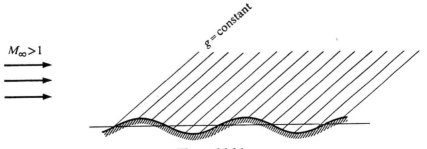

Figure 14.14

Therefore,

$$\left[\frac{\partial g}{\partial(x - \sqrt{M_\infty^2 - 1}\,y)}\right]_{y=0} = \frac{U_\infty}{\sqrt{M_\infty^2 - 1}}\frac{2\pi A}{\lambda}\cos\frac{2\pi x}{\lambda}$$

or, at any y,

$$\frac{\partial g}{\partial(x - \sqrt{M_\infty^2 - 1}\,y)} = \frac{U_\infty}{\sqrt{M_\infty^2 - 1}}\frac{2\pi A}{\lambda}\cos\left[\frac{2\pi(x - \sqrt{M_\infty^2 - 1}\,y)}{\lambda}\right]$$

$$= \frac{dg}{d(x - \sqrt{M_\infty^2 - 1}\,y)}$$

(We have written the preceding as a total derivative, since g is a function only of $x - \sqrt{M_\infty^2 - 1}\,y$.) Integrating, we obtain

$$g = \frac{U_\infty}{\sqrt{M_\infty^2 - 1}}\frac{2\pi A}{\lambda}\frac{\lambda}{2\pi}\sin\frac{2\pi(x - \sqrt{M_\infty^2 - 1}\,y)}{\lambda} + \text{constant}$$

$$= \frac{U_\infty A}{\sqrt{M_\infty^2 - 1}}\sin\left[\frac{2\pi}{\lambda}(x - \sqrt{M_\infty^2 - 1}\,y)\right] + \text{constant}$$

$$= \phi_p \tag{14.50}$$

Note that the boundary conditions at infinity, as given in Section 14.3, are also satisfied.

The perturbation velocities, u_p and v_p, are obtained by differentiation.

$$u_p = -\frac{\partial\phi_p}{\partial x}$$

$$= -\frac{U_\infty A}{\sqrt{M_\infty^2 - 1}}\frac{2\pi}{\lambda}\cos\frac{2\pi}{\lambda}(x - \sqrt{M_\infty^2 - 1}\,y) \tag{14.51}$$

$$v_p = -\frac{\partial\phi_p}{\partial y}$$

$$= + U_\infty A\frac{2\pi}{\lambda}\cos\left[\frac{2\pi}{\lambda}(x - \sqrt{M_\infty^2 - 1}\,y)\right] \tag{14.52}$$

Also, it has been shown that, for small-perturbation flow,

$$C_p = -\frac{2u_p}{U_\infty}$$

so that, for supersonic flow over a wave-shaped wall,

$$C_p = \frac{4A\pi}{\sqrt{M_\infty^2 - 1}\lambda}\cos\frac{2\pi}{\lambda}(x - \sqrt{M_\infty^2 - 1}\,y) \tag{14.53}$$

At the wall, y is approximately 0, so

$$C_{p_{\text{wall}}} = \frac{4\pi A}{\sqrt{M_\infty^2 - 1}\lambda}\cos\frac{2\pi x}{\lambda} \tag{14.54}$$

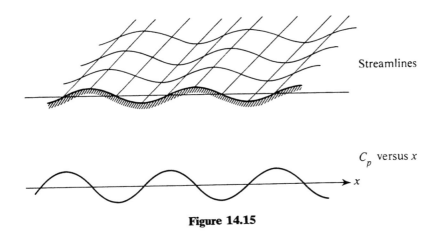

Streamlines

C_p versus x

Figure 14.15

Streamlines and pressure distribution on the wall are shown in Figure 14.15 for uniform supersonic flow over a wavy wall.

It can be seen from Eqs. (14.51), (14.52), and (14.53), that along lines of $x - \sqrt{M_\infty^2 - 1}\,y$, the perturbation velocities and pressure coefficient are constant. In other words, the disturbance introduced into the uniform flow by the presence of the wall is felt equally (out to infinity) along these lines. Thus the presence of a disturbance in supersonic linearized flow is propagated along the Mach lines or waves.

Once again, it can be seen why lines of f = constant were ruled out of the solution. These lines would be inclined to the flow as shown in Figure 14.16. For such lines to exist as Mach waves, it would be possible for the presence of the wall to be propagated back upstream into the supersonic flow; the supersonic flow would be able to sense the presence of the wall before reaching it. But from the discussion in Chapter 2, we know that this is physically impossible; so for this situation only lines of g = constant are physically present.

Several of the basic differences between supersonic and subsonic flow become evident from the example of flow over a wavy wall. In subsonic flow, the perturbation velocities die out as y gets very large. In supersonic flow, the disturbance of the wall is felt along Mach lines inclined to the flow direction.

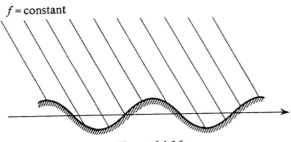

f = constant

Figure 14.16

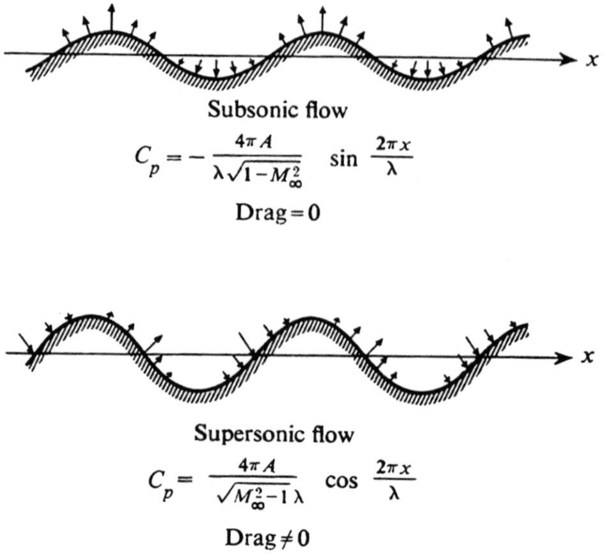

Subsonic flow

$$C_p = -\frac{4\pi A}{\lambda\sqrt{1-M_\infty^2}} \sin \frac{2\pi x}{\lambda}$$

Drag $= 0$

Supersonic flow

$$C_p = \frac{4\pi A}{\sqrt{M_\infty^2-1}\,\lambda} \cos \frac{2\pi x}{\lambda}$$

Drag $\neq 0$

Figure 14.17

The disturbance, or perturbation, does not die out, but remains constant, even as y approaches infinity. Thus, for supersonic flow, the Mach lines or waves have great significance; such Mach lines do not exist for subsonic flow.

In subsonic flow, the pressure coefficient is in phase with the wall shape, so there is no drag force on the wall. In supersonic flow, the pressure coefficient is out of phase with the wall shape, so there is a drag force on the wall (see Figure 14.17).

14.6 Thin Airfoils in Supersonic Flow

The general solution of the linearized potential equation for isentropic, irrotational supersonic flow has been shown to be

$$\phi_p = f(x + \sqrt{M_\infty^2 - 1}\,y) + g(x - \sqrt{M_\infty^2 - 1}\,y) \tag{14.45}$$

where f and g are arbitrary functions dependent on the boundary conditions. Let us apply this result to supersonic flow over a thin, two-dimensional airfoil (see Figure 14.18). Since a disturbance cannot be propagated upstream into a supersonic flow, the solution over the top of the foil will consist of lines of constant g, the solution under the foil will consist of lines of constant f. For y > 0,

$$\phi_p = g(x - \sqrt{M_\infty^2 - 1}\,y)$$

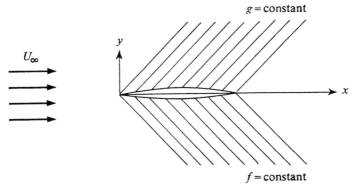

Figure 14.18

For $y < 0$,

$$\phi_p = f(x + \sqrt{M_\infty^2 - 1}\,y) \tag{14.55}$$

The boundary condition on the upper foil surface is

$$\left(\frac{dy}{dx}\right)_{\text{upper}} = \frac{v_{\text{upper}}}{U_\infty} = -\frac{1}{U_\infty}\frac{\partial \phi_p}{\partial y}$$

so

$$-\left(\frac{\partial \phi_p}{\partial y}\right)_{y>0} = \left[\frac{d\phi_p}{d(x - \sqrt{M_\infty^2 - 1}\,y)}\right]\sqrt{M_\infty^2 - 1} = U_\infty\left(\frac{dy}{dx}\right)_{\text{upper}} \tag{14.56}$$

The pressure coefficient on the upper surface of the foil, C_{p_u}, is given by

$$C_{p_u} = -\frac{2u_{\text{upper}}}{U_\infty} = -\frac{2}{U_\infty}\left(\frac{\partial \phi_p}{\partial x}\right)_{y=0}$$

$$= \frac{2}{U_\infty}\frac{d\phi_p}{d(x - \sqrt{M_\infty^2 - 1}\,y)} \tag{14.57}$$

Substituting Eq. (14.56) into Eq. (14.57), we obtain

$$C_{p_u} = \frac{2}{U_\infty}\frac{U_\infty\left(\frac{dy}{dx}\right)_{\text{upper}}}{\sqrt{M_\infty^2 - 1}}$$

$$= \frac{2}{\sqrt{M_\infty^2 - 1}}\left(\frac{dy}{dx}\right)_{\text{upper}} \tag{14.58}$$

Similarly, for the pressure coefficient on the lower surface C_{p_L}, we have

$$C_{p_L} = -\frac{2u_{\text{Lower}}}{U_\infty} = \frac{2}{U_\infty}\left(\frac{d\phi_p}{d(x + \sqrt{M_\infty^2 - 1}\,y)}\right)_{y=0}$$

where

$$\left[\frac{d\phi_p}{d(x + \sqrt{M_\infty^2 - 1}y)}\right]_{y=0} \sqrt{M_\infty^2 - 1} = -U_\infty\left(\frac{dy}{dx}\right)_{\text{lower}}$$

$$C_{p_L} = -\frac{2}{\sqrt{M_\infty^2 - 1}}\left(\frac{dy}{dx}\right)_{\text{lower}} \tag{14.59}$$

From these expressions for pressure coefficients, it is possible to calculate lift and drag coefficients for a thin airfoil in uniform supersonic flow. The slope dy/dx at a point on the surface of an airfoil is dependent on the angle of attack of the foil and the thickness and camber at the point. It is instructive to separate these airfoil properties in the way shown in Figure 14.19.

Consider a differential element of length ds on the upper airfoil surface, as shown in Figure 14.20.

$$dF = p \, ds$$

$$= C_{p_u}\tfrac{1}{2}\gamma p_\infty M_\infty^2 \, ds + p_\infty \, ds$$

The differential lift dL is given by

$$dL = -C_{p_u}\tfrac{1}{2}\gamma p_\infty M_\infty^2 \, dx - p_\infty \, dx$$

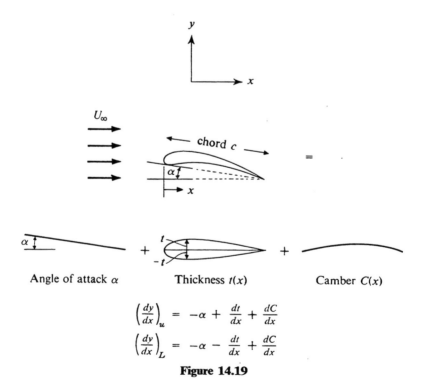

Angle of attack α Thickness $t(x)$ Camber $C(x)$

$$\left(\frac{dy}{dx}\right)_u = -\alpha + \frac{dt}{dx} + \frac{dC}{dx}$$

$$\left(\frac{dy}{dx}\right)_L = -\alpha - \frac{dt}{dx} + \frac{dC}{dx}$$

Figure 14.19

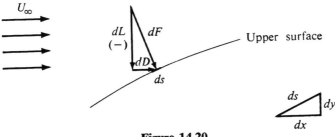

Figure 14.20

where lift is defined to be positive upward. The differential drag is

$$dD = C_{p_u}\tfrac{1}{2}\gamma p_\infty M_\infty^2 \left(\frac{dy}{dx}\right)_u dx + p_\infty\left(\frac{dy}{dx}\right)_u dx$$

where drag is positive in the flow direction.

For an element on the lower surface, as shown in Figure 14.21, we have

$$dL = C_{p_L}\tfrac{1}{2}\gamma p_\infty M_\infty^2\, dx + p_\infty\, dx$$

$$dD = -C_{p_L}\tfrac{1}{2}\gamma p_\infty M_\infty^2 \left(\frac{dy}{dx}\right)_L dx - p_\infty\left(\frac{dy}{dx}\right)_L dx$$

We can now add the two contributions to lift and drag and integrate to obtain the lift and drag for the airfoil.

$$L = \frac{1}{2}\gamma p_\infty M_\infty^2 \int_0^c (C_{p_L} - C_{p_u})\, dx \qquad\qquad \textbf{(14.60)}$$

$$D = \frac{1}{2}\gamma p_\infty M_\infty^2 \int_0^c \left[C_{p_u}\left(\frac{dy}{dx}\right)_u - C_{p_L}\left(\frac{dy}{dx}\right)_L \right] dx$$

$$+ p_\infty \int_0^c \left[\left(\frac{dy}{dx}\right)_u - \left(\frac{dy}{dx}\right)_L \right] dx \qquad\qquad \textbf{(14.61)}$$

The upper limit of the integration has been taken as $x = c$, where c is the chord length (see Figure 14.19). Actually, the upper limit should be $x = c \cos \alpha$. However, the small-perturbation theory can only be used for small angles of attack, so $\cos \alpha \approx 1$.

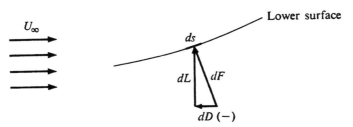

Figure 14.21

We now substitute Eqs. (14.58) and (14.59) into Eq. (14.60):

$$L = \frac{1}{2}\gamma p_\infty M_\infty^2 \int_0^c \left(-\frac{2}{\sqrt{M_\infty^2 - 1}}\right)\left\{\left(\frac{dy}{dx}\right)_L + \left(\frac{dy}{dx}\right)_u\right\} dx$$

$$= -\frac{\gamma p_\infty M_\infty^2}{\sqrt{M_\infty^2 - 1}}\left[\int_0^c (-2\alpha)\, dx + \int_0^c \left(\frac{dt}{dx} + \frac{dC}{dx}\right)_{upper} dx - \int_0^c \left(\frac{dt}{dx} - \frac{dC}{dx}\right)_L dx\right]$$

For a foil with sharp or only slightly rounded leading and trailing edges, the thickness and camber are both 0 at $x = 0$ and at $x = c$. For this case,

$$L = \frac{\gamma p_\infty M_\infty^2 2\alpha c}{\sqrt{M_\infty^2 - 1}} \tag{14.62}$$

Defining a lift coefficient as

$$C_L = \frac{L}{\frac{1}{2}\gamma p_\infty M_\infty^2}$$

we get

$$C_L = \frac{4\alpha}{\sqrt{M_\infty^2 - 1}} \tag{14.63}$$

For a thin airfoil in supersonic linearized flow, with sharp leading and trailing edges, the lift coefficient depends only on angle of attack, not on the camber or thickness of the foil.

An expression for the drag coefficient can be found in a similar fashion. Using Eqs. (14.58), (14.59) and (14.61), we get

$$D = \frac{1}{2}\gamma p_\infty M_\infty^2 \int_0^c \frac{2}{\sqrt{M_\infty^2 - 1}}\left[\left(\frac{dy}{dx}\right)_u^2 + \left(\frac{dy}{dx}\right)_L^2\right] dx + p_\infty \int_0^c \left[\left(\frac{dy}{dx}\right)_u - \left(\frac{dy}{dx}\right)_L\right] dx$$

$$= \frac{\gamma p_\infty M_\infty^2}{\sqrt{M_\infty^2 - 1}} \int_0^c \left\{\left[-\alpha + \frac{dt}{dx} + \frac{dC}{dx}\right]_u^2 + \left[-\alpha - \frac{dt}{dx} + \frac{dC}{dx}\right]_L^2\right\} dx$$

$$\left(\text{Note that } \int_0^c \left[\left(\frac{dy}{dx}\right)_u - \left(\frac{dy}{dx}\right)_L\right] dx = 0\right)$$

Expanding this expression for drag, we obtain,

$$D = \frac{\gamma p_\infty M_\infty^2}{\sqrt{M_\infty^2 - 1}}\left\{\int_0^c \left[2\alpha^2 + 2\left(\frac{dt}{dx}\right)^2 + 2\left(\frac{dC}{dx}\right)^2\right] dx - \int_0^c 4\alpha \frac{dC}{dx}\, dx\right\}$$

But

$$\int_0^c \alpha \frac{dC}{dx}\, dx = 0$$

for a foil with sharp or only slightly rounded leading and trailing edges. Also, we can write

$$\int_0^c \left(\frac{dt}{dx}\right)^2 dx = \overline{\left(\frac{dt}{dx}\right)^2} c \quad \text{and} \quad \int_0^c \left(\frac{dC}{dx}\right)^2 dx = \overline{\left(\frac{dC}{dx}\right)^2} c$$

so that

$$D = \frac{2\gamma p_\infty M_\infty^2}{\sqrt{M_\infty^2 - 1}} \left[\alpha^2 c + \overline{\left(\frac{dt}{dx}\right)^2} c + \overline{\left(\frac{dC}{dx}\right)^2} c \right]$$

or

$$C_D = \frac{4}{\sqrt{M_\infty^2 - 1}} \left[\alpha^2 + \overline{\left(\frac{dt}{dx}\right)^2} + \overline{\left(\frac{dC}{dx}\right)^2} \right]$$

There is thus a contribution to drag from the angle of attack, thickness, and camber.

Example 14.4

Using the linearized theory, calculate the lift and drag coefficients for a flat-plate airfoil at an angle of attack of $10°$ and $M_\infty = 2.5$. Compare the result with the exact solution obtained from oblique shock theory in Example 8.5.

Solution

$$C_L = \frac{4\alpha}{\sqrt{M_\infty^2 - 1}}$$

with α expressed in radians. Thus

$$C_L = \frac{4(0.1745)}{\sqrt{(2.5)^2 - 1}} = \underline{0.3046}$$

Since a flat-plate airfoil has no thickness and no camber,

$$C_D = \frac{4}{\sqrt{M_\infty^2 - 1}} \alpha^2 = \frac{4(0.1745)^2}{\sqrt{(2.5)^2 - 1}} = \underline{0.0532}$$

The exact solution in Example 8.5 yielded

$$C_L = 0.311, \qquad C_D = 0.0550$$

It can be seen that, even at angles of attack as large as $10°$, the linearized theory gives results within 5 percent of the exact solution.

Example 14.5

Using the linearized theory, compute the lift and drag coefficients for the symmetrical, diamond-shaped airfoil shown in Figure 14.22. Assume $M_\infty = 2.0$, with an angle of attack of $7°$.

Figure 14.22

Solution

The lift coefficient can be obtained from Eq. (14.63):

$$C_L = \frac{4\alpha}{\sqrt{M_\infty^2 - 1}} = \underline{0.2821}$$

The diamond-shaped foil has thickness but no camber, with

$$\overline{\left(\frac{dt}{dx}\right)^2} = (\tan 2.5°)^2$$

$$= 0.04366^2$$

$$= 0.001906$$

Thus

$$C_D = \frac{4}{\sqrt{M_\infty^2 - 1}}\left[\alpha^2 + \overline{\left(\frac{dt}{dx}\right)^2}\right]$$

$$= \underline{0.03887}$$

14.7 Summary

In Chapter 14 we have presented certain approximate methods for handling problems in two-dimensional compressible flow. Again, it is very important to recognize the nature of the approximation that we were forced to make in order to linearize the partial differential equation, that is, the body introduced into the flow causes only small perturbations in the flow.

Certain results from this analysis should stand out in the student's mind: the charge in the nature of the partial differential equation upon going from subsonic to supersonic flow, the possibility of deriving similarity relationships between subsonic compressible flow and the corresponding incompressible flow, and the importance of Mach waves in a supersonic solution.

REFERENCES

1. CHEERS, F., *Elements of Compressible Flow*, New York, John Wiley & Sons, Inc., 1963.

2. LIEPMANN, H., AND ROSHKO, A., *Elements of Gas Dynamics*, New York, John
   Wiley & Sons, Inc., 1957.
3. SHAPIRO, A., *The Dynamics and Thermodynamics of Compressible Fluid Flow*, Vol.
   1 New York, Ronald Press Company, 1953.
4. MILNE-THOMSON, L. M., *Theoretical Aerodynamics*, New York, Macmillan, Inc.,
   1952.

PROBLEMS

1. The lift coefficient versus angle of attack for an airfoil, as measured in a low-speed
   wind tunnel, is given in Figure P14.1. Sketch this curve for the same airfoil at a
   Mach number of 0.45.

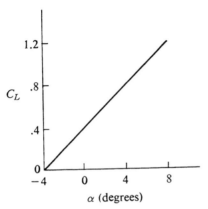

Figure P14.1

2. Using the potential equation

$$\frac{\partial^2 \phi_p}{\partial x^2}(1 - M_\infty^2) + \frac{\partial^2 \phi_p}{\partial y^2} + \frac{\partial^2 \phi_p}{\partial z^2} = 0$$

develop the Goethert similarity rules for three-dimensional potential subsonic flow.

3. Tests run at $M_\infty = 0.3$ show that the lift coefficient versus angle of attack for an
   airfoil is given by $C_L = 0.1(\alpha + 1)$ with α in degrees. Using the appropriate
   similarity laws, derive an expression for C_L versus α for this airfoil at $M_\infty = 0.50$.

4. For the airfoil of Problem 3, plot C_L versus M_∞ from $M_\infty = 0$ to $M_\infty = 0.60$ at
   angles of attack of 0°, 2°, and 4°.

3. During the testing of a two-dimensional, streamlined shape, it is found that sonic
   flow first occurs on the surface for $M_\infty = 0.70$. Calculate the pressure coefficient at
   this point and also the minimum pressure coefficient for this shape in incompressible flow.

6. Two-dimensional subsonic linearized potential flow takes place between two wavy
   walls as shown Figure P14.6. Solve for ϕ_p and determine the pressure distribution
   along the centerline.

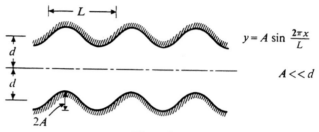

$$y = A \sin \frac{2\pi x}{L}$$

$$A << d$$

Figure P14.6

7. Consider two-dimensional, supersonic, linearized flow under a wavy wall, as shown in Figure P14.7. Solve for the velocity potential of the flow and pressure coefficient along the wall. Derive an expression for the lift and drag per wave length.

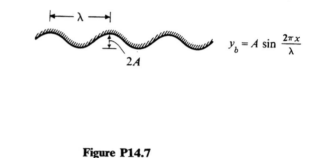

$$y_b = A \sin \frac{2\pi x}{\lambda}$$

Figure P14.7

8. A wing has the shape of a sine wave, as shown in Figure P14.8. Compute the lift and drag for supersonic flow. Assume linearized, two-dimensional flow above and below the foil.

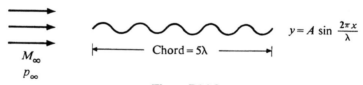

$$y = A \sin \frac{2\pi x}{\lambda}$$

Chord = 5λ

Figure P14.8

9. Using thin airfoil theory, find C_L and C_D for a two-dimensional, flat-plate airfoil with deflected flap in supersonic flow of Mach number M_∞. Plot C_L versus α for various δ for $F = 0.25$ (Figure P14.9).

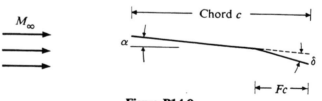

Chord c

Fc

Figure P14.9

10. Consider uniform supersonic flow over a wall in which there exists a bump, as shown in Figure P14.10. Assuming linearized, two-dimensional potential flow, calculate the vertical and horizontal components of the force on the bump. Assume $M_\infty = 2.0$, with $p_\infty = 50$ kPa.

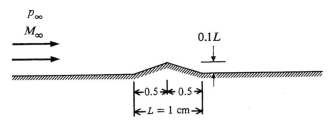

Figure P14.10

11. A supersonic airfoil consists of a circular arc, as shown in Figure P14.11. Compute the lift and drag coefficients of the foil versus angle of attack.

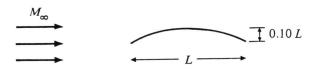

Figure P14.11

12. For the airfoil shown in Figure P14.12, determine C_L and C_D versus angle of attack in supersonic flow.

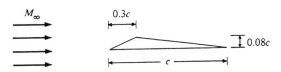

Figure P14.12

15

METHOD
OF CHARACTERISTICS

15.1 Introduction

In Chapter 13, we derived the partial differential equations for the velocity potential for the case of steady, irrotational, and isentropic flow, in the absence of gravity, viscous, electric, and magnetic forces. Linearization of these equations in Chapter 14 allowed approximate solutions to be worked out for several types of flow problems. Linearized solutions were found for both subsonic and supersonic flows, with the nature of the solutions being quite different for these two types of flow.

In certain cases, however, approximate solutions obtained by linearization are not adequate, but greater accuracy is necessary. Instead of dropping terms, as was done in Chapter 14, it is often desirable to consider the complete nonlinear equation. Fortunately, the very nature of supersonic flow allows a numerical method of solution of the complete differential equation for problems in two-dimensional flow, axisymmetric flow, and one-dimensional unsteady flow. The procedure used is called the method of characteristics. In this chapter, we shall first discuss the physical nature of supersonic flow, as already covered in previous chapters. Then, using the equations of motion developed in Chapter 13, the method of characteristics will be demonstrated for the case of two-dimensional flow.

Several examples will demonstrate the numerical procedures used for solution of problems in supersonic flow. It is emphasized that in this chapter we shall be dealing with two-dimensional, supersonic, steady flow of a perfect gas with constant specific heats. We shall assume that the flow is irrotational and isentropic, so that, for example, no shock waves are present.

15.2 Nature of Supersonic Flow

It was shown in Chapter 2 that when a small disturbance is introduced into a subsonic flow the effect of the disturbance is felt throughout the entire flow field. However, when a small disturbance is introduced into a supersonic flow, the effect of the disturbance is confined to that portion of the flow field contained within a Mach cone, defined by the Mach lines or waves. Likewise, for linearized supersonic flow over a wavy wall, discussed in Chapter 14, the effect of the wall was propagated to infinity along straight lines inclined at the Mach angle to the flow. The existence of these Mach waves or lines, which as we shall see are also called characteristics, is a property of the solution of the equations of supersonic flow. The solution of supersonic flow problems will be shown to depend on a determination of the Mach waves or characteristics.

For linearized two-dimensional flow, the Mach waves are straight lines and relatively easy to determine. However, if the nonlinear terms in the equations of motion are taken into account, a determination of the Mach waves and characteristic lines becomes considerably more complex—they are no longer straight lines. Before returning to the complete nonlinear, partial differential equation, let us examine some of the properties of Mach waves that have been discussed in previous chapters. From Chapter 7, Prandtl Meyer flow, suppose a weak disturbance is introduced into a supersonic, two-dimensional flow, by a small change in the direction of the wall, as shown in Figure 15.1. The effect of the disturbance is felt in the flow behind the Mach wave, with uniform conditions prevailing before and after the wave. For isentropic flow, with no finite shock waves possible, there can be no finite discontinuity in the velocity components across the wave, so the first derivatives of the velocity potential are continuous. However, the change in wall direction $d\theta$ does bring about an infinitesimal expansion and resultant infinitesimal increase in velocity to $V + dV$. This infinitesimal velocity increase, occurring normal to a Mach wave of

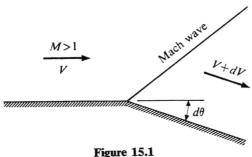

Figure 15.1

infinitesimal thickness, indicates the presence of a finite-velocity gradient, $\partial V/\partial x$. In the flow field ahead of the wave, the spatial derivatives of velocity are zero, since the flow is uniform. In the flow field after the wave, again the velocity derivatives are zero, since the flow is uniform. The presence of a finite derivative at the wave itself, however, indicates that there is a discontinuity in the velocity derivative across a Mach wave. We shall use this property as a definition of a characteristic curve, a line on which the velocity is continuous, although the first derivative of velocity is discontinuous. Having established the existence of characteristics on physical grounds, via a simplified model, we wish now to extend this discussion to the complete nonlinear, partial differential equation and provide a more substantial mathematical foundation for the existence of characteristics.

15.3 Theory of Characteristics

Starting with the complete partial differential equation developed in Chapter 13, we shall first establish the conditions under which there exist characteristic lines, defined as lines on which there exists a discontinuity of velocity gradient. Then we shall determine procedures for calculating the shape of such lines and, finally, show how they can be used to solve problems in supersonic flow.

The potential equation for isentropic, irrotational, steady flow of a perfect gas with constant specific heats was found in Chapter 13:

$$\frac{\partial^2 \phi}{\partial x^2} + \frac{\partial^2 \phi}{\partial y^2} + \frac{\partial^2 \phi}{\partial z^2} - \frac{1}{a^2}\left[\left(\frac{\partial \phi}{\partial x}\right)^2 \frac{\partial^2 \phi}{\partial x^2} + \left(\frac{\partial \phi}{\partial y}\right)^2 \frac{\partial^2 \phi}{\partial y^2} + \left(\frac{\partial \phi}{\partial z}\right)^2 \frac{\partial^2 \phi}{\partial z^2}\right]$$

$$-\frac{2}{a^2}\left[\frac{\partial \phi}{\partial x}\frac{\partial \phi}{\partial y}\frac{\partial^2 \phi}{\partial x \partial y} + \frac{\partial \phi}{\partial y}\frac{\partial \phi}{\partial z}\frac{\partial^2 \phi}{\partial y \partial z} + \frac{\partial \phi}{\partial z}\frac{\partial \phi}{\partial x}\frac{\partial^2 \phi}{\partial z \partial x}\right] = 0 \qquad \textbf{(13.30)}$$

For two-dimensional flow, this reduces to

$$\frac{\partial^2 \phi}{\partial x^2} + \frac{\partial^2 \phi}{\partial y^2} - \frac{1}{a^2}\left[\left(\frac{\partial \phi}{\partial x}\right)^2 \frac{\partial^2 \phi}{\partial x^2} + \left(\frac{\partial \phi}{\partial y}\right)^2 \frac{\partial^2 \phi}{\partial y^2}\right] - \frac{2}{a^2}\left[\frac{\partial \phi}{\partial x}\frac{\partial \phi}{\partial y}\frac{\partial^2 \phi}{\partial x \partial y}\right] = 0 \quad \textbf{(15.1)}$$

or

$$\left[a^2 - \left(\frac{\partial \phi}{\partial x}\right)^2\right]\frac{\partial^2 \phi}{\partial x^2} - 2\frac{\partial \phi}{\partial x}\frac{\partial \phi}{\partial y}\frac{\partial^2 \phi}{\partial x \partial y} + \left[a^2 - \left(\frac{\partial \phi}{\partial y}\right)^2\right]\frac{\partial^2 \phi}{\partial y^2} = 0 \quad \textbf{(15.2)}$$

This partial differential equation is of the general form

$$C_1 \frac{\partial^2 \phi}{\partial x^2} + 2C_2 \frac{\partial^2 \phi}{\partial x \partial y} + C_3 \frac{\partial^2 \phi}{\partial y^2} = 0 \qquad \textbf{(15.3)}$$

where $C_1, C_2,$ and C_3 depend on $\partial \phi/\partial x$ and $\partial \phi/\partial y$. We shall assume, as discussed previously, that for this isentropic flow, the velocity components, and hence the derivatives $\partial \phi/\partial x$ and $\partial \phi/\partial y$, are continuous. Thus we can write, for the differential changes in velocity components,

$$d\frac{\partial\phi}{\partial x} = \frac{\partial^2\phi}{\partial x^2}dx + \frac{\partial^2\phi}{\partial x\,\partial y}dy \qquad (15.4)$$

and

$$d\frac{\partial\phi}{\partial y} = \frac{\partial^2\phi}{\partial y\,\partial x}dx + \frac{\partial^2\phi}{\partial y^2}dy \qquad (15.5)$$

From Eqs. (15.3), (15.4), and (15.5), we obtain three linear equations for determining

$$\frac{\partial^2\phi}{\partial x^2}, \quad \frac{\partial^2\phi}{\partial y^2}, \quad \text{and} \quad \frac{\partial^2\phi}{\partial x\,\partial y}$$

According to Cramer's rule[1], if there exists a set of n linear equations in the unknowns $x_1, x_2, \ldots x_n$,

$$a_{11}x_1 + a_{12}x_2 + \cdots + a_{1n}x_n = k_1$$

$$a_{21}x_1 + a_{22}x_2 + \cdots + a_{2n}x_n = k_2$$

$$\vdots$$

$$a_{n1}x_1 + a_{n2}x_2 + \cdots + a_{nn}x_n = k_n$$

then the system has a unique solution if the determinant of the system

$$|A| = \begin{vmatrix} a_{11}a_{12}\ldots a_{1n} \\ \vdots \\ a_{n1}a_{n2}\ldots a_{nn} \end{vmatrix}$$

is not equal to zero. The solution is given by

$$x_1 = \frac{|A_1|}{|A|}$$

where $|A_1|$, for example, is the determinant formed by replacing the elements $a_{11}, a_{21}, \ldots, a_{n1}$ in the first column of A by $k_1, k_2, \ldots, k_n$.

For our case, we have the following set of three linear equations in the unknowns

$$\frac{\partial^2\phi}{\partial x^2}, \quad \frac{\partial^2\phi}{\partial y^2}, \quad \text{and} \quad \frac{\partial^2\phi}{\partial x\,\partial y}$$

$$d\frac{\partial\phi}{\partial x} = dx\frac{\partial^2\phi}{\partial x^2} + dy\frac{\partial^2\phi}{\partial x\,\partial y} + 0\frac{\partial^2\phi}{\partial y^2}$$

$$d\frac{\partial\phi}{\partial y} = 0\frac{\partial^2\phi}{\partial x^2} + dx\frac{\partial^2\phi}{\partial x\,\partial y} + dy\frac{\partial^2\phi}{\partial y^2} \qquad (15.6)$$

$$0 = C_1\frac{\partial^2\phi}{\partial x^2} + 2C_2\frac{\partial^2\phi}{\partial x\,\partial y} + C_3\frac{\partial^2\phi}{\partial y^2}$$

According to the discussion of the previous section, we are interested in obtaining expressions for the characteristics, lines across which there is a discontinuity in the velocity derivative (or, expressing this another way, lines on which there is not a unique solution for the velocity derivative). Furthermore, we are interested in determining under what conditions such characteristics exist.

The determinant $|A|$ of the set of equations (15.6) is equal to

$$C_3 \, dx^2 + C_1 \, dy^2 - 2C_2 \, dx \, dy$$

If there is not to be a unique solution for velocity derivative ($\partial^2 \phi/\partial x^2, \partial^2 \phi/\partial y^2$, or $\partial^2 \phi/\partial x \, \partial y$), then $|A|$ must equal zero. Therefore, the slope of a characteristic line in the xy plane, dy/dx, is given by

$$C_1 \left(\frac{dy}{dx}\right)^2 - 2C_2 \frac{dy}{dx} + C_3 = 0$$

Solving this quadratic, we obtain

$$\frac{dy}{dx} = \frac{2C_2 \pm \sqrt{4C_2^2 - 4C_1 C_3}}{2C_1}$$

$$= \frac{C_2 \pm \sqrt{C_2^2 - C_1 C_3}}{C_1} \tag{15.7}$$

For $C_2^2 - C_1 C_3 < 0$, dy/dx will be imaginary, and for $C_2^2 - C_1 C_3 > 0$, dy/dx will be real.

Comparing with Eq. (15.2), we see

$$C_1 = a^2 - \left(\frac{\partial \phi}{\partial x}\right)^2 = a^2 - u^2$$

$$C_2 = -\frac{\partial \phi}{\partial x} \frac{\partial \phi}{\partial y} = -uv$$

$$C_3 = a^2 - \left(\frac{\partial \phi}{\partial y}\right)^2 = a^2 - v^2$$

so that

$$\frac{dy}{dx} = \frac{-uv \pm \sqrt{a^2 u^2 + a^2 v^2 - a^4}}{a^2 - u^2} \tag{15.8}$$

Thus, at each point in the flow, there exist two characteristics, I and II, with slopes given by

$$\left(\frac{dy}{dx}\right)_I = \frac{-(uv/a^2) + \sqrt{M^2 - 1}}{1 - (u^2/a^2)} \tag{15.9}$$

and

$$\left(\frac{dy}{dx}\right)_{II} = \frac{-(uv/a^2) - \sqrt{M^2 - 1}}{1 - (u^2/a^2)} \tag{15.10}$$

We can see that, for $M > 1$, there exist characteristic lines along which a unique value of velocity derivative does not exist. For $M < 1$, these lines are imaginary. According to the mathematical theory of partial differential equations, if $\sqrt{C_2^2 - C_1 C_3}$ is real, the equation is termed of the hyperbolic type; if $\sqrt{C_2^2 - C_1 C_3} = 0$, the equation is parabolic, and if $\sqrt{C_2^2 - C_1 C_3}$ is imaginary, the equation is elliptic. Thus, for supersonic flow, we are dealing with a hyperbolic partial differential equation.

Before proceeding, it is interesting to return to the simple linearized equation for velocity potential. For linearized flow, Eq. (15.3) reduces to

$$C_1 \frac{\partial^2 \phi}{\partial x^2} = C_3 \frac{\partial^2 \phi}{\partial y^2} = 0$$

with

$$C_1 = 1 - M_\infty^2$$
$$C_3 = 1$$

Now

$$\frac{dy}{dx} = \pm \frac{1}{\sqrt{M_\infty^2 - 1}}$$
$$= \pm \tan \mu$$

For the linearized equation, the characteristics are straight lines inclined at the Mach angle to the flow.

For the general case, we have determined the slope of the characteristic lines in the xy plane in terms of the velocity components u and v, or $\partial \phi / \partial x$ and $\partial \phi / \partial y$. To fully establish the characteristic lines, we must find how the velocity components vary with x and y. To this end, let us now return to the set of three linear equations (15.6) and develop an expression for one of the velocity derivatives. For example,

$$\frac{\partial^2 \phi}{\partial x^2} = \frac{\begin{vmatrix} du & dy & 0 \\ dv & dx & dy \\ 0 & 2C_2 & C_3 \end{vmatrix}}{\begin{vmatrix} dx & dy & 0 \\ 0 & dx & dy \\ C_1 & 2C_2 & C_3 \end{vmatrix}} \tag{15.11}$$

It will now be assumed that the velocity derivatives, $\partial^2 \phi / \partial x^2$, $\partial^2 \phi / \partial y^2$, and $\partial^2 \phi / \partial x \, \partial y$, with no shocks present, must be finite throughout the entire flow field. Since the denominator of Eq. (15.11) is zero, the determinant is also zero.

$$\begin{vmatrix} du & dy & 0 \\ dv & dx & dy \\ 0 & 2C_2 & C_3 \end{vmatrix} = 0$$

We obtain

$$du\, dx\, C_3 - du\, dy\, 2C_2 - dv\, dy\, C_3 = 0$$

or

$$\frac{dv}{du} = \frac{dx}{dy} - \frac{2C_2}{C_3} \tag{15.12}$$

Combining with Eq. (15.8), we obtain

$$\frac{dv}{du} = \frac{a^2 - u^2}{-uv \pm \sqrt{a^2u^2 + a^2v^2 - a^4}} + \frac{2uv}{a^2 - v^2}$$

or

$$\frac{dv}{du} = \frac{uv \mp \sqrt{a^2(u^2 + v^2) - a^4}}{a^2 - v^2} \tag{15.13}$$

Thus, for characteristic I,

$$\left(\frac{dv}{du}\right)_I = \frac{-uv + a\sqrt{u^2 + v^2 - a^2}}{v^2 - a^2} \tag{15.14}$$

and, for characteristic II,

$$\left(\frac{dv}{du}\right)_{II} = \frac{-uv - a\sqrt{u^2 + v^2 - a^2}}{v^2 - a^2} \tag{15.15}$$

Equations (15.14) and (15.15) present the slopes of the characteristics in the hodograph or velocity plane (u versus v). Given initial values of velocity, u, v, and a, at some point in the flow field, Eqs. (15.9), (15.10), (15.14), and (15.15) can be simultaneously numerically integrated to obtain the characteristic curves in the physical (x, y) plane and in the hodograph (u, v) plane. Note that, since dy/dx has been shown to be a function of u and v, solution for the characteristics in the (x, y) plane is dependent on the simultaneous solution for the characteristic curves in the (u, v) plane.

It will be found more useful to express the slope of the characteristic lines in polar coordinates. Using the system shown in Figure 15.2, with positive θ counterclockwise, we see that $u = V \cos \theta$, $v = V \sin \theta$, and $\tan \theta = v/u$. We have shown in Chapter 2 that $\sin \mu = 1/M = a/V$ (where $\mu = $ Mach angle),

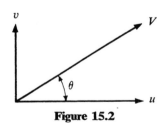

Figure 15.2

or $\tan \mu = 1/\sqrt{M^2 - 1}$. By algebraic manipulation, it can be shown that

$$\frac{dy}{dx} = \frac{-uv \pm a\sqrt{u^2 + v^2 - a^2}}{a^2 - u^2} = \frac{\dfrac{u}{v} \mp \dfrac{a}{\sqrt{u^2 + v^2 - a^2}}}{1 + \dfrac{a}{\sqrt{u^2 + v^2 - a^2}}\dfrac{v}{u}} \qquad (15.16)$$

In polar coordinates, Eq. (15.16) becomes

$$\frac{dy}{dx} = \frac{\tan \theta \mp \dfrac{1}{\sqrt{M^2 - 1}}}{1 + (\tan \theta)\dfrac{1}{\sqrt{M^2 - 1}}} = \frac{\tan \theta \mp \tan \mu}{1 + \tan \theta \tan \mu} \qquad (15.17)$$

Using the trigonometric identity

$$\tan (\theta \mp \mu) = \frac{\tan \theta \mp \tan \mu}{1 \pm \tan \mu \tan \theta} \qquad (15.18)$$

we arrive at

$$\left(\frac{dy}{dx}\right)_{\mathrm{I}} = \tan (\theta - \mu) \qquad (15.19)$$

$$\left(\frac{dy}{dx}\right)_{\mathrm{II}} = \tan (\theta + \mu) \qquad (15.20)$$

In other words, at each point in the flow, the characteristic lines are inclined at the Mach angle to the flow direction, as shown in Figure 15.3, with characteristics of type I making a negative acute angle with the flow direction, and characteristics of type II making a positive acute angle with the flow direction. Using the mathematical approach, we have so far demonstrated the existence of characteristics for supersonic flow and have shown their equivalence with the Mach waves. The student is cautioned that, whereas the mathematics shows the existence of two characteristics at each point in a supersonic flow field, there

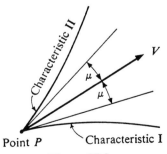

Point P Characteristic I

Figure 15.3

may exist only one Mach wave, since from physical considerations a disturbance in a supersonic flow cannot propagate back upstream. This was shown in
Chapter 14, for example, for the flow over a wave-shaped wall, where lines of
f = constant, corresponding to Mach lines making a negative angle with the
flow direction, were ruled out.

To demonstrate the utility of the characteristics for the solution of flow
problems, we must obtain a workable expression for velocity or Mach number,
along a characteristic. Let us write an expression for the slope of the characteristic in the hodograph plane in polar coordinates.

We have shown

$$\frac{dv}{du} = \frac{-uv \pm a\sqrt{u^2 + v^2 - a^2}}{v^2 - a^2} \tag{15.13}$$

Multiply numerator and denominator by $uv \pm a\sqrt{u^2 + v^2 - a^2}$ and simplify to
obtain

$$\left(\frac{dv}{du}\right)_{I} = \frac{1 - \dfrac{u^2}{a^2}}{\dfrac{uv}{a^2} + \sqrt{M^2 - 1}} \tag{15.21}$$

$$\left(\frac{dv}{du}\right)_{II} = \frac{1 - \dfrac{u^2}{a^2}}{\dfrac{uv}{a^2} - \sqrt{M^2 - 1}} \tag{15.22}$$

Comparing with Eqs. (15.9) and (15.10), we see that

$$\left(\frac{dv}{du}\right)_{I} = -\frac{1}{\left(\dfrac{dy}{dx}\right)_{II}} \quad \text{and} \quad \left(\frac{dv}{du}\right)_{II} = -\frac{1}{\left(\dfrac{dy}{dx}\right)_{I}} \tag{15.23}$$

or, from Eqs. (15.19) and (15.20), there results

$$\left(\frac{dv}{du}\right)_{I} = -\text{ctn}\,(\theta + \mu) \tag{15.24}$$

and

$$\left(\frac{dv}{du}\right)_{II} = -\text{ctn}\,(\theta - \mu) \tag{15.25}$$

Since $u = V\cos\theta$ and $v = V\sin\theta$, then $du = \cos\theta\,dV - V\sin\theta\,d\theta$ and
$dv = \sin\theta\,dV + V\cos\theta\,d\theta$, and we can write

$$\frac{dv}{du} = \frac{\sin\theta\,dV + V\cos\theta\,d\theta}{\cos\theta\,dV - V\sin\theta\,d\theta} \tag{15.26}$$

Combining Eq. (15.26) with Eqs. (15.24) and (15.25), we obtain

$$d\theta = \frac{dV \tan \theta \tan (\theta \pm \mu) + 1}{V \; \tan \theta - \tan (\theta \pm \mu)} \tag{15.27}$$

Using the trigonometric identity, Eq. (15.18), in Eq. (15.27), there results

$$\frac{1}{V} \frac{dV}{d\theta} = \mp \tan \mu = \mp \frac{1}{\sqrt{M^2 - 1}} \tag{15.28}$$

In other words, along characteristic I, we can write

$$\frac{1}{V} \frac{dV}{d\theta} = -\frac{1}{\sqrt{M^2 - 1}} \tag{15.29}$$

and along characteristic II,

$$\frac{1}{V} \frac{dV}{d\theta} = +\frac{1}{\sqrt{M^2 - 1}} \tag{15.30}$$

We now wish to integrate Eqs. (15.29) and (15.30) to obtain Mach number as a function of θ *along* the characteristics. Fortunately, we have already come across a similar integral in dealing with Prandtl Meyer flow in Chapter 7, where it was shown that, *across* a Mach wave,

$$\frac{dV}{V} = \frac{1}{\sqrt{M^2 - 1}} \, dv \tag{7.2}$$

or

$$\Delta v = \int \sqrt{M^2 - 1} \frac{dV}{V}$$

It is very important to appreciate the difference between Eq. (7.2) and Eqs. (15.29) and (15.30). In Chapter 7, we integrated *across* the waves, or characteristics, to find changes taking place normal to the waves. In Eqs. (15.29) and (15.30), we are integrating *along* the characteristic, not across, or normal to it. The reason that the resultant equation for the two cases is similar will be discussed later in this chapter, when we derive expressions for changes taking place across the characteristics. Returning to the integration, we have

$$\int_a^b d\theta = -\int_a^b \sqrt{M^2 - 1} \frac{dV}{V} \qquad \text{along a characteristic of type I}$$

$$\int_b^b d\theta = +\int_a^b \sqrt{M^2 - 1} \frac{dV}{V} \qquad \text{along a characteristic of type II}$$

where a and b are points on the characteristics. Using the results from Chapter 7, for the type I characteristic,

$$\theta_b - \theta_a = -(v_b - v_a) \tag{15.31}$$

Let a be a reference state, so that $\theta - \theta_{\text{ref}} = -(v - v_{\text{ref}})$, or, along a characteristic of type I, we have

$$\theta + v = \text{constant} = C_1 \tag{15.32}$$

Similarly, along a characteristic of type II,

$$v - \theta = \text{constant} = C_{\text{II}} \tag{15.33}$$

15.4 Calculation Procedure

We are now in a position to demonstrate how the method of characteristics can be used to solve for a two-dimensional flow field. Suppose that, along a curve LM, we know the flow conditions, that is, velocity and its direction, pressure, temperature, and so on. This information represents a boundary condition for the problem. Now select points A and B on the curve, as shown in Figure 15.4.

With velocity (magnitude and direction) and temperature known at A and B, v and θ can be found, so that the characteristic values C_I and C_{II} can be calculated from Eqs. (15.32) and (15.33). Characteristic I from A intersects characteristic II from B at point P (Figure 15.4). Since $v + \theta = $ constant along a characteristic of type I and $v - \theta = $ constant along a characteristic of type II, it follows that, at P

$$C_I = v_A + \theta_A = v_P + \theta_P$$
$$C_{II} = v_B - \theta_B = v_P - \theta_P$$

so that

$$v_P = \frac{C_I + C_{II}}{2}$$

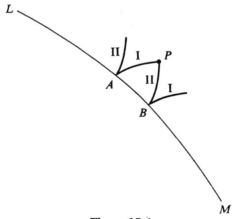

Figure 15.4

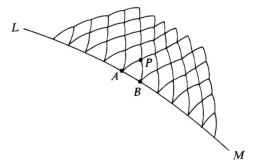

Figure 15.5

and

$$\theta_P = \frac{C_I - C_{II}}{2} \tag{15.34}$$

Therefore, from Eq. (15.34), the Mach number and velocity direction at P can be found. For this isentropic flow, the temperature and pressure at P can be calculated from the temperature and pressure at A or B, with T_t and p_t equal to constants.

Working outward from the boundary curve LM, it is then possible to build up a net or mesh, with points connected by the characteristics (see Figure 15.5). The only problem left is the determination of the shape of the characteristics in the (x, y) plane. Since Mach number varies throughout the flow field, the characteristic lines AP or BP are not necessarily straight, but rather are curved. However, if the size of the mesh is kept small enough, the lines AP or BP can be approximated by straight lines (Figure 15.6) inclined at the Mach angle to the flow direction. Naturally, the accuracy of this procedure depends on the mesh size: the smaller the mesh, the more accurate the solution. In other words, along the curve LM, points should be closely spaced for a high degree of accuracy in the resultant solution. Example 15.1 will illustrate the computational procedure.

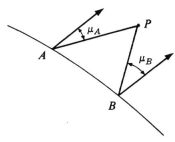

Figure 15.6

Example 15.1

Uniform radial flow at Mach 2.0 enters a two-dimensional diverging channel with straight walls. Compute the variation of Mach number in this radial flow field, assuming isentropic, steady flow. The walls are inclined at a total angle of 12°, as shown in Figure 15.7.

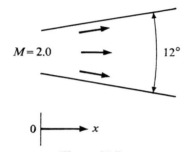

Figure 15.7

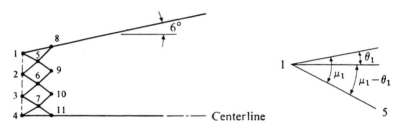

Figure 15.8

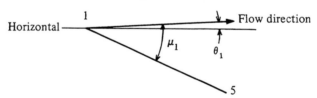

Angle of 1–5 with horizontal $= \theta_1 - \mu_1 = -24°$

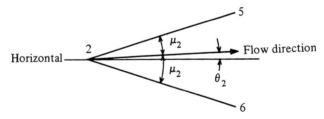

Angle of 2–5 with horizontal $= \theta_2 + \mu_2 = 34°$

Angle of 2–6 with horizontal $= \theta_2 - \mu_2 = -26°$

Figure 15.9

Solution

From symmetry, it is only necessary to consider one-half of the flow field (see Figure 15.8). At $x = 0$, the flow is known, so we can work outward from this line. Select four equally spaced points, such that $\theta_4 = 0°$, $\theta_3 = 2°$, $\theta_2 = 4°$, and $\theta_1 = 6°$. $M_1 = M_2 = M_3 = M_4 = 2.0$, so from Appendix D, $\nu_1 = \nu_2 = \nu_3 = \nu_4 = 26.38°$. Along the characteristic line 2–5, $C_{II} = $ constant $= \nu - \theta = 22.38°$. Along the characteristic line 1–5, $C_I = $ constant $= \nu + \theta = 32.38°$. Now we have found C_I and C_{II} at point 5, so that, from Eq. (15.34), $\nu_5 = 27.38°$ and $\theta_5 = 5°$. From Appendix D, for $\nu = 27.38°$, $M_5 = 2.04$. For this numerical procedure, the line 1–5 is assumed straight, as shown in Figure 15.9, making an angle $-\mu_1$ with

Table 15.1   Computations for Example 15.1

Point	M	μ	ν	θ	C_I	C_{II}	$\theta + \mu$	$\theta - \mu$
1	2.00	30.00	26.38	6	32.38	20.38	36.00	−24.00
2	2.00	30.00	26.38	4	30.38	22.38	34.00	−26.00
3	2.00	30.00	26.38	2	28.38	24.38	32.00	−28.00
4	2.00	30.00	26.38	0	26.38	26.38	30.00	−30.00
5	2.04	29.35	27.38	5	32.38	22.38	34.35	−24.35
6	2.04	29.35	27.38	3	30.38	24.38	32.35	−26.35
7	2.04	29.35	27.38	1	28.38	26.38	30.35	−28.35
8	2.07	28.89	28.38	6	34.38	22.38	34.89	−22.89
9	2.07	28.89	28.38	4	32.38	24.38	32.89	−24.89
10	2.07	28.89	28.38	2	30.38	26.38	30.89	−26.89
11	2.07	28.89	28.38	0	28.38	28.38	28.89	−28.89
12	2.11	28.29	29.38	5	34.38	24.38	33.29	−23.29
13	2.11	28.29	29.38	3	32.38	26.38	31.29	−25.29
14	2.11	28.29	29.38	1	30.38	28.38	29.29	−27.29
15	2.15	27.72	30.38	6	36.38	24.38	33.72	−21.72
16	2.15	27.72	30.38	4	34.38	26.38	31.72	−23.72
17	2.15	27.72	30.38	2	32.38	28.38	29.72	−25.72
18	2.15	27.72	30.38	0	30.38	30.38	27.72	−27.72
19	2.19	27.17	31.38	5	36.38	26.38	32.17	−22.17
20	2.19	27.17	31.38	3	34.38	28.38	30.17	−24.17
21	2.19	27.17	31.38	1	32.38	30.38	28.17	−26.17
22	2.23	26.64	32.38	6	38.38	26.38	32.64	−20.64
23	2.23	26.64	32.38	4	36.38	28.38	30.64	−22.64
24	2.23	26.64	32.38	2	34.38	30.38	28.64	−24.64
25	2.23	26.64	32.38	0	32.38	32.38	26.64	−26.64
26	2.26	26.26	33.38	5	38.38	28.38	31.26	−21.26
27	2.26	26.26	33.38	3	36.38	30.38	29.26	−23.26
28	2.26	26.26	33.38	1	34.38	32.38	27.26	−25.26
29	2.30	25.77	34.38	6	40.38	28.38	31.77	−19.77
30	2.30	25.77	34.38	4	38.38	30.38	29.77	−21.77
31	2.30	25.77	34.38	2	36.38	32.38	27.77	−23.77
32	2.30	25.77	34.38	0	34.38	34.38	25.77	−25.77

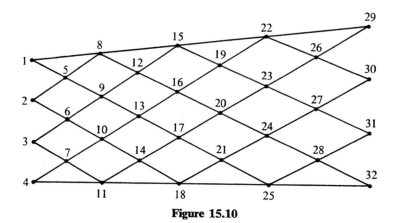

Figure 15.10

Table 15.2 Example 15.1–comparison with one-dimensional flow

Point	Area	A/A^*	M (Chapter 3)	M (Method of Characteristics)
4	A_4	1.688	2.00	2.00
11	$1.07A_4$	1.806	2.08	2.07
18	$1.13A_4$	1.907	2.14	2.15
25	$1.21A_4$	2.042	2.22	2.23
32	$1.28A_4$	2.181	2.28	2.30

the flow direction at 1, or $\theta - \mu$ with the horizontal. μ can be found from $\sin \mu = 1/M$ (μ is given in Appendix D). For lines such as 2–5, which make a positive (counterclockwise) acute angle with the flow direction, the angle between these lines and the horizontal is $\theta + \mu$. For lines such as 2–6, which make a negative acute angle with the flow direction, the angle these lines make with the horizontal is $\theta - \mu$ (see Figure 15.9). At points such as 8 and 11, it is only possible to find one characteristic value (C_{II} at 8, C_I at 11). However, at these points, we have additional boundary conditions, since the flow must follow the walls. At point 8, θ is equal to 6°; at 11, θ is equal to 0°. With one characteristic value and θ, ν and the other characteristic value can be calculated from Eq. (15.34). Table 15.1 and Figure 15.10 show the results of the numerical procedure.

To make a rough check on the computations, we can compare the Mach number along the centerline of the channel with the results from the simple one-dimensional flow methods of Chapter 3. (Over the range considered, it can be seen that flow in the channel is close to one-dimensional.) The areas given in Table 15.2 have been measured directly from Figure 15.10.

15.5 Another Computational Procedure: The Region-to-Region Method

The procedure outlined in Section 15.4 has involved calculating the flow properties at points in the flow field, points determined by the intersection of

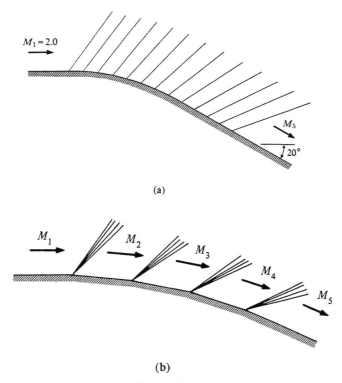

(a)

(b)

Figure 15.11

characteristic lines. In many examples, it is easier to use a slightly different procedure in which properties are found in the regions bounded by characteristics. For example, suppose we have a continuously varying flow about a curved wall (See Figure 15.11a). Several Mach waves (characteristics of type II) have been shown; for this flow, the characteristics of type I are not Mach waves, since, as has been discussed previously, a disturbance cannot propagate back upstream in a supersonic flow. If the wall is smooth, there will be an infinite number of Mach waves between M_1 and M_5. In order to analyze the flow, let us break the continuous curved wall into segments, as shown in Figure 15.11b. To carry this procedure one step further, let us replace each expansion fan in Figure 15.11b by a single wave, as shown in Figure 15.12. These waves (a, b, c, d) are not truly characteristics or Mach waves, since they separate

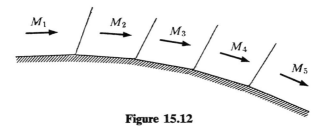

Figure 15.12

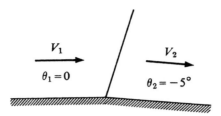

Figure 15.13

regions between which small but finite changes take place. However, let us assume that the wall is divided into enough segments that the waves can be approximated as characteristics. (Strictly speaking, only if the wall were divided into an infinite number of segments would the waves be characteristics.) The four waves (a, b, c, d) divide up the flow into five regions, in each of which Mach number and flow direction are uniform.

We wish to determine the changes in v and θ on going from one region to another across the waves. Consider regions 1 and 2, shown in Figure 15.13. It is evident from the results of Chapter 7 that, across this wave, $\Delta v = -\Delta \theta$, where again θ is defined to be positive counterclockwise. This result can be seen also by sketching in the characteristics of type I (see Figure 15.14). Notice that, in order to go from region 1 to region 2, we must go *across* a characteristic of type II. However, by proceeding *along* a characteristic of type I we can also go from region 1 to region 2. Thus, when we talk about changes taking place *across* characteristics of type II, we are also talking about changes that take place *along* characteristics of type I. We have shown from Eq. (15.32) that *along* type I characteristics $\Delta v = -\Delta \theta$. *Across* type II characteristics, likewise

$$\Delta v = -\Delta \theta \tag{15.35}$$

From Eq. (15.33), *along* a characteristic of type II, $\Delta v = \Delta \theta$, so that *across* a type I characteristic, we have

$$\Delta v = \Delta \theta \tag{15.36}$$

It thus becomes possible to work out a procedure involving the computation of a supersonic flow from region to region across characteristics, rather

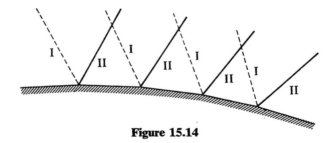

Figure 15.14

than from point to point along the characteristics. Before going through an example of this procedure, it is well to consider the nature of the approximation made here and for the previous method.

In the region-to-region method, the approximation involves dividing a continuous flow into regions of uniform flow divided by weak but finite waves, with these weak waves assumed to have the properties of characteristics. As the number of regions into which the flow is divided is increased, the approximate solution becomes more and more exact.

For the point-to-point method, the approximation involves treating the characteristics as straight lines; again, this result comes closer and closer to the exact solution as more and more points are taken and the size of the mesh reduced. Both methods, of course, assume isentropic flow, so that no shocks are present.

Example 15.2

Flow at the exit of a Mach 1.5 supersonic nozzle is expanded from an exit-plane pressure of 200 kPa to a back pressure of 100 kPa. Determine the flow just downstream of the nozzle exit.

Solution

From symmetry, we need only consider one-half of the flow. First divide the Prandtl Meyer expansion fan at the nozzle exit into four segments (Figure 15.15), with $p_4 = 100$ kPa. For isentropic flow,

$$\frac{p_4}{p_t} = \frac{p_4}{p_1}\frac{p_1}{p_t} = \frac{100}{200}0.2724 = 0.1362$$

so that

$$M_4 = 1.959$$

$$\theta_4 = \nu_4 - \nu_1 = 25.243 - 11.905 = 13.338$$

yielding

$$\theta_2 = 4.446, \qquad \theta_3 = 8.892$$

From regions 1 to 2, we are crossing a wave or characteristic of type I so that

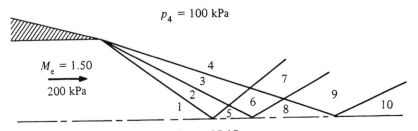

$p_4 = 100$ kPa

$M_e = 1.50$

200 kPa

Figure 15.15

$\Delta v = \Delta \theta$. Since $\theta_2 = 4.446°$, $\Delta \theta = \theta_2 - \theta_1 = 4.446°$ and $\Delta v = v_2 - v_1 = 4.446°$; hence, we obtain $v_2 = 11.905 + 4.446 = 16.351°$. From regions 2 to 5, we cross a wave or characteristic of type II, so that across this wave, $\Delta v = -\Delta \theta$. From the boundary condition at the centerline, we have that $\theta_5 = 0°$, so that from 2 to 5, $\Delta \theta = -4.446°$ and $\Delta v = 4.446°$. Therefore, $v_5 = v_2 + \Delta v = 16.351° + 4.446° = 20.797°$. On going from 5 to 6, $\Delta v = \Delta \theta$, so that $v_6 - v_5 = \theta_6 - \theta_5$, or $v_6 - \theta_6 = v_5 - \theta_5$. On going from 3 to 6, $\Delta v = -\Delta \theta$, so that $v_6 - v_3 = \theta_3 - \theta_6$, or $v_6 + \theta_6 = v_3 + \theta_3$.

Combining, we can solve for v_6:

$$v_6 = \frac{(v_5 - \theta_5) + (v_3 + \theta_3)}{2}$$

$$= \frac{20.797 + 29.689}{2}$$

$$= 25.243°$$

and

$$\theta_6 = v_6 - (v_5 - \theta_5) = 4.446°$$

To establish the physical boundaries of the regions, we must determine the wave angles. Since the single wave between 1 and 2 represents, in actuality, a

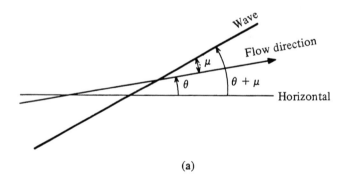

(a)

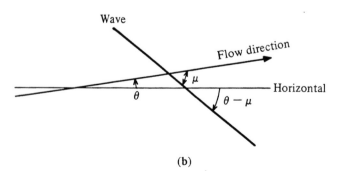

(b)

Figure 15.16

Table 15.3   Computations for Example 15.2

Region	θ	ν	M	μ	$\theta + \mu$	$\theta - \mu$
1	0	11.905	1.500	41.81	41.81	−41.81
2	4.446	16.351	1.650	37.31	41.76	−32.86
3	8.892	20.797	1.802	33.71	42.60	−24.82
4	13.338	25.243	1.959	30.70	44.04	−17.36
5	0	20.797	1.802	33.71	33.71	−33.71
6	4.446	25.243	1.959	30.70	35.15	−26.25
7	8.892	29.689	2.122	28.11	37.00	−19.22
8	0	29.689	2.122	28.11	28.11	−28.11
9	4.446	34.135	2.294	25.84	30.29	−21.39
10	0	38.581	2.477	23.81	23.81	−23.81

Prandtl Meyer fan, approximate the angle that the single wave makes with the horizontal by averaging the angles that the first and last waves of the fan make with the horizontal. For a wave that makes a positive acute angle with the flow direction (Figure 15.16a), the angle that the wave makes with the horizontal is $\theta + \mu$. For a wave that makes a negative acute angle with the flow direction, the angle that the wave makes with the horizontal is $\theta - \mu$ (see Figure 15.16b). Draw the wave between regions 2 and 5 at an angle of

$$\frac{41.76 + 33.71}{2} = 37.74°$$

with the horizontal; draw the wave between regions 1 and 2 at an angle of

$$\frac{-41.81 + (-32.86)}{2} = -37.34°$$

with the horizontal. The remainder of the computations are shown in Table 15.3, with the graphical construction in Figure 15.17.

From the preceding example, a property of the waves becomes apparent: the strength of the wave, as measured by $\Delta\theta$ across the wave, is undiminished by intersection with other weak waves or by reflection. For example, the initial wave separating regions 1 and 2 is reflected from the centerline, with the

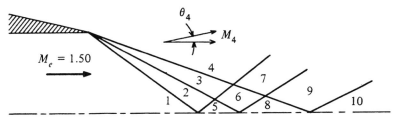

Figure 15.17

reflected wave intersected by two other waves. Yet $\theta_2 - \theta_1 = \theta_4 - \theta_7$. This result really follows from our assumption that the waves are very weak, that they behave as characteristics.

15.6 Supersonic Nozzle Design

One application of the method of characteristics is the design of the contour of supersonic nozzles. For a wind tunnel nozzle, it is essential that the flow at the nozzle exit (test section) be parallel and wave free at the desired Mach number. The presence of waves is undesirable, in that these weak waves can coalesce to form a finite shock and prevent uniform flow in the test section.

Let us now examine how a nozzle contour can be designed so as to cancel an incident wave. If a weak wave of turning angle $\Delta\theta$ is incident upon a plane surface, as shown in Figure 15.18, a reflected wave of turning angle $\Delta\theta$ must be present so as to satisfy the boundary conditions at the wall.

To cancel the incident wave, let the wall turn through an angle $\Delta\theta$ at the point of impingement of the incident wave (see Figure 15.19). There is no reflected wave, since the boundary condition at the wall is satisfied without it; the incident wave is effectively canceled.

Now let us turn our attention to the converging–diverging supersonic nozzle. We shall assume parallel, sonic flow at the nozzle throat. From symmetry, we need only consider one-half of the flow; since there can be no flow across the centerline, we can represent the centerline as a plane wall. The flow first expands from q to a. Four of the expansion waves (type I

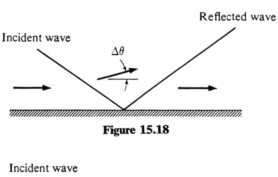

Figure 15.18

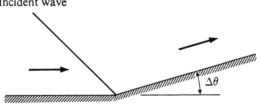

Figure 15.19

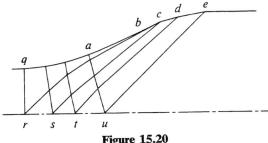

Figure 15.20

characteristics) have been sketched in Figure 15.20. They intersect the center-line at r, s, t, and u and are reflected, as characteristics of type II, impinging on the nozzle wall at b, c, d, and e. Thus the waves reflected from the centerline turn the flow back to the horizontal at e. The wall contours at b, c, d, and e must turn the flow through the same $\Delta\theta$ as the incident waves, so as to cancel the incident waves. Naturally, the more waves considered, the smoother and more exact the resultant contour. We are only interested here in wave cancellations, so the initial expansion (from q to a) is arbitrary.

Example 15.3

Design the diverging section of a supersonic nozzle to produce uniform Mach 1.8 flow. Assume that the length of the nozzle is to be kept to a minimum.

Solution

For the shortest possible nozzle, the expansion from q to a can be assumed to take place as a Prandtl Meyer expansion fan. We shall represent the expansion as four waves of equal strength and design the nozzle from these waves (see Figure 15.21). The waves separating regions q and s, s and r, r and p, and p and a are of type I, so across these waves $\Delta\nu = \Delta\theta$. The waves separating a and b, b and c, c and d, and d and e are of type II, so $\Delta\nu = -\Delta\theta$. The overall $\Delta\nu = \nu_e - \nu_q = 20.72$ for

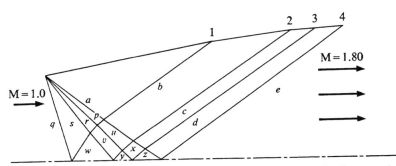

Note: As in Example 15-2, the angle that the wave between q and s, for example, makes with the horizontal is the average value of $\mu - \theta$ in the two regions.

Figure 15.21

Table 15.4   Supersonic nozzle design

Region	θ	ν	M	μ	$\theta + \mu$	$\theta - \mu$
q	0	0	1.0	90	+90	−90
s	2.59	2.59	1.16	59.55	+62.14	−56.96
r	5.18	5.18	1.26	52.53	+57.71	−47.35
p	7.77	7.77	1.36	47.33	+55.10	−39.56
a	10.36	10.36	1.45	43.60	+53.96	−33.24
b	7.77	12.95	1.54	40.49	+48.26	−32.72
c	5.18	15.54	1.62	38.12	+43.30	−32.94
d	2.59	18.13	1.71	35.79	+38.38	−33.20
e	0	20.72	1.80	33.75	+33.75	−33.75
u	5.18	10.36	1.45	43.60	+48.78	−38.42
v	2.59	7.77	1.36	47.33	+49.92	−44.74
w	0	5.18	1.26	52.53	+52.53	−52.53
x	2.59	12.95	1.54	40.49	+43.08	−37.90
y	0	10.36	1.45	43.60	+43.60	−43.60
z	0	15.54	1.62	38.12	+38.12	−38.12

$M_e = 1.80$. The waves of the Prandtl Meyer fan are assumed of equal strength (equal $\Delta\theta$ or $\Delta\nu$). Since the strength of the waves is undiminished by reflection or intersection with other waves, it follows that the strength of the waves separating a and b, b and c, c and d, and d and e are equal to those of the Prandtl Meyer fan. Therefore, for each of these waves, $\Delta\nu = 20.72/8 = 2.59°$, since the flow crosses a total of eight waves between q and e. Note that the flow is horizontal at q and at e, so $\theta_q = \theta_e = 0$. The flow angle θ increases (counterclockwise) across the Prandtl Meyer fan, and then decreases to the exit. At each of the points 1, 2, 3, and 4, where the reflected waves impinge on the wall, the wall must be turned through 2.59°. The remainder of the construction, shown in Table 15.4 and Figure 15.21, is the same as that of Example 15.2. The required contour is 0–1–2–3–4.

Since uniform, one-dimensional flow is present at q and e, with isentropic flow in between, A/A^* from the construction should agree with that from Appendix A, that is, at $M = 1.80$, $A/A^* = 1.439$. Naturally, the more waves considered, the smoother the resultant nozzle contour, and the more adequate the design.

15.7 Summary

The method of characteristics has been presented as a method for solving problems in two-dimensional isentropic, supersonic flow. Characteristics are defined as lines along which there exist discontinuities in velocity derivatives. By treating the complete equations of motion for isentropic, two-dimensional flow, expressions were derived for the characteristics in the x, y plane and in the hodograph or u, v plane. It was shown that characteristics only exist in the real plane for supersonic flow, and that the characteristics are inclined at the local Mach angle to the flow direction, thereby establishing a relationship between the characteristics and Mach waves. Having found expressions for the velocity vector along the characteristics, it is possible to proceed from point to point in the flow field and effect a solution for a supersonic flow.

An equivalent method was also discussed, in which regions of the flow are separated by Mach waves or characteristics. Knowing the necessary relationships for $\Delta\theta$ and $\Delta\nu$ across such waves, it is then possible to work from region to region to solve a problem in two-dimensional supersonic flow.

If nothing else, this coverage of the method of characteristics should impress the student with the very real differences between subsonic flow and supersonic flow. Since, from experience, we are somewhat more accustomed to dealing with low-velocity flows, it follows that some of the relationships between the various properties may be more easily understood for subsonic flow. However, it should also be realized that the solution of a general two-dimensional, compressible flow problem for supersonic flow, for which the method of characteristics is available, is often easier than the solution of a similar problem in subsonic flow, where no such procedure exists.

REFERENCES

Specific References

1. RABENSTEIN, A.L., *Introduction to Ordinary Differential Equations*, New York, Academic Press, 1972, pp. 489–493.

General References

2. LIEPMANN, H.W., AND ROSHKO, A., *Elements of Gasdynamics*, New York, John Wiley & Sons, Inc., 1957.
3. CHEERS, F., *Elements of Compressible Flow*, New York, John Wiley & Sons, Inc., 1963.
4. SHAPIRO, A., *The Dynamics and Thermodynamics of Compressible Fluid Flow*, New York, Ronald Press Company, 1953.
5. OWCZAREK, J.A., *Fundamentals of Gas Dynamics*, Scranton, Pa., International Textbook Company, 1964.
6. ZUCROW, M.J., AND HOFFMAN, J.D., *Gas Dynamics*, Vol. 1. New York. John Wiley & Sons, Inc., 1976.

PROBLEMS

1. A two-dimensional supersonic nozzle uses circular arc walls as shown in Figure P15.1. Using the method of characteristics, determine the distribution of Mach number and pressure throughout the flow. The inlet Mach number is 1.8, and inlet

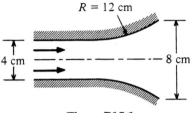

Figure P15.1

pressure is 50 kPa. *Hint*: Break the smooth wall into a series of segments, and use the region-to-region method.

2. A thin airfoil has the form of a circular arc, as shown in Figure P15.2. Determine the lift and drag coefficients for the foil at a Mach number of 2.0, and compare with the linearized theory.

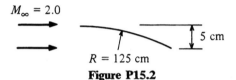

Figure P15.2

3. Derive Eq. 15.16.
4. Continue Example 15.2 until the series of waves has its second intersection with the centerline.
5. A supersonic flow at Mach 1.8 enters the channel shown in Figure P15.5. Using the method of characteristics, determine the Mach number distribution throughout the flow and the pressure distribution along the walls.

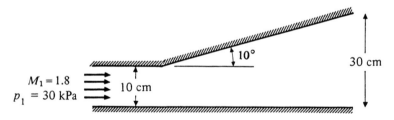

Figure P15.5

6. Design the contour of a two-dimensional plug nozzle (see Chapter 8) so as to produce cancellation of the waves incident on the plug (see Figure P15.6). The

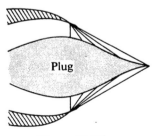

Figure P15.6

nozzle is to provide a flow of 1 kg/s of air at Mach 2.0. Assume a stagnation pressure of 0.5 MPa and a stagnation temperature of 500 K.

7. Using the method of characteristics, design the contour of the shortest possible converging–diverging nozzle to fit the specifications of Problem 6.

8. A converging–diverging nozzle discharges a uniform supersonic flow at Mach 2.2 and static pressure of 1 atm two dimensionally into a back pressure region of 0.8 atm. Sketch the flow pattern and calculate the pressure along the axis as far as the region after the intersection of the first Prandtl Meyer expansion fans.

9. Repeat Problem 1 using the point-to-point method. Compare the results.

16

RAREFIED GAS DYNAMICS

16.1 Introduction

The previous 15 chapters have dealt exclusively with continuum flow, in which the number of molecules per unit volume has been assumed large enough so that, in general, the fluid properties could be assumed to vary continuously from point to point throughout a region. In reality, of course, a volume of fluid is not a continuum, but consists of a finite number of molecules. However, at 1 atm and 20°C there are approximately 2×10^{19} molecules in $1 \, cm^3$ of air, with the mean free path (mean distance between molecular collisions) only 6.35×10^{-6} cm. Under these conditions, the smallest volume we are interested in will contain enough molecules so that we can effectively average over the molecules present and use a macroscopic approach; we do not need to consider the motions of individual molecules.

The assumption of a continuum, however, breaks down at very low pressures, where the mean free path becomes comparable or greater in length than the length of a body or other object under consideration. For example, at a pressure of 10^{-6} atm, the mean free path increases to 6.35 cm. At this pressure, obviously, we can no longer speak of the fluid as a continuum, but rather must consider, for example, the interaction between individual molecules and the surface under examination.

It is the intent of this chapter to show some of the effects that take place in rarefied flows, when the macroscopic approach is not entirely valid. Due to the great complexity of the physical problem, we shall try to reduce the number of equations to a minimum and put the emphasis on a description of some of the phenomena that occur in rarefied flows.

16.2 Knudsen Number

A dimensionless parameter called *Knudsen number, Kn,* relates the mean free path $\bar{l}$ in a flow to the characteristic length L of a body immersed in the flow.

$$Kn = \frac{\bar{l}}{L} \qquad (16.1)$$

For continuum flows, Kn is very small; however, as has been mentioned, for rarefied flows, in which $\bar{l}$ takes on appreciable values, the Knudsen number will be greater than 1 and will be a significant parameter to be used in describing the resultant flow. It is of interest to express Knudsen number in terms of Mach number and Reynolds number.

$$Re = \frac{\rho V L}{\mu}$$

$$M = \frac{V}{a}$$

From kinetic theory[1], the viscosity of a gas can be expressed as

$$\mu = \frac{1}{3} \rho \bar{c} \bar{l} \qquad (16.2)$$

where $\bar{c}$ = mean molecular velocity. Also, $\bar{c}$ can be related to the velocity of sound by

$$\bar{c} = \sqrt{\frac{8}{\pi \gamma}}\, a \qquad (16.3)$$

so that

$$Kn = \frac{\bar{l}}{L}$$

$$= \frac{3\mu}{\rho a L \sqrt{8/\pi\gamma}} \qquad (16.4)$$

For flow at low Reynolds numbers, for example Stokes flow, a significant length is a body dimension L_b so that

$$Kn \sim \frac{M}{Re_{L_b}} \qquad (16.5)$$

where

$$Re_{L_b} = \frac{\rho V L_b}{\mu}$$

For flow at high Reynolds numbers, a significant dimension is boundary layer thickness. For a laminar boundary layer, $\delta \sim 1/\sqrt{Re_{L_b}}$, so that

$$Kn_\delta = \frac{\bar{l}}{\delta}$$

$$\sim \frac{M}{\sqrt{Re_{L_b}}} \qquad \text{for high Reynolds numbers} \qquad (16.6)$$

It can be seen that rarefied gas flow can occur at high Mach number, for cases in which low Reynolds numbers are present, for example, high velocity flow at high altitude. As might be expected, the line of demarcation between rarefied flow and continuum flow is not sharply defined, but represents really a gradual transition. Nevertheless, it is possible to set rough boundaries between the two types of flow. For flow at appreciable Reynolds numbers (Re greater than 1), we have continuum flow for $M/\sqrt{Re} < 0.01$, whereas for $Re < 1$, continuum flow exists for $M/Re < 0.01$[2].

As an example, suppose a missile 3.0 m long is traveling at Mach 10 at 50 km where, from Appendix H, $T = 270.7$ K, $\rho = 0.00103$ kg/m^3, and $\mu = 1.70 \times 10^{-5}$ N $\cdot$ s/m^2. The vehicle velocity is $10\sqrt{\gamma RT} = 3298$ m/s, so

$$Re = \frac{(0.00103 \text{ kg/m}^3)(3298 \text{ m/s})(3.0 \text{ m})}{1.7 \times 10^{-5} \text{ N} \cdot \text{s/m}^2}$$

$$= 5.99 \times 10^5$$

For this case, $M/\sqrt{Re} = 0.0129$, which puts the situation in the range where rarefied gas effects must be considered.

It is important to realize that the study of rarefied flows is more than an academic curiosity; such widely diverse problems as the calculation of drag on a space satellite and the determination of molecular flows in an ultrahigh vacuum system require a knowledge of the phenomena peculiar to this flow.

16.3 Molecular Model of Flow Near a Surface

The flow of a gas in the continuum regime can be thought of as the combination of a random molecular motion superposed on an ordered directional flow. The distribution of molecular velocities for a gas in equilibrium has been calculated by Maxwell; according to the Maxwellian distribution[3], the number of molecules per unit speed range is given by

$$\frac{dN}{dc} = 4\pi N\alpha^3 c^2 e^{-\beta^2 c^2} \qquad (16.7)$$

where α and β are constants.

$$N = \text{total number of molecules}$$
$$c = \text{molecular speed}$$

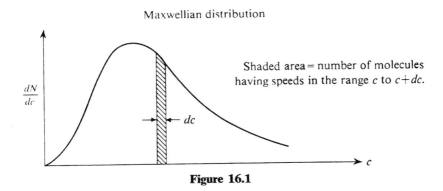

Figure 16.1

This equation expresses the number of molecules dN possessing speeds in the range c to $c + dc$, as shown in Figure 16.1.

The Maxwellian distribution was derived for a gas at equilibrium, in which molecular collisions are perfectly elastic; the molecules exert no force on one another—they are uniformly distributed and all directions of molecular velocities are equally probable. For this distribution, the most probable molecular speed is

$$c_{mp} = \sqrt{\frac{2kT}{m}} \tag{16.8}$$

and the average speed is

$$\bar{c} = \sqrt{\frac{8kT}{\pi m}} \tag{16.9}$$

where k = Boltzmann's constant and m = mass of molecule.

For a gas flow, we are usually interested in what occurs near a surface. There are two possible ways in which a molecule may reflect from a stationary solid surface. If the solid surface is perfectly smooth, the reflection is called *specular*. For a specular reflection, the angle of incidence is equal to the angle of reflection, with no change in the velocity component tangential to the surface and the normal component changed in direction, but not magnitude. For a specular reflection, if the incident gas obeys a Maxwellian distribution, so will the reflected gas. In the real case, however, a surface is not smooth to an incident molecule, since the molecule itself is extremely small in diameter, on the order of 10^{-8} cm. Thus the surface presents a rough boundary to the molecule. In this case, the reflection is called *diffuse*, and the molecule may move away from the wall at any angle to the surface (see Figure 16.2). For the diffuse reflection, then, the molecules essentially lose, on the average, their tangential velocity component. In the continuum regime, the molecular flow near a surface results from the interactions between incident and reflected molecules. If the mean free path of the molecules is very small, a great number of collisions occurs in the vicinity of the wall. For example, molecules colliding

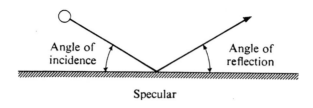

Specular

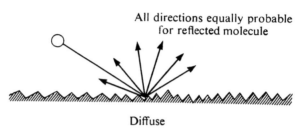

Diffuse

Figure 16.2

with those that have been reflected from the wall will, on the average, lose some of their tangential momentum. A continuous velocity distribution is built up in a thin layer near a surface; this layer is called the *boundary layer*. The velocity distribution in a boundary layer is shown in Figure 16.3. The random velocity distribution of the molecules in the boundary layer is not Maxwellian, since velocities in all directions are not equally probable.

At the other extreme is highly rarefied gas flow. If the mean free path becomes very large, intermolecular collisions become infrequent. Molecules arriving at a surface from the free stream have not had a chance to be slowed down by collisions with other molecules; rather, they arrive at the surface with free stream velocity. For very rarefied flows, the incident and reflected streams do not interact; rather, the reflected molecules must travel a long distance before collision, so that the incident molecules are unaffected by the presence of the body. For this flow, the velocity distribution is shown in Figure 16.4.

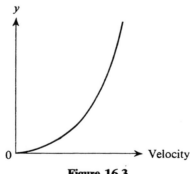

Figure 16.3

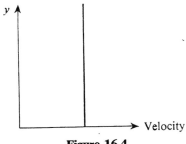

Figure 16.4

For highly rarefied flows, then, no boundary layer exists. The gas velocity at the surface is not zero, but rather the flow "slips" over the surface with an appreciable velocity. Maxwellian distributions can be assumed for the entire flow field.

One important difference between rarefied flow and continuum flow is the boundary condition at a stationary surface. Let us derive a relationship for this boundary condition. Suppose we have parallel flow over a flat plate aligned to the flow in the x direction (see Figure 16.5). Adjacent to the plate the gas consists of molecules, one-half of which have reflected from the plate, the other half coming, on the average, from a layer of gas a mean free path away from the plate. The velocity of the gas a mean free path from the plate can be written as a Taylor's expansion:

$$u_{y=\bar{l}} = u_{y=0} + \bar{l}\left(\frac{\partial u}{\partial y}\right)_{y=0} \qquad (16.10)$$

with u = velocity in the x direction.

The molecules reflected diffusely from the wall have no net tangential velocity. Those reflected specularly have the same tangential velocity as those a mean free path from the plate. We shall let d = fraction of molecules that are diffusely reflected.

The average velocity at the plate surface is equal to one-half the velocity of the molecules reflected from the surface plus one-half the velocity of the molecules incident on the surface, the latter coming, on the average, from a layer a mean free path from the plate.

$$\bar{u}_{y=0} = \frac{1}{2}\left\{\left[(1-d)\left(\bar{u}_{y=0} + \bar{l}\left(\frac{\partial \bar{u}}{\partial y}\right)_{y=0}\right)\right] + (d)0\right\} + \frac{1}{2}\left\{\bar{u}_{y=0} + \bar{l}\left(\frac{\partial \bar{u}}{\partial y}\right)_{y=0}\right\}$$

$$(16.11)$$

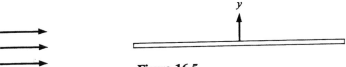

Figure 16.5

Table 16.1 Diffuse reflection coefficient d (Ref. 4)

Gas and surface	d
Air on machined brass	1.00
Air on oil	0.895
Air on glass	0.89
Helium on oil	0.87
CO_2 on oil	0.92

Therefore, we obtain

$$\bar{u}_{y=0} = \frac{2-d}{d}\,\bar{l}\left(\frac{\partial \bar{u}}{\partial y}\right)_{y=0} \tag{16.12}$$

We find that, even for continuum flow and $d = 1$, a slip velocity $(\bar{u}_{y=0}>0)$ exists. However, for continuum flow, $\bar{l}$ is small enough so that $\bar{u}_{y=0} \approx 0$. As $\bar{l}$ gets larger, though, and we get into the rarefied-gas flow regime, $\bar{u}_{y=0}$ can assume appreciable values.

The preceding derivation only holds for relatively small values of $\bar{l}$; however, it does establish the existence of a "slip" velocity. Values of d are given in Table 16.1.

Rarefied gas flow can itself be broken down into several regimes, depending on the degree of rarefaction. For appreciable Reynolds numbers and $M/\sqrt{Re}$ close to 0.01, the regime is called *slip flow*, in which the slip condition is present at the boundary surface, yet the remainder of the flow can be treated by the continuum flow equations. For much greater values of $M/\sqrt{Re}$, there is the regime of free molecular flow in which the Maxwellian velocity distribution can be assumed throughout the entire field, irrespective of the presence of a body.

In between the regimes of slip flow and free molecular flow is a third rarefied-flow regime, called the transition region. In the next section, we shall discuss briefly the characteristics of the flow in the various regimes and try to define the limits of the regimes.

16.4 Slip Flow

The *slip flow* regime is the flow regime of slight rarefaction. The transition from continuum flow to slip flow is gradual; however, on the basis of experimental evidence[5] it is appropriate to define the limits of the slip flow regime as

$$0.01 < \frac{M}{\sqrt{Re_L}} < 0.1 \qquad \text{for } Re_L > 1$$

and

$$0.01 < \frac{M}{Re_L} < 0.1 \qquad \text{for } Re_L < 1$$

(16.13)

In this regime, boundary layers will in general be laminar and quite thick, in some instances so thick that boundary-layer theory is not strictly applicable. Because of difficulties such as those associated with the interaction between the boundary layer and the external flow, few solutions are available for specific flow situations. Best results seem to be obtained by using the continuum equations of motion (*Navier Stokes equations*) plus the slip boundary conditions. Such a technique will be illustrated by the following example, discussed in Reference 6.

Example 16.1

Derive the velocity distribution for Poiseuille flow in a circular tube in the slip-flow regime. See Figure 16.6.

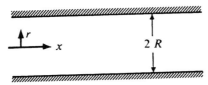

Figure 16.6

Solution

We shall apply the usual momentum equation for Poiseuille flow [see, for example, Reference 7] with the slip boundary condition at the pipe wall. For laminar flow in a circular tube, the momentum equation is given by

$$\nu_k\left(\frac{d^2u}{dr^2} + \frac{1}{r}\frac{du}{dr}\right) = \frac{1}{r}\frac{dp}{dx} = \text{constant } C_1 = \nu_k\left(\frac{1}{r}\frac{d}{dr}r\frac{du}{dr}\right) \qquad \textbf{(16.14)}$$

where ν_k = kinematic viscosity = μ/p.
    At the wall ($r = R$ or $y = 0$) we have from Eq. (16.12), with $d = 1$,

$$u = \bar{l}\left(\frac{du}{dy}\right)_{y=0}$$

or

$$u = -\bar{l}\left(\frac{du}{dr}\right)_{r=R}$$

Also, we have at $r = 0$, $du/dr = 0$ from symmetry. Integrating Eq. (16.14), we obtain

$$r\frac{du}{dr} = \frac{C_1}{\nu_k}\frac{r^2}{2} + C_2$$

From the second boundary condition, $C_2 = 0$, so that

$$r\frac{du}{dr} = \frac{C_1}{\nu_k}\frac{r^2}{2}$$

Integrating again yields

$$u = \frac{C_1}{\nu_k} \frac{r^2}{4} + C_3$$

Using the first boundary condition, we obtain

$$-\bar{l}\left(\frac{du}{dr}\right)_{r=R} = \frac{C_1}{\nu_k} \frac{R^2}{4} + C_3$$

so that

$$u = \frac{1}{\mu} \frac{dp}{dx}\left(\frac{r^2 - R^2}{4}\right) - \bar{l}\left(\frac{du}{dr}\right)_{r=R}$$

But

$$\left(\frac{du}{dr}\right)_{r=R} = \frac{1}{\mu} \frac{dp}{dx} \frac{R}{2}$$

so that

$$u = \frac{1}{4\mu} \frac{dp}{dx}(r^2 - R^2 - 2\bar{l}R)$$

or

$$u = \frac{1}{4\mu} \frac{dp}{dx}[r^2 - R^2(1 - Kn)] \qquad (16.15)$$

where

$$Kn = \frac{\bar{l}}{2R}$$

Note that, for low values of Knudsen number, the flow approaches the continuum flow velocity distribution.

16.5 Free Molecular Flow

For free molecular flow, the mean free path between molecular collisions is very much larger than a characteristic dimension of a body or surface. Free molecular flow is generally assumed to exist for $M/Re > 3^{[7]}$. It is meaningless to specify a range of $M/\sqrt{Re}$ since no boundary layer exists for this flow. For free molecular flow, the gas stream incident on a surface can be treated completely independently of the reflected stream, and vice versa, with both flows usually assumed to obey a Maxwellian velocity distribution.

This will be illustrated by Example 16.2.

Example 16.2

From kinetic theory[8] it can be shown that the number of molecules striking a

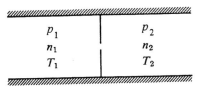

Orifice area $= A_0$ $p_1 > p_2$

Figure 16.7

unit area of surface per unit time, on the average, is given by

$$\frac{\dot{N}}{A} = \frac{1}{4} n \bar{c}$$

where

n = number of molecules per unit volume

$\bar{c}$ = average molecular speed

Use this result to obtain an expression for the mass flow rate of a gas through a very small orifice in a pipe, assuming that the orifice diameter is very much less than the mean free path on either side of the orifice (see Figure 16.7).

Solution

For free molecular flow, it can be assumed that the passage of gas molecules from 1 to 2 can be calculated independently of that from 2 to 1. This is to be contrasted with continuum flow, in which the orifice diameter would be very much greater than the mean free path. In continuum flow, there would be many intermolecular collisions, and the molecular flow from 1 to 2 would interact with that from 2 to 1. Returning to free molecular flow, it will further be assumed that the states on either side of the orifice remain in equilibrium, irrespective of the small movement of molecules from one side of the orifice to the other.

The rate at which molecules from side 1 strike the orifice area and pass into region 2 is given by

$$\frac{1}{4} n_1 \bar{c}_1 A_0$$

The rate at which molecules from 2 pass into 1 is given by

$$\frac{1}{4} n_2 \bar{c}_2 A_0$$

Each molecule has a mass m equal to $\bar{M}/N_A$, with N_A = Avogadro's number and $\bar{M}$ = molecular mass. But $n = pN_A/\bar{R}T$ and, for a Maxwellian velocity distribution,

$$\bar{c} = \sqrt{8kT/\pi m} = \sqrt{8kTN_A/\pi\bar{M}},$$

so that the net mass flow rate from 1 to 2 is given by

$$\dot{m} = \left(\frac{1}{4} A_0 N_A \frac{p_1 - p_2}{\bar{R}T} \sqrt{\frac{8kTN_A}{\pi\bar{M}}}\right) \frac{\bar{M}}{N_A}$$

It can be shown that $kN_A = \bar{R}$, so that

$$\dot{m} = \frac{1}{4} A_0 \frac{p_1 - p_2}{\sqrt{T}} \sqrt{\frac{8\bar{M}}{\pi\bar{R}}}$$

$$= \frac{A_0(p_1 - p_2)}{\sqrt{2\pi RT}} \quad \text{with } R = \frac{\bar{R}}{\bar{M}} \qquad (16.16)$$

For continuum flow, with A_0 much greater than $\bar{l}$, the mass flow of a gas across a pipe orifice can be shown to be[9]

$$\dot{m} = \frac{C_0 A_0 Y \sqrt{2\rho_1(p_1 - p_2)}}{\sqrt{1 - (A_0/A_1)^2}} \qquad (16.17)$$

with Y = compressibility factor and C_0 = orifice coefficient. There can be seen to be a completely different relationship for flow across an orifice for the two types of flow.

The third regime of rarefied flows, between slip flow and free molecular flow, is called the *transition regime*. In this region, the mean free path is of the same order as a body dimension, so that characteristics of both slip flow and free molecular flow are present. The resultant analysis is extremely complex, and the information available is mostly empirical.

16.6 Summary

When the mean free path between molecular collisions becomes of the same order of magnitude as a body dimension, the assumption of continuum flow starts to break down. An important dimensionless parameter for rarefied flow is the Knudsen number, the ratio of mean free path to a characteristic dimension. With infrequent collisions between molecules, the concept of a boundary layer must be altered. Molecules near a surface are not slowed down by collisions with other, slower-moving molecules, but rather possess appreciable velocities in the vicinity of a surface. Therefore, a slip condition exists at a surface in rarefied gas flow.

Rarefied gas dynamics can be broken down into three regimes, depending on the degree of rarefaction: slightly rarefied flow or slip flow, highly rarefied flow or free molecular flow, and transition flow. In slip flow, the usual equations of continuum flow can be used with the slip boundary condition at a body surface. For free molecular flow, motion of individual molecules must be considered; it is necessary to assume a molecular velocity distribution and analyze the individual molecule–surface collisions.

REFERENCES

Specific References

1. LEE, J., SEARS, F.W., and TURCOTTE, D.L., *Statistical Thermodynamics*, Reading, Mass., Addison-Wesley Publishing Co., Inc., 1963, p. 81.

2. SCHAAF, S.A., and CHAMBRE, P.L., "Flow of Rarefied Gases," Section H in *Fundamentals of Gas Dynamics*, H.W. EMMONS, ed., Princeton, N.J., Princeton University Press, 1958, p. 689.

3. See Reference 1, p. 48.

4. See Reference 2, p. 695.

5. See Reference 2, p. 689.

6. ROHSENOW, W.M., and CHOI, H.Y., *Heat, Mass and Momentum Transfer*, Englewood Cliffs, N.J., Prentice-Hall, Inc., 1961, p. 287.

7. See Reference 2, p. 689.

8. TIEN, C.L., and LIENHARD, J.H., *Statistical Thermodynamics*, New York, Holt, Rinehart and Winston, Inc., 1971, p. 51.

9. JOHN, J.E.A., and HABERMAN, W.L., *Introduction to Fluid Mechanics*, 2nd ed., Englewood Cliffs, N.J. Prentice-Hall, Inc., 1980, p. 487.

General References

10. PATTERSON, G.N., *Molecular Flow of Gases*, New York, John Wiley & Sons, Inc., 1956.

11. See Reference 2.

PROBLEMS

1. A missile 1 m long travels at Mach 15, at altitudes of 40 km and 60 km. Determine whether, at these altitudes, the missile is in continuum flow or rarefied flow and, if rarefied, specify whether the flow is slip, transition, or free molecular.

2. Repeat Problem 1 for Mach 5.0.

3. A vacuum bell jar is 1 m in diameter by 1 m high, as shown in Figure P16.3. The pressure inside the bell jar is maintained at 10^{-10} atm. With valve V closed, isolating the pumping system from the chamber, a leak develops in the bell jar. Assuming the leak to be equivalent to a hole 10^{-6} cm in diameter, determine the time required for the pressure inside the bell jar to rise to 10^{-6} atm. Assume an air temperature of 20°C.

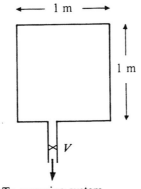

To pumping system

Figure P16.3

4. Consider the system shown in Figure P16.4. Two chambers are connected by an orifice, with the orifice size small enough that free molecule flow can be assumed in calculating the flow through the orifice. Initially, the pressures in the two compartments are the same, but the temperatures are different. Give an expression for the initial mass flow rate between the compartments. As time proceeds, the pressures in the chambers change. Give an expression for the pressures in the chambers at equilibrium, when the mass flow is zero.

T_1	T_2	
P_1	P_2	

Initially, $T_1 > T_2$

$$p_1 = p_2$$

T_1 and T_2 are constant, independent of time.

Figure P16.4

5. With the mean free path between molecular collisions equal to 6.35×10^{-6} cm, at standard atmospheric conditions, determine the magnitude of $\bar{u}_{y=0}$ from Eq. (16.12) for continuum laminar flow boundary layer flow of air over a flat plate. Use the Blasius profile to find $(\partial \bar{u}/\partial y)_{y=0}$. Assume $d = 1$, and express your answer in terms of free-stream velocity U_∞.

6. Consider flow from a large reservoir tank through a small opening in the wall of the tank. (Figure P16.6). It is desired to find the mass flow rate $\dot{m}$ through the hole. Using the methods of continuum flow from Chapter 3, develop an expression for $\dot{m}$. Assuming free molecule flow and utilizing the material in this chapter, develop an expression for $\dot{m}$. Are the two the same? Under what conditions should each of these expressions be used?

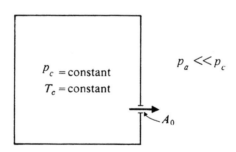

p_c = constant
T_c = constant

$p_a \ll p_c$

A_0

Figure P16.6

17

MEASUREMENTS IN COMPRESSIBLE FLOW

17.1 Introduction

In this chapter we shall discuss some of the methods used for measuring the properties of a compressible flow. Important variables that require measurement are pressure, temperature, velocity, flow direction, and density. We shall restrict ourselves to the more conventional techniques used in continuum flow measurements. Several of the devices to be mentioned here should be familiar to the student from a study of incompressible flow. It is important to keep in mind, then, the differences due to compressibility effects, both in subsonic and supersonic flow.

17.2 Pressure Measurement

Static pressure is the pressure indicated by a measuring device moving at the flow velocity, or, in other words, by a device that introduces no disturbance or velocity change to the flow. The usual method for measuring the static pressure of a flow along a wall is to drill a small hole normal to the surface of the wall and connect this to a manometer, pressure gauge, or other similar device (see Figure 17.1). The hole must be small with respect to the boundary-layer thickness and must be free from roughness or burrs that might disturb the flow. Since there is no pressure change through a boundary layer in a direction

M_∞, p_∞

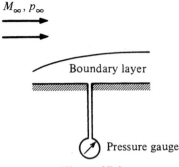

Boundary layer

Pressure gauge

Figure 17.1

normal to the wall, the pressure indicated is a true measure of the free-stream static pressure p_∞.

If there is no wall present in the flow, static pressure can be measured in a similar fashion by introducing a probe, which, in a sense, creates a wall in the flow. Such a probe must be very thin and aligned with the flow direction so as to introduce a minimum disturbance into the flow. A typical static pressure probe is shown in Figure 17.2. The probe generally has a sharp, conical nose, with the pressure tap located far enough downstream to be out of the influence of the disturbance introduced by the nose (for example, 10 to 20 probe diameters). For supersonic flow, there will be an attached shock at the nose, as shown in Figure 17.3. However, there will also be an expansion where the nose joins the cylindrical section of the probe at S, so the pressure measurement at the tap at P, at least for a weak attached shock, will be very close to the free-stream static pressure p_∞. The probe is very sensitive to flow alignment; it can be seen that the more the probe is misaligned, the greater will be the disturbance introduced into the flow by the probe. This sensitivity can be reduced by drilling several holes around the circumference of the probe at p,

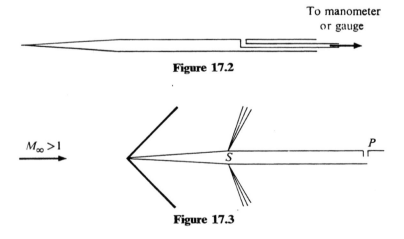

To manometer or gauge

Figure 17.2

$M_\infty > 1$

S

P

Figure 17.3

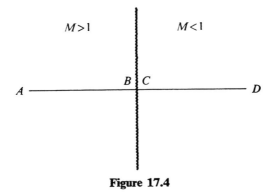

Figure 17.4

the resultant measured pressure then being an average pressure. However, even with this, if accuracy to within 1 percent is desired, the misalignment cannot exceed 5°[1].

Care must be exercised in the interpretation of the results of static pressure probe measurements in a supersonic flow field in which a shock is present. Consider such a supersonic flow field in which there occurs a normal shock, as shown in Figure 17.4. If we move the probe along the line *ABCD*, we should get a static pressure curve as shown in Figure 17.5. Actually, however, a boundary layer builds up on the probe, so the actual velocity at the fixed probe surface is zero. In the subsonic part of the boundary layer, with no shock present, the effect of the pressure change occurring outside the boundary layer can be felt upstream. In other words, in the presence of a shock, the pressure changes that occur in the boundary layer are not the same as the pressure changes that occur across the shock outside the boundary layer. Instead of a sudden pressure rise, the pressure probe indicates a more gradual pressure rise, as shown in Figure 17.6. In the presence of a shock, the static pressure indicated by the probe can only be taken as a true reading when the probe is either far upstream or downstream of the shock.

The stagnation pressure is the pressure measured by an instrument that brings the flow isentropically to rest. Such a device is a simple pitot tube, an

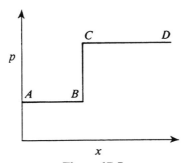

Figure 17.5

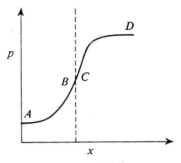

Figure 17.6

open-ended tube facing directly into the flow, as shown in Figure 17.7. In supersonic flow, since the probe is essentially a blunt-nosed body, there will be a detached normal shock in front of the probe (see Figure 17.8). In this case, the stagnation pressure indicated on the manometer or gauge will be the stagnation pressure after a shock that is normal to the flow direction. Whereas, for subsonic flow, the stagnation pressure measured by the pitot tube is the true stagnation pressure,

$$p_{t_{\text{pitot}}} = p_{t_\infty} = p_\infty \left(1 + \frac{\gamma - 1}{2} M_\infty^2\right)^{\gamma/(\gamma-1)} \tag{17.1}$$

for supersonic flow, the stagnation pressure indicated by the pitot tube is the stagnation pressure after a normal shock occurring at the free-stream Mach number. Therefore, for supersonic flow,

$$p_{t_{\text{pitot}}} = p_{t_2} = \frac{p_{t_2}}{p_{t_1}} \frac{p_{t_\infty}}{p_\infty} p_\infty \tag{17.2}$$

Since p_{t_2}/p_{t_1} is a function of M_∞ (see Appendix B), and p_{t_∞}/p_∞ is a function of M_∞ (see Appendix A), it follows that M_∞ can be determined from $p_{t_{\text{pitot}}}$ and p_∞ for supersonic flow. For convenience, the ratio p_1/p_{t_2} has been given in Appendix B for supersonic flow. The stagnation pressure probe is not as sensitive to misalignment as the static probe; 1 percent accuracy can be

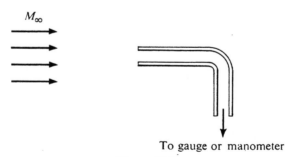

To gauge or manometer

Figure 17.7

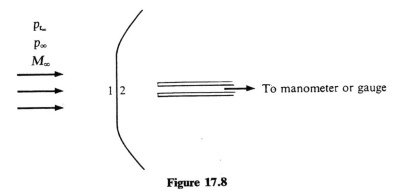

Figure 17.8

achieved for angles up to $20^{\circ(2)}$. The choice of the size of the orifice (tube diameter) is dependent on the fineness of the measurement to be taken; for stagnation pressure measurements in a boundary layer, a probe with a diameter as small as 0.5 mm might be used.

Example 17.1

The Mach number of a compressible flow is to be determined from static probe and pitot tube measurements. If the static probe indicates 20 kPa and the pitot tube 32 kPa, determine the flow Mach number. Repeat for a pitot pressure of 80 kPa.

Solution

For the first case, the ratio $p_\infty/p_{t_{pitot}}$ is greater than that for Mach 1, so the free-stream Mach number is subsonic and Eq. (17.1) is applicable. From Appendix A, at $p/p_\infty = 20/32 = 0.625$, $\underline{M_\infty = 0.48}$. For the second case, the ratio $p_\infty/p_{t_{pitot}}$ is less than that for Mach 1, the free-stream Mach number is supersonic, and Eq. (17.2) is applicable. From Appendix B, at $p_1/p_{t_2} = 20/80 = 0.25$, $\underline{M_\infty =}$ $\underline{1.647}$.

17.3 Temperature Measurement

The direct measurement of the static temperature of a gas flow would have to be made by a device that does not disturb the flow, in other words, by a device that would travel at the velocity of the flow. Unfortunately, just as with static pressure measurement, this is impractical. The first alternative that occurs is to perform the temperature measurement just as was done with the static pressure measurement; that is, locate a thermocouple or other thermometric device at the wall surface (see Figure 17.9). However, this locates the thermocouple inside the boundary layer; at a fixed wall, the flow velocity is equal to zero, and we would expect the measured wall temperature to be closer to the free-stream

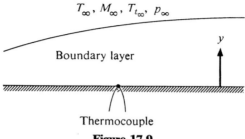

$$T_\infty,\ M_\infty,\ T_{t_\infty},\ p_\infty$$

Boundary layer

y

Thermocouple
Figure 17.9

stagnation temperature T_{t_∞} than T_∞. This differs from the discussion of static pressure measurement previously considered, in that the static pressure does not vary through the boundary layer (in the y direction).

A temperature distribution in a compressible-flow boundary layer is given in Figure 17.10. The wall temperature for the case shown, in which the wall is insulated so that $\partial T/\partial y = 0$ at $y = 0$, is called the adiabatic wall temperature T_{aw}. Since the fluid near the wall is at a higher temperature than the fluid near the outer edge of the boundary layer, there will be a heat flow in the y direction due to conduction. However, due to the flow of a high velocity gas near a surface, there can be appreciable frictional heating of the fluid. In other words, the faster moving layers in the boundary layer do work on the slower moving layers. If the heat loss due to conduction and energy gain from viscous heating cancel, the flow can be considered to be brought to rest adiabatically in the boundary layer, and $T_{aw} = T_{t_\infty}$. A measure of the relative importance of heat conduction and viscous heating is given by the Prandtl number

$$Pr = \frac{c_p \mu}{k} \qquad (17.3)$$

In other words, for a fluid with $Pr = 1$, the adiabatic wall temperature is equal to the free-stream stagnation temperature. (If $Pr < 1$, then $T_{aw} < T_{t_\infty}$). This

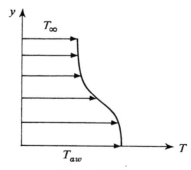

y

T_∞

T_{aw}

T

Figure 17.10

can be summarized by defining a recovery factor $\mathscr{R}$, where

$$\mathscr{R} = \frac{T_{aw} - T_\infty}{T_{t_\infty} - T_\infty} \tag{17.4}$$

Since

$$T_{t_\infty} = T_\infty \left(1 + \frac{\gamma - 1}{2} M_\infty^2 \right)$$

it follows that

$$T_{aw} = T_\infty \left(1 + \mathscr{R}\frac{\gamma - 1}{2} M_\infty^2 \right) \tag{17.5}$$

It can be shown[3] that for the laminar compressible boundary layer, $\mathscr{R} = \sqrt{Pr}$, whereas for a turbulent boundary layer $\mathscr{R}$ is approximately equal to $\sqrt[3]{Pr}$. For air, up to moderately high temperature, $Pr = 0.72$, so that $\mathscr{R} \approx 1$ for a turbulent boundary layer.

We have shown then that, whereas a direct measurement of free-stream static temperature is not possible, a measurement of the adiabatic wall temperature can be used to determine T_{t_∞}; then, by measuring p_{t_∞} and p_∞ with the methods of the previous section, M_∞ and T_∞ can be calculated.

In the absence of a wall, a stagnation temperature probe, as shown in Figure 17.11, can be used to determine T_{t_∞}. The measurement of T_{t_∞} is unaffected by the presence of the detached shock in front of the probe, since the shock is an adiabatic process. The flow is brought to rest in the tube. Vent holes are provided in the sides of the probe to allow for proper ventilation of the space inside the probe; if the air were allowed to stagnate, it would cool and yield a false reading. It is necessary that the flow be slowed down to zero velocity at the thermocouple with no gain or loss of heat. Shields have been provided to prevent radiation heat loss from the thermocouple; also, the thermocouple lead wires must be made as thin as possible so as to minimize heat flow by conduction back along the wires. Nevertheless, it is appropriate to define a correction factor K such that

$$K = \frac{T_{t_{indicated}} - T_\infty}{T_t - T_\infty} \tag{17.6}$$

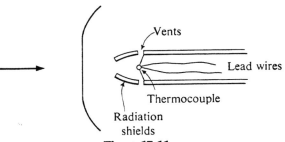

Vents

Lead wires

Thermocouple

Radiation shields

Figure 17.11

where $T_{t_{\text{indicated}}}$ is the temperature indicated by the probe. By suitable design, K can be made very close to 1 for air; in any case, such a probe must be calibrated to define K as a function of Reynolds number, Prandtl number, and Mach number.

Example 17.2

A total temperature probe is found to have $K = 0.95$ over a wide range of operating conditions. A pitot tube inserted into a supersonic flow indicates a pressure of 300 kPa; a static probe indicates 60 kPa. If the indicated temperature of the total temperature probe is 600 K when inserted into the same supersonic flow, determine T_{∞}.

Solution

From Appendix B, for $p_x/p_{t_2} = 60/300$, $M_{\infty} = 1.871$. From Appendix A, at $M_{\infty} = 1.871$, $T_{\infty}/T_{t_x} = 0.5881$.

$$K = 0.95 = \frac{T_{t_{\text{ind}}} - T_{\infty}}{T_t - T_{\infty}}$$

$$= \frac{T_{t_{\text{ind}}} - T_{\infty}}{\dfrac{T_{\infty}}{0.5881} - T_{\infty}}$$

$$= \frac{600 - T_{\infty}}{0.7004 T_{\infty}}$$

Solving, we obtain $\underline{T_{\infty} = 360.3 \text{ K}}$.

17.4 Velocity and Direction

With pressure and temperature measured as shown in Sections 17.2 and 17.3, the magnitude of velocity is calculable:

$$V = M\sqrt{\gamma RT}$$

We are also interested in finding the direction of the velocity vector. At

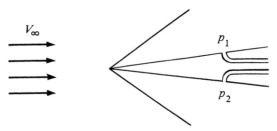

Figure 17.12

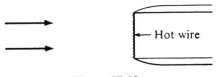

Figure 17.13

supersonic speeds, this can be done with a wedge or cone, as shown in Figure 17.12. For a uniform two-dimensional flow passing over a symmetrical wedge, the angle of attack of the wedge with respect to the flow can be determined from a measurement of the pressure difference, $p_1 - p_2$. Alternatively, the wedge can be rotated until the pressure difference is zero, at which condition the centerline of the wedge is aligned with the flow.

The hot wire probe can be used for determining local velocities in a gas flow. Very briefly, the hot wire probe consists of a thin wire suspended in the gas stream (Figure 17.13), with the wire heated by the passing of electric current through it. In other words, at equilibrium, the convective cooling of the wire is balanced by electrical energy input:

$$I^2R = hA(T_w - T_\infty) \tag{17.7}$$

with

I = current
R = electrical resistance of the wire
A = wire surface area exposed to the flow
T_w = wire surface temperature
h = convective heat transfer coefficient at the wire surface

By measuring R, I, and T and knowing the wire temperature as a function of resistance, h can be calculated from Eq. (17.7). A dimensionless parameter Nusselt number, hL/k, is a function of Re, M, and Pr:

$$Nu = f(Re, M, Pr)$$

By determining the nature of this function f by extensive calibration, h can be related to local velocity; thus V can be determined from the measurement of h.

Due to its small size, the hot wire probe has been particularly useful in measuring fluctuating flow quantities in a turbulent flow. Greater details can be found in Reference 4.

17.5 Density

The actual measurement of density, at this point, seems superfluous. Certainly, by measuring or determining pressure and temperature, density can be calculated from the perfect gas law or other equation of state. However, the optical methods to be discussed here, which depend on the variation of density or its derivatives throughout the flow field, are extremely important techniques for the investigation of compressible flows. The use of any one of the three optical methods to be discussed, the schlieren, shadowgraph, or interferometer, allows one to "see" phenomena such as shock waves without introducing any object or disturbance into the flow.

The three optical methods depend on the principle that the speed of light in a medium is dependent on the density of that medium. The index of refraction relates the speed of light in vacuum c_0 to the speed of light in a medium c.

$$n = \frac{c_0}{c} \tag{17.8}$$

For gases, the index of refraction is very close to 1, with

$$n = 1 + K_1\rho, \qquad K_1 \text{ a positive constant} \tag{17.9}$$

As is well known, if light rays pass from one medium into another of different density, the rays are turned or refracted.

Consider a series of light rays passing from air into water, as shown in Figure 17.14. The plane wave front OA has been shown at the instant the light is incident on the surface. A short time later, the wave front reaches the position $O'A'$. Since the velocity of light in water is less than that in air, the distance OO' is less than AA', so the wave front turns, with

$$\frac{\sin i}{\sin r} = \frac{c_{\text{air}}}{c_{\text{water}}} = \frac{c_0/c_{\text{water}}}{c/c_{\text{air}}} = \frac{n_{\text{water}}}{n_{\text{air}}} \tag{17.10}$$

Now suppose that, instead of a sudden change in n, the light passes

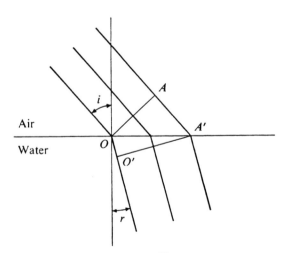

Figure 17.14

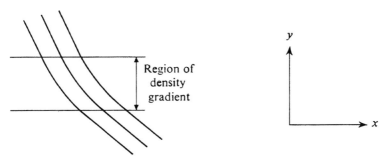

Figure 17.15

through a region in which there exists a gradual change of index of refraction. (see Figure 17.15). In this case, there is a smooth turning of the light rays, with the angle through which the rays have been turned clearly dependent on the gradient of n which, in turn, is dependent on the density gradient in the y direction.

Next consider a two-dimensional flow in the z direction, as shown in Figure 17.16, with a density gradient (perhaps caused by a shock) in the y direction. Light is to be passed through section AA in the x direction. For this two-dimensional flow, properties do not vary in the x direction. We shall assume that the density decreases with y, so the velocity of light for ray 1 is greater than that of ray 2. This causes a turning of the wave front and of the rays, as shown in Figure 17.17. Assume the density changes are relatively small so that the rays can be treated as straight lines. Let the velocity of light for ray 2 equal c, so the velocity of light for ray 1 will be $c + (dc/dy) \Delta y$. The angular deflection of the wave front and hence the angle through which the light rays

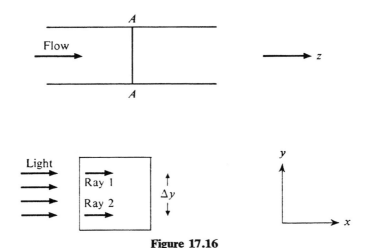

Figure 17.16

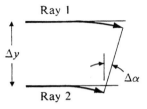

Figure 17.17

are turned is given by

$$\Delta\alpha = \frac{\Delta t \dfrac{dc}{dy} \Delta y}{\Delta y}$$

where

$$\Delta t = \frac{L}{c}$$

Expanding this, we obtain

$$\Delta\alpha = \frac{dc}{dy}\frac{L}{c} \tag{17.11}$$

Substitution from Eqs. (17.8) and (17.9) yields

$$\Delta\alpha = -\frac{L}{n}K_1\frac{d\rho}{dy} \tag{17.12}$$

For a three-dimensional flow, the angular deflection is dependent on the density gradient in both x and y directions. The important result is that the angular deflection of the light rays is dependent on the first derivative of density.

Having established the basic optical principles, we are now in a position to consider the details of the schlieren, shadowgraph, and interferometer systems.

The basic *schlieren* system is shown in Figure 17.18. Light from a source *ab* (for example, a filament) is collimated by lens L_1, providing a parallel light beam through the test section. Shown in Figure 17.18 are two extreme light rays from *a* and *b*, rays from each point passing parallel through the test section. After passing through the test section, the light rays are focused by lens L_2 at $a'b'$, an inverted image of the light source. The light is then focused on the screen or photographic plate by lens L_3, providing an inverted image of the test section. Now place a knife edge K at $a'b'$. Clearly, if this knife edge is raised too far, it will obstruct all the light and the screen will be dark. Instead, let the knife edge obstruct approximately half the incident light. In this case, the image of the test section on the screen will be darker than before, but still uniformly illuminated; that is, the ray $rb'r'$ will reach the screen, but the ray

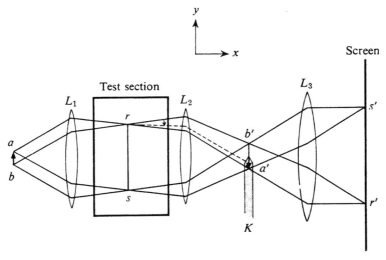

Schlieren apparatus

Figure 17.18

Figure 17.19 Schlieren photograph of flow about a model in a shock tube at Mach 2.0. (By permission of AVCO Corporation, Wilmington, Mass.)

$ra'r'$ will not, so the image of the point r will actually appear only half as bright as it would without the knife edge.

Let there be a density gradient in the test section in the plane of r (same y coordinate as r). In this case, the rays from r will be deflected so that, for example, the ray $rb'r'$ will be turned downward, as shown, to the dashed line. Now this ray will be intercepted by the knife edge, so the image of r on the screen will be darker than the rest of the image of the test section. If the density gradient had the opposite sign, the rays from r would be bent upward, so some rays that had been obstructed by the knife edge would now pass over

Figure 17.20 Schlieren photograph of 20° cone travelling at 3800 m/s. (By permission of AVCO Corporation, Wilmington, Mass.)

it and give a brighter image of r on the screen. Thus it is possible to use a schlieren optical system to visualize, for example, a shock wave by observing the light or dark areas on the screen. It is emphasized that the schlieren system involves the visualization of density gradients in a flow, unlike the shadow-graph and interferometer devices. In actual practice, it is found beneficial to use mirrors rather than lenses. Typical schlieren photographs are provided in Figures 17.19 and 17.20.

A basic *shadowgraph* system is shown in Figure 17.21. A parallel light beam passes through the test section, as shown, and illuminates a screen. If a density gradient exists in the test section, the light rays will be deflected, as shown. In region R_1, the light rays have diverged, so there is a decrease in illumination of the screen in this region. In region R_2, the rays have converged and there is a resultant increase in illumination. Thus a shock wave can be seen on the screen as a dark region together with a light region. Figure 17.22 provides a typical shadowgraph of a shock wave.

It is important to realize that the shadowgraph really depicts the second derivative of density. If there were to be a uniform density gradient in the test section (Figure 17.23), all the rays would be deflected through the same angle and the screen would still be uniformly illuminated. It is only with a change in density gradient that the dark and light regions appear on the screen. The shadowgraph, indicating the second derivative of density, is more suitable for use in flows with large, sudden changes in density (e.g., strong shock waves). For flows with slowly varying density, the schlieren system should be employed. The shadowgraph has the advantages of being simpler, less expensive, and easier to operate than the schlieren, but it is not as precise an instrument.

The third optical system is the *interferometer*, shown in Figure 17.24. The apparatus consists of two mirrors, M_1 and M_2, and two splitter plates, SP_1 and SP_2, the latter being half-silvered mirrors that allow half the light to pass through and reflect the other half. Thus monochromatic light from the source at A has two possible paths, one from SP_1 to M_1 to SP_2 to S, and the other from SP_1 to M_2, through the test section to SP_2 and to S. If the light waves

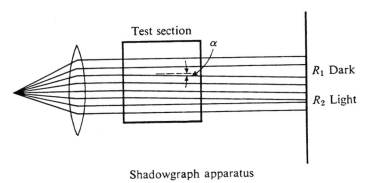

Shadowgraph apparatus
Figure 17.21

Figure 17.22 Shadowgraph of 10° cone travelling at 3800 m/s. (By permission of AVCO Corporation, Wilmington, Mass.)

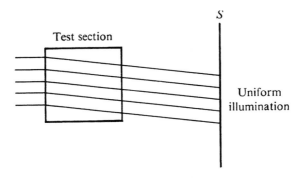

Figure 17.23

from each path arrive at S in phase, they will reinforce one another and brighten the screen. If the waves are one half-wavelength out of phase, they will cancel one another and the screen will be dark. Suppose the system is set (mirror angles adjusted) so that, with no flow in the test section, the reinforcements and cancellations of the waves due to the geometry of the system yield a regular, uniform pattern of fringes, as shown at the outer edges of Figure 17.25, away from the object, and also in Figure 17.26. Now introduce flow and density changes to the test section, thus altering the speed with which the rays traverse the lower path. By increasing test-section density, the effective path of the lower rays is increased; that is, the time for the rays to traverse the lower path is increased by

$$\Delta t = +\frac{L}{c_f} - \frac{L}{c_i}$$

with

c_i = velocity of light through test section with no flow

c_f = velocity of light through test section with flow and higher density

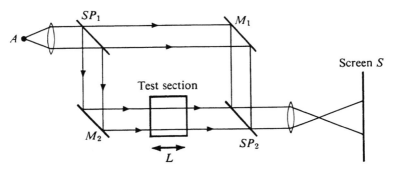

Interferometer

Figure 17.24

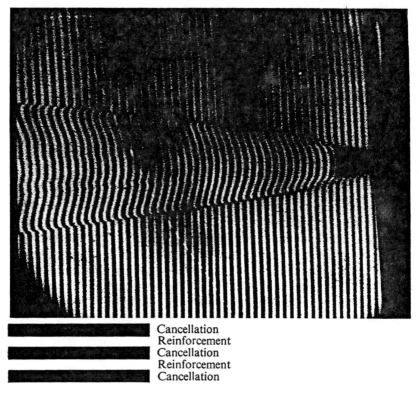

Cancellation
Reinforcement
Cancellation
Reinforcement
Cancellation

Figure 17.25   Interferogram of 20° cone travelling at 3000 m/s. (By permission of AVCO Corporation, Wilmington, Mass.)

This causes an effective increase in the optical path of

$$\Delta L = c\Delta t$$

or a fringe for monochromatic light (single frequency) of

$$N = \frac{\Delta L}{\lambda} \text{ fringes}$$

where λ is the wavelength of light

By observing the fringe shift at different points throughout the region of interest, the density at these points can be determined. It can be seen that the uniform, regular pattern of fringes for a background helps in counting the fringe shift.

Of the three optical methods, the interferometer is the most precise, yet also the most costly, delicate, and difficult to set up. Notice that it gives a direct indication of density, unlike the schlieren or shadowgraph. The desire for a monochromatic, small, coherent light source makes the laser suitable for

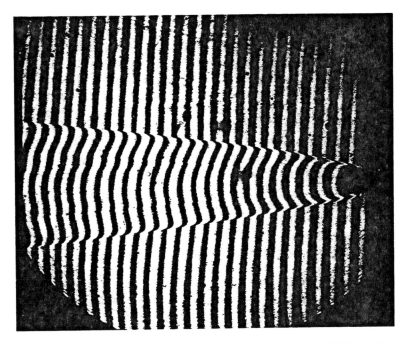

Figure 17.26 Laser interferogram of 20° cone travelling at 4570 m/s. (By permission of AVCO Corporation, Wilmington, Mass.)

application to the interferometer. In Figure 17.26, a laser was employed as the source of monochromatic light.

REFERENCES

Specific References

1. LIEPMANN, H.W., AND ROSHKO, A., *Elements of Gasdynamics*, New York, John Wiley & Sons, Inc., 1957, p. 145.
2. See Reference 1, p. 148.
3. HOLMAN, J.P., *Heat Transfer*, 5th ed., New York, McGraw-Hill Book Company, 1981, pp. 212, 213.
4. HINZE, J.O., *Turbulence*, 2nd ed., New York, McGraw-Hill Book Company, 1975, Chapter 2.

PROBLEMS

1. A pitot tube is placed in a uniform air flow of Mach 2.5. If the pitot tube indicates a pressure of 500 kPa, find the static pressure of the flow.
2. A pitot tube is placed in a uniform helium flow. If the pitot tube indicates a

pressure of 280 kPa and the static pressure of the flow is measured to be 20 kPa, find the Mach number.

3. A uniform flow of air at Mach 2.0 passes over an insulated wall. The static temperature and pressure in the free stream outside the boundary layer are, respectively, 250 K and 20 kPa. Determine the free-stream stagnation temperature, adiabatic wall temperature, and static pressure at the wall surface.

4. A total temperature probe is inserted into the flow of Problem 3. If the probe has K equal to 0.97, what temperature will be indicated by the probe?

5. Sketch a plot of p/p_t versus M for isentropic flow. On the same coordinates, plot p_∞/p_{pitot} versus M_∞. Take $\gamma = 1.40$.

6. Why is it desirable to have a monochromatic light source for use with an interferometer?

7. For the schlieren system shown in Figure 17.18, is density increasing or decreasing in the y direction? (Refer to the dashed ray.)

8. Repeat Problem 7 for the shadowgraph shown in Figure 17.21. Sketch plots of density and its first and second derivatives versus y for Figure 17.21.

9. A symmetrical wedge of 10° total included angle is placed in a uniform Mach 2.0 flow of static pressure of 60 kPa. If the axis of the wedge is misaligned with the flow direction by 3°, determine the static pressure difference between the top and bottom surfaces of the wedge.

10. A conical probe is aligned with a Mach 3.0 supersonic flow. Determine the static pressure on the cone surface if the free-stream static pressure is 20 kPa. Assume $\gamma = 1.40$.

11. Refer to Figure 17.20. Use the conical shock charts of Appendix C to estimate the wave angle and compare with the wave angle measured from the figure.

APPENDICES

Appendix A
Table A.1 Isentropic flow tables ($\gamma = 1.4$)
Table A.2 Isentropic flow tables ($\gamma = 1.3$)
Table A.3 Isentropic flow tables ($\gamma = 5/3$)

Appendix B
Table B.1 Normal shock tables ($\gamma = 1.4$)
Table B.2 Normal shock tables ($\gamma = 1.3$)
Table B.3 Normal shock tables ($\gamma = 5/3$)

Appendix C
Oblique shock charts ($\gamma = 1.4$)
Conical shock charts ($\gamma = 1.405$)

Appendix D
Prandtl Meyer functions ($\gamma = 1.4$)

Appendix E
Fanno line flow ($\gamma = 1.4$)

Appendix F
Rayleigh line flow ($\gamma = 1.4$)

Appendix G
Table G.1 Ideal gas properties of various gases
Table G.2 Ideal gas specific heat ratio γ of several gases as function of temperature
Table G.3 Thermodynamic properties of air at 1 atmosphere (R for air $= 0.2870 \text{ kJ/kg} \cdot \text{K}$)
Table G.4 Thermodynamic properties of air at 10 atmospheres (R for air $= 0.2870 \text{ kJ/kg} \cdot \text{K}$)
Table G.5 Thermodynamic properties of nitrogen at 1 atmosphere (R for nitrogen $= 0.2968 \text{ kJ/kg} \cdot \text{K}$)
Table G.6 Thermodynamic properties of nitrogen at 10 atmospheres (R for nitrogen $= 0.2968 \text{ kJ/kg} \cdot \text{K}$)

Appendix H
Standard atmosphere

Appendix I
Conversion factors

Table A.1 Isentropic flow tables ($\gamma = 1.4$)

M	p/p_t	T/T_t	A/A^*	M	p/p_t	T/T_t	A/A^*
0	1.0000	1.0000	∞	.60	.7840	.9328	1.1882
.01	.9999	1.0000	57.8738	.61	.7778	.9307	1.1767
.02	.9997	.9999	28.9421	.62	.7716	.9286	1.1657
.03	.9994	.9998	19.3005	.63	.7654	.9265	1.1552
.04	.9989	.9997	14.4815	.64	.7591	.9243	1.1452
.05	.9983	.9995	11.5914	.65	.7528	.9221	1.1356
.06	.9975	.9993	9.6659	.66	.7465	.9199	1.1265
.07	.9966	.9990	8.2915	.67	.7401	.9176	1.1179
.08	.9955	.9987	7.2616	.68	.7338	.9153	1.1097
.09	.9944	.9984	6.4613	.69	.7274	.9131	1.1018
.10	.9930	.9980	5.8218	.70	.7209	.9107	1.0944
.11	.9916	.9976	5.2992	.71	.7145	.9084	1.0873
.12	.9900	.9971	4.8643	.72	.7080	.9061	1.0806
.13	.9883	.9966	4.4969	.73	.7016	.9037	1.0742
.14	.9864	.9961	4.1824	.74	.6951	.9013	1.0681
.15	.9844	.9955	3.9103	.75	.6886	.8989	1.0624
.16	.9823	.9949	3.6727	.76	.6821	.8964	1.0570
.17	.9800	.9943	3.4635	.77	.6756	.8940	1.0519
.18	.9776	.9936	3.2779	.78	.6691	.8915	1.0471
.19	.9751	.9928	3.1123	.79	.6625	.8890	1.0425
.20	.9725	.9921	2.9635	.80	.6560	.8865	1.0382
.21	.9697	.9913	2.8293	.81	.6495	.8840	1.0342
.22	.9668	.9904	2.7076	.82	.6430	.8815	1.0305
.23	.9638	.9895	2.5968	.83	.6365	.8789	1.0270
.24	.9607	.9886	2.4956	.84	.6300	.8763	1.0237
.25	.9575	.9877	2.4027	.85	.6235	.8737	1.0207
.26	.9541	.9867	2.3173	.86	.6170	.8711	1.0179
.27	.9506	.9856	2.2385	.87	.6106	.8685	1.0153
.28	.9470	.9846	2.1656	.88	.6041	.8659	1.0129
.29	.9433	.9835	2.0979	.89	.5977	.8632	1.0108
.30	.9395	.9823	2.0351	.90	.5913	.8606	1.0089
.31	.9355	.9811	1.9765	.91	.5849	.8579	1.0071
.32	.9315	.9799	1.9219	.92	.5785	.8552	1.0056
.33	.9274	.9787	1.8707	.93	.5721	.8525	1.0043
.34	.9231	.9774	1.8229	.94	.5658	.8498	1.0031
.35	.9188	.9761	1.7780	.95	.5595	.8471	1.0022
.36	.9143	.9747	1.7358	.96	.5532	.8444	1.0014
.37	.9098	.9733	1.6961	.97	.5469	.8416	1.0008
.38	.9052	.9719	1.6587	.98	.5407	.8389	1.0003
.39	.9004	.9705	1.6234	.99	.5345	.8361	1.0001
.40	.8956	.9690	1.5901	1.00	.5283	.8333	1.000
.41	.8907	.9675	1.5587	1.01	.5221	.8306	1.000
.42	.8857	.9659	1.5289	1.02	.5160	.8278	1.000
.43	.8807	.9643	1.5007	1.03	.5099	.8250	1.001
.44	.8755	.9627	1.4740	1.04	.5039	.8222	1.001
.45	.8703	.9611	1.4487	1.05	.4979	.8193	1.002
.46	.8650	.9594	1.4246	1.06	.4919	.8165	1.003
.47	.8596	.9577	1.4018	1.07	.4860	.8137	1.004
.48	.8541	.9560	1.3801	1.08	.4800	.8108	1.005
.49	.8486	.9542	1.3595	1.09	.4742	.8080	1.006
.50	.8430	.9524	1.3398	1.10	.4684	.8052	1.008
.51	.8374	.9506	1.3212	1.11	.4626	.8023	1.010
.52	.8317	.9487	1.3034	1.12	.4568	.7994	1.011
.53	.8259	.9468	1.2865	1.13	.4511	.7966	1.013
.54	.8201	.9449	1.2703	1.14	.4455	.7937	1.015
.55	.8142	.9430	1.2550	1.15	.4398	.7908	1.017
.56	.8082	.9410	1.2403	1.16	.4343	.7879	1.020
.57	.8022	.9390	1.2263	1.17	.4287	.7851	1.022
.58	.7962	.9370	1.2130	1.18	.4232	.7822	1.025
.59	.7901	.9349	1.2003	1.19	.4178	.7793	1.026

M	p/p_t	T/T_t	A/A^*	M	p/p_t	T/T_t	A/A^*
1.20	.4124	.7764	1.030	1.85	.1612	.5936	1.495
1.21	.4070	.7735	1.033	1.86	.1587	.5910	1.507
1.22	.4017	.7706	1.037	1.87	.1563	.5884	1.519
1.23	.3964	.7677	1.040	1.88	.1539	.5859	1.531
1.24	.3912	.7648	1.043	1.89	.1516	.5833	1.543
1.25	.3861	.7619	1.047	1.90	.1492	.5807	1.555
1.26	.3809	.7590	1.050	1.91	.1470	.5782	1.568
1.27	.3759	.7561	1.054	1.92	.1447	.5756	1.580
1.28	.3708	.7532	1.058	1.93	.1425	.5731	1.593
1.29	.3658	.7503	1.062	1.94	.1403	.5705	1.606
1.30	.3609	.7474	1.066	1.95	.1381	.5680	1.619
1.31	.3560	.7445	1.071	1.96	.1360	.5655	1.633
1.32	.3512	.7416	1.075	1.97	.1339	.5630	1.646
1.33	.3464	.7387	1.080	1.98	.1318	.5605	1.660
1.34	.3417	.7358	1.084	1.99	.1298	.5580	1.674
1.35	.3370	.7329	1.089	2.00	.1278	.5556	1.688
1.36	.3323	.7300	1.094	2.01	.1258	.5531	1.702
1.37	.3277	.7271	1.099	2.02	.1239	.5506	1.716
1.38	.3232	.7242	1.104	2.03	.1220	.5482	1.730
1.39	.3187	.7213	1.109	2.04	.1201	.5458	1.745
1.40	.3142	.7184	1.115	2.05	.1182	.5433	1.760
1.41	.3098	.7155	1.120	2.06	.1164	.5409	1.775
1.42	.3055	.7126	1.126	2.07	.1146	.5385	1.790
1.43	.3012	.7097	1.132	2.08	.1128	.5361	1.806
1.44	.2969	.7069	1.138	2.09	.1111	.5337	1.821
1.45	.2927	.7040	1.144	2.10	.1094	.5313	1.837
1.46	.2886	.7011	1.150	2.11	.1077	.5290	1.853
1.47	.2845	.6982	1.156	2.12	.1060	.5266	1.869
1.48	.2804	.6954	1.163	2.13	.1043	.5243	1.885
1.49	.2764	.6925	1.169	2.14	.1027	.5219	1.902
1.50	.2724	.6897	1.176	2.15	.1011	.5196	1.919
1.51	.2685	.6868	1.183	2.16	.9956 $^{-1}$	.5173	1.935
1.52	.2646	.6840	1.190	2.17	.9802 $^{-1}$	.5150	1.953
1.53	.2608	.6811	1.197	2.18	.9649 $^{-1}$	.5127	1.970
1.54	.2570	.6783	1.204	2.19	.9500 $^{-1}$	.5104	1.987
1.55	.2533	.6754	1.212	2.20	.9352 $^{-1}$	.5081	2.005
1.56	.2496	.6726	1.219	2.21	.9207 $^{-1}$	.5059	2.023
1.57	.2459	.6698	1.227	2.22	.9064 $^{-1}$	.5036	2.041
1.58	.2423	.6670	1.234	2.23	.8923 $^{-1}$	.5014	2.059
1.59	.2388	.6642	1.242	2.24	.8785 $^{-1}$	.4991	2.078
1.60	.2353	.6614	1.250	2.25	.8648 $^{-1}$	.4969	2.096
1.61	.2318	.6586	1.258	2.26	.8514 $^{-1}$	.4947	2.115
1.62	.2284	.6558	1.267	2.27	.8382 $^{-1}$	.4925	2.134
1.63	.2250	.6530	1.275	2.28	.8251 $^{-1}$	.4903	2.154
1.64	.2217	.6502	1.284	2.29	.8123 $^{-1}$	.4881	2.173
1.65	.2184	.6475	1.292	2.30	.7997 $^{-1}$	.4859	2.193
1.66	.2151	.6447	1.301	2.31	.7873 $^{-1}$	.4837	2.213
1.67	.2119	.6419	1.310	2.32	.7751 $^{-1}$	.4816	2.233
1.68	.2088	.6392	1.319	2.33	.7631 $^{-1}$	.4793	2.254
1.69	.2057	.6364	1.328	2.34	.7512 $^{-1}$	.4773	2.274
1.70	.2026	.6337	1.338	2.35	.7396 $^{-1}$	.4752	2.295
1.71	.1996	.6310	1.347	2.36	.7281 $^{-1}$	.4731	2.316
1.72	.1966	.6283	1.357	2.37	.7168 $^{-1}$	.4709	2.338
1.73	.1936	.6256	1.367	2.38	.7057 $^{-1}$	.4688	2.359
1.74	.1907	.6229	1.376	2.39	.6948 $^{-1}$	.4668	2.381
1.75	.1878	.6202	1.386	2.40	.6840 $^{-1}$	.4647	2.403
1.76	.1850	.6175	1.397	2.41	.6734 $^{-1}$	.4626	2.425
1.77	.1822	.6148	1.407	2.42	.6630 $^{-1}$	.4606	2.448
1.78	.1794	.6121	1.418	2.43	.6527 $^{-1}$	.4585	2.471
1.79	.1767	.6095	1.428	2.44	.6426 $^{-1}$	.4565	2.494
1.80	.1740	.6068	1.439	2.45	.6327 $^{-1}$	.4544	2.517
1.81	.1714	.6041	1.450	2.46	.6229 $^{-1}$	.4524	2.540
1.82	.1688	.6015	1.461	2.47	.6133 $^{-1}$	.4504	2.564
1.83	.1662	.5989	1.472	2.48	.6038 $^{-1}$	.4484	2.588
1.84	.1637	.5963	1.484	2.49	.5945 $^{-1}$	.4464	2.612

Table A.1 (continued) Isentropic flow tables ($\gamma = 1.4$)

M	p/p_t	T/T_t	A/A^*	M	p/p_t	T/T_t	A/A^*
2.50	$.5853^{-1}$	.4444	2.637	3.15	$.2177^{-1}$	.3351	4.884
2.51	$.5762^{-1}$	.4425	2.661	3.16	$.2146^{-1}$	.3337	4.930
2.52	$.5674^{-1}$	.4405	2.686	3.17	$.2114^{-1}$	.3323	4.977
2.53	$.5586^{-1}$	.4386	2.712	3.18	$.2083^{-1}$	.3309	5.025
2.54	$.5500^{-1}$	.4366	2.737	3.19	$.2053^{-1}$	.3295	5.073
2.55	$.5415^{-1}$	.4347	2.763	3.20	$.2023^{-1}$	.3281	5.121
2.56	$.5332^{-1}$	.4328	2.789	3.21	$.1993^{-1}$	.3267	5.170
2.57	$.5250^{-1}$	.4309	2.815	3.22	$.1964^{-1}$	.3253	5.219
2.58	$.5169^{-1}$	.4289	2.842	3.23	$.1936^{-1}$	.3240	5.268
2.59	$.5090^{-1}$	.4271	2.869	3.24	$.1908^{-1}$	.3226	5.319
2.60	$.5012^{-1}$	.4252	2.896	3.25	$.1880^{-1}$	.3213	5.369
2.61	$.4935^{-1}$	.4233	2.923	3.26	$.1853^{-1}$	.3199	5.420
2.62	$.4859^{-1}$	.4214	2.951	3.27	$.1826^{-1}$	.3186	5.472
2.63	$.4784^{-1}$	.4196	2.979	3.28	$.1799^{-1}$	.3173	5.523
2.64	$.4711^{-1}$	.4177	3.007	3.29	$.1773^{-1}$	.3160	5.576
2.65	$.4639^{-1}$	.4159	3.036	3.30	$.1748^{-1}$	.3147	5.629
2.66	$.4568^{-1}$	.4141	3.065	3.31	$.1722^{-1}$	.3134	5.682
2.67	$.4498^{-1}$	.4122	3.094	3.32	$.1698^{-1}$	.3121	5.736
2.68	$.4429^{-1}$	.4104	3.123	3.33	$.1673^{-1}$	.3108	5.790
2.69	$.4362^{-1}$	.4086	3.153	3.34	$.1649^{-1}$	.3095	5.845
2.70	$.4295^{-1}$	.4068	3.183	3.35	$.1625^{-1}$	.3082	5.900
2.71	$.4229^{-1}$	.4051	3.213	3.36	$.1602^{-1}$	.3069	5.956
2.72	$.4165^{-1}$	.4033	3.244	3.37	$.1579^{-1}$	.3057	6.012
2.73	$.4102^{-1}$	.4015	3.275	3.38	$.1557^{-1}$	.3044	6.069
2.74	$.4039^{-1}$	.3998	3.306	3.39	$.1534^{-1}$	.3032	6.126
2.75	$.3978^{-1}$	.3980	3.338	3.40	$.1512^{-1}$	.3019	6.184
2.76	$.3917^{-1}$	.3963	3.370	3.41	$.1491^{-1}$	.3007	6.242
2.77	$.3858^{-1}$	.3945	3.402	3.42	$.1470^{-1}$	.2995	6.301
2.78	$.3799^{-1}$	.3928	3.434	3.43	$.1449^{-1}$	.2982	6.360
2.79	$.3742^{-1}$	.3911	3.467	3.44	$.1428^{-1}$	.2970	6.420
2.80	$.3685^{-1}$	.3894	3.500	3.45	$.1408^{-1}$	.2958	6.480
2.81	$.3629^{-1}$	.3877	3.534	3.46	$.1388^{-1}$	.2946	6.541
2.82	$.3574^{-1}$	.3860	3.567	3.47	$.1368^{-1}$	.2934	6.602
2.83	$.3520^{-1}$	.3844	3.601	3.48	$.1349^{-1}$	.2922	6.664
2.84	$.3467^{-1}$	.3827	3.636	3.49	$.1330^{-1}$	.2910	6.727
2.85	$.3415^{-1}$	.3810	3.671	3.50	$.1311^{-1}$	.2899	6.790
2.86	$.3363^{-1}$	.3794	3.706	3.51	$.1293^{-1}$	.2887	6.853
2.87	$.3312^{-1}$	.3777	3.741	3.52	$.1274^{-1}$	.2875	6.917
2.88	$.3263^{-1}$	.3761	3.777	3.53	$.1256^{-1}$	.2864	6.982
2.89	$.3213^{-1}$	.3745	3.813	3.54	$.1239^{-1}$	.2852	7.047
2.90	$.3165^{-1}$	.3729	3.850	3.55	$.1221^{-1}$	.2841	7.113
2.91	$.3118^{-1}$	.3712	3.887	3.56	$.1204^{-1}$	.2829	7.179
2.92	$.3071^{-1}$	.3696	3.924	3.57	$.1188^{-1}$	.2818	7.246
2.93	$.3025^{-1}$	.3681	3.961	3.58	$.1171^{-1}$	.2806	7.313
2.94	$.2980^{-1}$	.3665	3.999	3.59	$.1155^{-1}$	.2795	7.382
2.95	$.2935^{-1}$	.3649	4.038	3.60	$.1138^{-1}$	.2784	7.450
2.96	$.2891^{-1}$	.3633	4.076	3.61	$.1123^{-1}$	.2773	7.519
2.97	$.2848^{-1}$	.3618	4.115	3.62	$.1107^{-1}$	.2762	7.589
2.98	$.2805^{-1}$	.3602	4.155	3.63	$.1092^{-1}$	.2751	7.659
2.99	$.2764^{-1}$	.3587	4.194	3.64	$.1076^{-1}$	.2740	7.730
3.00	$.2722^{-1}$	.3571	4.235	3.65	$.1062^{-1}$	.2729	7.802
3.01	$.2682^{-1}$	.3556	4.275	3.66	$.1047^{-1}$	.2718	7.874
3.02	$.2642^{-1}$	.3541	4.316	3.67	$.1032^{-1}$	.2707	7.947
3.03	$.2603^{-1}$	.3526	4.357	3.68	$.1018^{-1}$	.2697	8.020
3.04	$.2564^{-1}$	.3511	4.399	3.69	$.1004^{-1}$	.2686	8.094
3.05	$.2526^{-1}$	.3496	4.441	3.70	$.9903^{-2}$	.2675	8.169
3.06	$.2489^{-1}$	.3481	4.483	3.71	$.9767^{-2}$	.2665	8.244
3.07	$.2452^{-1}$	.3466	4.526	3.72	$.9633^{-2}$	.2654	8.320
3.08	$.2416^{-1}$	.3452	4.570	3.73	$.9500^{-2}$	.2644	8.397
3.09	$.2380^{-1}$	.3437	4.613	3.74	$.9370^{-2}$	.2632	8.474
3.10	$.2345^{-1}$	.3422	4.657	3.75	$.9242^{-2}$	.2623	8.552
3.11	$.2310^{-1}$	.3408	4.702	3.76	$.9116^{-2}$	.2613	8.630
3.12	$.2276^{-1}$	.3393	4.747	3.77	$.8991^{-2}$	.2602	8.709
3.13	$.2243^{-1}$	.3379	4.792	3.78	$.8869^{-2}$	.2592	8.789
3.14	$.2210^{-1}$	.3365	4.838	3.79	$.8748^{-2}$	.2582	8.870

Table A.1 (continued) Isentropic flow tables ($\gamma = 1.4$)

M	p/p_t	T/T_t	A/A^*	M	p/p_t	T/T_t	A/A^*
3.80	$.8629^{-2}$	.2572	8.951	4.45	$.3678^{-2}$	.2016	15.87
3.81	$.8512^{-2}$	.2562	9.032	4.46	$.3633^{-2}$	.2009	16.01
3.82	$.8396^{-2}$	.2582	9.115	4.47	$.3587^{-2}$	.2002	16.15
3.83	$.8283^{-2}$	.2542	9.198	4.48	$.3543^{-2}$	.1994	16.28
3.84	$.8171^{-2}$	.2532	9.282	4.49	$.3499^{-2}$	.1987	16.42
3.85	$.8000^{-2}$	.2522	9.366	4.50	$.3455^{-2}$	.1980	16.56
3.86	$.7951^{-2}$	.2513	9.451	4.51	$.3412^{-2}$	.1973	16.70
3.87	$.7844^{-2}$	.2503	9.537	4.52	$.3370^{-2}$	.1966	16.84
3.88	$.7730^{-2}$	.2493	9.624	4.53	$.3329^{-2}$	.1959	16.90
3.89	$.7635^{-2}$	.2434	9.711	4.54	$.3288^{-2}$	.1952	17.13
3.90	$.7532^{-2}$	.2474	9.799	4.55	$.3247^{-2}$	.1945	17.28
3.91	$.7431^{-2}$	.2464	9.888	4.56	$.3207^{-2}$	.1938	17.42
3.92	$.7332^{-2}$	.2455	9.977	4.57	$.3108^{-2}$	.1932	17.57
3.93	$.7233^{-2}$	.2446	10.07	4.58	$.3129^{-2}$	.1925	17.72
3.94	$.7137^{-2}$	.2436	10.16	4.59	$.3090^{-2}$	.1918	17.87
3.95	$.7042^{-2}$	.2427	10.25	4.60	$.3053^{-2}$	.1911	18.03
3.96	$.6948^{-2}$	.2418	10.34	4.61	$.3015^{-2}$	.1905	16.17
3.97	$.6855^{-2}$	.2408	10.44	4.62	$.2978^{-2}$	.1898	18.33
3.98	$.6764^{-2}$	.2399	10.53	4.63	$.2912^{-2}$	.1891	18.48
3.99	$.6675^{-2}$	.2390	10.62	4.64	$.2906^{-2}$	.1885	18.63
4.00	$.6586^{-2}$	.2381	10.72	4.65	$.2871^{-2}$	.1878	18.79
4.01	$.6499^{-2}$	.2372	10.81	4.66	$.2836^{-2}$	.1872	18.94
4.02	$.6413^{-2}$	.2363	10.91	4.67	$.2802^{-2}$	.1865	19.10
4.03	$.6328^{-2}$	.2354	11.01	4.68	$.2768^{-2}$	.1859	19.26
4.04	$.6245^{-2}$	.2345	11.11	4.69	$.3734^{-2}$	.1852	19.42
4.05	$.6163^{-2}$	.2326	11.21	4.70	$.2701^{-2}$	.1846	19.58
4.06	$.6082^{-2}$	.2337	11.31	4.71	$.2669^{-2}$	.1839	19.75
4.07	$.6002^{-2}$	.2319	11.41	4.72	$.2637^{-2}$	.1833	19.91
4.08	$.5923^{-2}$	.2310	11.51	4.73	$.2605^{-2}$	.1827	20.07
4.09	$.5845^{-2}$	.2301	11.61	4.74	$.2573^{-2}$	.1820	20.24
4.10	$.5769^{-2}$	.2293	11.71	4.75	$.2543^{-2}$	.1814	20.41
4.11	$.5694^{-2}$	.2284	11.82	4.76	$.2512^{-2}$	.1809	20.58
4.12	$.5619^{-2}$	.2275	11.92	4.77	$.2482^{-2}$	.1802	20.75
4.13	$.5546^{-2}$	.2267	12.03	4.78	$.2452^{-2}$	.1795	20.92
4.14	$.5474^{-2}$	.2258	12.14	4.79	$.2423^{-2}$	.1789	21.09
4.15	$.5403^{-2}$	.2250	12.24	4.80	$.2394^{-2}$	.1783	21.26
4.16	$.5333^{-2}$	.2242	12.35	4.81	$.2366^{-2}$	.1777	21.44
4.17	$.5264^{-2}$	.2233	12.46	4.82	$.2338^{-2}$	.1771	21.61
4.18	$.5195^{-2}$	.2225	12.57	4.83	$.2310^{-2}$	.1765	21.79
4.19	$.5128^{-2}$	.2217	12.68	4.84	$.2283^{-2}$	.1759	21.97
4.20	$.5062^{-2}$	.2208	12.79	4.85	$.2255^{-2}$	.1753	22.15
4.21	$.4997^{-2}$	.2500	12.90	4.86	$.2229^{-2}$	.1747	22.33
4.22	$.4932^{-2}$	.2192	13.02	4.87	$.2202^{-2}$	.1741	22.51
4.23	$.4869^{-2}$	.2184	13.13	4.88	$.2177^{-2}$	.1735	22.70
4.24	$.4806^{-2}$	.2176	13.25	4.89	$.2151^{-2}$	.1729	22.88
4.25	$.4745^{-2}$	.2168	13.36	4.90	$.2126^{-2}$	.1724	23.07
4.26	$.4684^{-2}$	.2160	13.48	4.91	$.2102^{-2}$	.1718	23.25
4.27	$.4624^{-2}$	.2152	13.60	4.92	$.2076^{-2}$	.1712	23.44
4.28	$.4565^{-2}$	.2144	13.72	4.93	$.2062^{-2}$	.1706	23.63
4.29	$.4507^{-2}$	.2136	13.83	4.94	$.2028^{-2}$	.1700	23.82
4.30	$.4449^{-2}$	.2129	13.95	4.95	$.2004^{-2}$	.1695	24.02
4.31	$.4393^{-2}$	.2121	14.08	4.96	$.1981^{-2}$	.1689	24.21
4.32	$.4337^{-2}$	.2113	14.20	4.97	$.1957^{-2}$	.1683	24.41
4.33	$.4282^{-2}$	.2105	14.32	4.98	$.1935^{-2}$	.1678	24.60
4.34	$.4228^{-2}$	.2098	14.45	4.99	$.1912^{-2}$	.1672	24.80
4.35	$.4174^{-2}$	.2090	14.57	5.00	$.1890^{-2}$	.1667	25.00
4.36	$.4121^{-2}$	.2083	14.70	6.00	$.6334^{-3}$	.1220	53.18
4.37	$.4069^{-2}$	.2075	14.82	7.00	$.2416^{-3}$	$.9259^{-1}$	104.1
4.38	$.4018^{-2}$	.2067	14.95	8.00	$.1024^{-3}$	$.7246^{-1}$	190.1
4.39	$.3968^{-2}$	.2060	15.08	9.00	$.4739^{-4}$	$.5814^{-1}$	327.2
4.40	$.3918^{-2}$	.2053	15.21	10.00	$.2356^{-4}$	$.4762^{-1}$	535.9
4.41	$.3868^{-2}$	.2045	15.34				
4.42	$.3820^{-2}$	.2038	15.47				
4.43	$.3772^{-2}$	.2030	15.61				
4.44	$.3725^{-2}$	.2023	15.74				

Material in Appendix A.1 has been adapted from NACA Report 1135, "Equations, Tables, and Charts for Compressible Flow," Ames Research Staff, 1953.

Table A.2 Isentropic flow tables ($\gamma = 1.3$)

M	p/p_t	T/T_t	A/A^*	M	p/p_t	T/T_t	A/A^*
0	1.0000	1.0000	∞	1.75	0.1944	0.6852	1.424
.05	.9984	.9996	11.7202	1.80	.1797	.6729	1.484
.10	.9936	.9985	5.8852	1.85	.1660	.6607	1.549
.15	.9855	.9966	3.9520	1.90	.1533	.6487	1.618
.20	.9744	.9940	2.9938	1.95	.1415	.6368	1.693
.25	.9603	.9907	2.4262	2.00	.1305	.6250	1.773
.30	.9435	.9867	2.0537	2.05	.1203	.6134	1.859
.35	.9241	.9820	1.7930	2.10	.1108	.6019	1.951
.40	.9023	.9766	1.6023	2.15	.1020	.5905	2.050
.45	.8784	.9705	1.4586	2.20	$.9393^{-1}$	.5794	2.156
.50	.8526	.9639	1.3479	2.25	$.8645^{-1}$	.5684	2.268
.55	.8251	.9566	1.2614	2.30	$.7955^{-1}$	.5576	2.388
.60	.6267	.9488	1.1932	2.35	$.7318^{-1}$	.5470	2.517
.65	.7662	.9404	1.1395	2.40	$.6731^{-1}$	.5365	2.654
.70	.7354	.9315	1.0972	2.45	$.6190^{-1}$	.5262	2.799
.75	.7724	.9222	1.0644	2.50	$.5692^{-1}$	.5161	2.954
.80	.6273	.9124	1.0395	2.55	$.5234^{-1}$	.5062	3.119
.85	.6403	.9022	1.0214	2.60	$.4813^{-1}$	.4965	3.295
.90	.6084	.8917	1.0092	2.65	$.4426^{-1}$	.4870	3.482
.95	.5768	.8808	1.0022	2.70	$.4070^{-1}$	.4777	3.681
1.00	.5457	.8696	1.0000	2.75	$.3743^{-1}$	.4686	3.892
1.05	.5152	.8581	1.002	2.80	$.3442^{-1}$	.4596	4.116
1.10	.4854	.8464	1.008	2.85	$.3166^{-1}$	.4508	4.354
1.15	.4565	.8345	1.018	2.90	$.2913^{-1}$	.4422	4.607
1.20	.4285	.8224	1.032	2.95	$.2680^{-1}$	.4338	4.875
1.25	.4015	.8101	1.049	3.00	$.2466^{-1}$	.4255	5.160
1.30	.3756	.7978	1.070	3.50	$.1090^{-1}$	.3524	9.110
1.35	.3509	.7853	1.095	4.00	$.4977^{-2}$	.2941	15.94
1.40	.3273	.7728	1.123	4.50	$.2363^{-2}$	.2477	27.39
1.45	.3049	.7603	1.154	5.00	$.1169^{-2}$	.2105	45.96
1.50	.2836	.7477	1.189	6.00	$.3120^{-3}$	.1563	120.1
1.55	.2635	.7351	1.228	7.00	$.1014^{-3}$	.1198	285.3
1.60	.2446	.7225	1.271	8.00	$.3606^{-4}$	$.9434^{-1}$	623.1
1.65	.2268	.7100	1.318	9.00	$.1417^{-4}$	$.7605^{-1}$	1265
1.70	.2100	.6976	1.369	10.00	$.6056^{-5}$	$.6250^{-1}$	2416

Table A.3 Isentropic flow tables ($\gamma = 5/3$)

M	p/p_t	T/T_t	A/A^*	M	p/p_t	T/T_t	A/A^*
0	1.0000	1.0000	∞	1.75	.1723	.4948	1.313
.05	.9979	.9992	11.2683	1.80	.1603	.4808	1.352
.10	.9917	.9967	5.6623	1.85	.1491	.4671	1.394
.15	.9815	.9926	3.8062	1.90	.1388	.4539	1.437
.20	.9674	.9868	2.8878	1.95	.1292	.4410	1.483
.25	.9498	.9796	2.3447	2.00	.1202	.4286	1.531
.30	.9288	.9709	1.9892	2.05	.1120	.4165	1.582
.35	.9048	.9608	1.7411	2.10	.1043	.4049	1.634
.40	.8782	.9494	1.5603	2.15	$.9718^{-1}$	.3936	1.689
.45	.8493	.9368	1.4244	2.20	$.9058^{-1}$	.3827	1.746
.50	.8186	.9231	1.3203	2.25	$.8446^{-1}$	.3721	1.806
.55	.7865	.9084	1.2394	2.30	$.7878^{-1}$	.3619	1.868
.60	.7533	.8929	1.1760	2.35	$.7352^{-1}$	.3520	1.932
.65	.7194	.8766	1.1263	2.40	$.6863^{-1}$	.3425	1.998
.70	.6851	.8596	1.0875	2.45	$.6411^{-1}$	.3332	2.067
.75	.6508	.8421	1.0576	2.50	$.5990^{-1}$	.3243	2.139
.80	.6167	.8242	1.0351	2.55	$.5600^{-1}$	.3157	2.213
.85	.5831	.8059	1.0189	2.60	$.5238^{-1}$	.3074	2.290
.90	.5502	.7874	1.0081	2.65	$.4902^{-1}$	.2993	2.369
.95	.5181	.7687	1.0019	2.70	$.4589^{-1}$	.2915	2.451
1.00	.4871	.7500	1.000	2.75	$.4299^{-1}$	.2840	2.536
1.05	.4573	.7313	1.002	2.80	$.4029^{-1}$	.2768	2.623
1.10	.4286	.7126	1.007	2.85	$.3778^{-1}$	.2697	2.713
1.15	.4013	.6940	1.015	2.90	$.3545^{-1}$	.2629	2.806
1.20	.3753	.6757	1.027	2.95	$.3327^{-1}$	.2564	2.901
1.25	.3506	.6575	1.041	3.00	$.3125^{-1}$	.2500	3.000
1.30	.3272	.6397	1.058	3.50	$.1716^{-1}$	.1967	4.153
1.35	.3052	.6221	1.077	4.00	$.9906^{-2}$	.1579	5.641
1.40	.2845	.6048	1.098	4.50	$.5981^{-2}$	.1290	7.508
1.45	.2650	.5879	1.122	5.00	$.3758^{-2}$	.1071	9.800
1.50	.2468	.5714	1.148	6.00	$.1641^{-2}$	$.7692^{-1}$	15.84
1.55	.2298	.5553	1.177	7.00	$.7995^{-3}$	$.5769^{-1}$	24.14
1.60	.2139	.5396	1.208	8.00	$.4243^{-3}$	$.4478^{-1}$	35.07
1.65	.1990	.5242	1.240	9.00	$.2410^{-3}$	$.3571^{-1}$	49.00
1.70	.1851	.5093	1.275	10.00	$.1448^{-3}$	$.2913^{-1}$	66.30

APPENDIX B

Table B.1　Normal shock tables ($\gamma = 1.4$)

M_1	M_2	p_2/p_1	ρ_2/ρ_1	T_2/T_1	p_{t2}/p_{t1}	p_1/p_{t2}
1.00	1.000	1.000	1.000	1.000	1.000	0.5283
1.01	.9901	1.023	1.017	1.007	1.000	.5221
1.02	.9805	1.047	1.033	1.013	1.000	.5160
1.03	.9712	1.071	1.050	1.020	1.000	.5100
1.04	.9620	1.095	1.067	1.026	.9999	.5039
1.05	.9531	1.120	1.084	1.033	.9999	.4980
1.06	.9444	1.144	1.101	1.039	.9997	.4920
1.07	.9360	1.169	1.118	1.046	.9996	.4861
1.08	.9277	1.194	1.135	1.052	.9994	.4803
1.09	.9196	1.219	1.152	1.059	.9992	.4746
1.10	.9118	1.245	1.169	1.065	.9989	.4689
1.11	.9041	1.271	1.186	1.071	.9986	.4632
1.12	.8966	1.297	1.203	1.078	.9982	.4576
1.13	.8892	1.323	1.221	1.084	.9978	.4521
1.14	.8820	1.350	1.238	1.090	.9973	.4467
1.15	.8750	1.376	1.255	1.097	.9967	.4413
1.16	.8682	1.403	1.272	1.103	.9961	.4360
1.17	.8615	1.430	1.290	1.109	.9953	.4307
1.18	.8549	1.458	1.307	1.115	.9946	.4255
1.19	.8485	1.485	1.324	1.122	.9937	.4204
1.20	.8422	1.513	1.342	1.128	.9928	.4154
1.21	.8360	1.541	1.359	1.134	.9918	.4104
1.22	.8300	1.570	1.376	1.141	.9907	.4055
1.23	.8241	1.598	1.394	1.147	.9896	.4006
1.24	.8183	1.627	1.411	1.153	.9884	.3958
1.25	.8126	1.656	1.429	1.159	.9871	.3911
1.26	.8071	1.686	1.446	1.166	.9857	.3865
1.27	.8016	1.715	1.463	1.172	.9842	.3819
1.28	.7963	1.745	1.481	1.178	.9827	.3774
1.29	.7911	1.775	1.498	1.185	.9811	.3729
1.30	.7860	1.805	1.516	1.191	.9794	.3685
1.31	.7809	1.835	1.533	1.197	.9776	.3642
1.32	.7760	1.866	1.551	1.204	.9758	.3599
1.33	.7712	1.897	1.568	1.210	.9738	.3557
1.34	.7664	1.928	1.585	1.216	.9718	.3516
1.35	.7618	1.960	1.603	1.223	.9697	.3475
1.36	.7572	1.991	1.620	1.229	.9676	.3435
1.37	.7527	2.023	1.638	1.235	.9653	.3395
1.38	.7483	2.055	1.655	1.242	.9630	.3356
1.39	.7440	2.087	1.672	1.248	.9607	.3317
1.40	.7397	2.120	1.690	1.255	.9582	.3280
1.41	.7355	2.153	1.707	1.261	.9557	.3242
1.42	.7314	2.186	1.724	1.268	.9531	.3205
1.43	.7274	2.219	1.742	1.274	.9504	.3169
1.44	.7235	2.253	1.759	1.281	.9476	.3133
1.45	.7196	2.286	1.776	1.287	.9448	.3098
1.46	.7157	2.320	1.793	1.294	.9420	.3063
1.47	.7120	2.354	1.811	1.300	.9390	.3029
1.48	.7083	2.389	1.828	1.307	.9360	.2996
1.49	.7047	2.423	1.845	1.314	.9329	.2962
1.50	.7011	2.458	1.862	1.320	.9298	.2930
1.51	.6976	2.493	1.879	1.327	.9266	.2898
1.52	.6941	2.529	1.896	1.334	.9233	.2866
1.53	.6907	2.564	1.913	1.340	.9200	.2835
1.54	.6874	2.600	1.930	1.347	.9166	.2804
1.55	.6841	2.636	1.947	1.354	.9132	.2773
1.56	.6809	2.673	1.964	1.361	.9097	.2744
1.57	.6777	2.709	1.981	1.367	.9061	.2714
1.58	.6746	2.746	1.998	1.374	.9026	.2685
1.59	.6715	2.783	2.015	1.381	.8989	.2656

M_1	M_2	p_2/p_1	ρ_2/ρ_1	T_2/T_1	p_{t2}/p_{t1}	p_1/p_{t2}
1.60	.6684	2.820	2.032	1.388	.8952	.2628
1.61	.6655	2.857	2.049	1.395	.8915	.2600
1.62	.6625	2.895	2.065	1.402	.8877	.2573
1.63	.6596	2.933	2.082	1.409	.8838	.2546
1.64	.6568	2.971	2.099	1.416	.8799	.2519
1.65	.6540	3.010	2.115	1.423	.8760	.2493
1.66	.6512	3.048	2.132	1.430	.8720	.2467
1.67	.6485	3.087	2.148	1.437	.8680	.2442
1.68	.6458	3.126	2.165	1.444	.8640	.2417
1.69	.6431	3.165	2.181	1.451	.8598	.2392
1.70	.6405	3.205	2.198	1.458	.8557	.2368
1.71	.6380	3.245	2.214	1.466	.8516	.2344
1.72	.6355	3.285	2.230	1.473	.8474	.2320
1.73	.6330	3.325	2.247	1.480	.8431	.2296
1.74	.6305	3.366	2.263	1.487	.8389	.2273
1.75	.6281	3.406	2.279	1.495	.8346	.2251
1.76	.6257	3.447	2.295	1.502	.8302	.2228
1.77	.6234	3.488	2.311	1.509	.8259	.2206
1.78	.6210	3.530	2.327	1.517	.8215	.2184
1.79	.6188	3.571	2.343	1.524	.8171	.2163
1.80	.6165	3.613	2.359	1.532	.8127	.2142
1.81	.6143	3.655	2.375	1.539	.8082	.2121
1.82	.6121	3.698	2.391	1.547	.8038	.2100
1.83	.6099	3.740	2.407	1.554	.7993	.2080
1.84	.6078	3.783	2.422	1.562	.7948	.2060
1.85	.6057	3.826	2.438	1.569	.7902	.2040
1.86	.6036	3.870	2.454	1.577	.7857	.2020
1.87	.6016	3.913	2.469	1.585	.7811	.2001
1.88	.5996	3.957	2.485	1.592	.7765	.1982
1.89	.5976	4.001	2.500	1.600	.7720	.1963
1.90	.5956	4.045	2.516	1.608	.7674	.1945
1.91	.5937	4.089	2.531	1.616	.7627	.1927
1.92	.5918	4.134	2.546	1.624	.7581	.1909
1.93	.5899	4.179	2.562	1.631	.7535	.1891
1.94	.5880	4.224	2.577	1.639	.7488	.1873
1.95	.5862	4.270	2.592	1.647	.7442	.1856
1.96	.5844	4.315	2.607	1.655	.7395	.1839
1.97	.5826	4.361	2.622	1.663	.7349	.1822
1.98	.5808	4.407	2.637	1.671	.7302	.1806
1.99	.5791	4.453	2.652	1.679	.7255	.1789
2.00	.5774	4.500	2.667	1.688	.7209	.1773
2.01	.5757	4.547	2.681	1.696	.7162	.1757
2.02	.5740	4.594	2.696	1.704	.7115	.1741
2.03	.5723	4.641	2.711	1.712	.7069	.1726
2.04	.5707	4.689	2.725	1.720	.7022	.1710
2.05	.5691	4.736	2.740	1.729	.6975	.1695
2.06	.5675	4.784	2.755	1.737	.6928	.1680
2.07	.5659	4.832	2.769	1.745	.6882	.1665
2.08	.5643	4.881	2.783	1.754	.6835	.1651
2.09	.5628	4.929	2.798	1.762	.6789	.1636
2.10	.5613	4.978	2.812	1.770	.6742	.1622
2.11	.5598	5.027	2.826	1.779	.6696	.1608
2.12	.5583	5.077	2.840	1.787	.6649	.1594
2.13	.5568	5.126	2.854	1.796	.6603	.1580
2.14	.5554	5.176	2.868	1.805	.6557	.1567
2.15	.5540	5.226	2.882	1.813	.6511	.1553
2.16	.5525	5.277	2.896	1.822	.6464	.1540
2.17	.5511	5.327	2.910	1.831	.6419	.1527
2.18	.5498	5.378	2.924	1.839	.6373	.1514
2.19	.5484	5.429	2.938	1.848	.6327	.1502
2.20	.5471	5.480	2.951	1.857	.6281	.1489
2.21	.5457	5.531	2.965	1.866	.6236	.1476
2.22	.5444	5.583	2.978	1.875	.6191	.1464
2.23	.5431	5.636	2.992	1.883	.6145	.1452
2.24	.5418	5.687	3.005	1.892	.6100	.1440

Table B.1 (continued)   Normal shock tables ($\gamma = 1.4$)

M_1	M_2	p_2/p_1	ρ_2/ρ_1	T_2/T_1	p_{t2}/p_{t1}	p_1/p_{t2}
2.25	.5406	5.740	3.019	1.901	.6055	.1428
2.26	.5393	5.792	3.032	1.910	.6011	.1417
2.27	.5381	5.845	3.045	1.919	.5966	.1405
2.28	.5368	5.898	3.058	1.929	.5921	.1394
2.29	.5356	5.951	3.071	1.938	.5877	.1382
2.30	.5344	6.005	3.085	1.947	.5833	.1371
2.31	.5332	6.059	3.098	1.956	.5789	.1360
2.32	.5321	6.113	3.110	1.965	.5745	.1349
2.33	.5309	6.167	3.123	1.974	.5702	.1338
2.34	.5297	6.222	3.136	1.984	.5658	.1328
2.35	.5286	6.276	3.149	1.993	.5615	.1317
2.36	.5275	6.331	3.162	2.002	.5572	.1307
2.37	.5264	6.386	3.174	2.012	.5529	.1297
2.38	.5253	6.442	3.187	2.021	.5486	.1286
2.39	.5242	6.497	3.199	2.031	.5444	.1276
2.40	.5231	6.553	3.212	2.040	.5401	.1266
2.41	.5221	6.609	3.224	2.050	.5359	.1257
2.42	.5210	6.666	3.237	2.059	.5317	.1247
2.43	.5200	6.722	3.249	2.069	.5276	.1237
2.44	.5189	6.779	3.261	2.079	.5234	.1228
2.45	.5179	6.836	3.273	2.088	.5193	.1218
2.46	.5169	6.894	3.285	2.098	.5152	.1209
2.47	.5159	6.951	3.298	2.108	.5111	.1200
2.48	.5149	7.009	3.310	2.118	.5071	.1191
2.49	.5140	7.067	3.321	2.128	.5030	.1182
2.50	.5130	7.125	3.333	2.138	.4990	.1173
2.51	.5120	7.183	3.345	2.147	.4950	.1164
2.52	.5111	7.242	3.357	2.157	.4911	.1155
2.53	.5102	7.301	3.369	2.167	.4871	.1147
2.54	.5092	7.360	3.380	2.177	.4832	.1138
2.55	.5083	7.420	3.392	2.187	.4793	.1130
2.56	.5074	7.479	3.403	2.198	.4754	.1122
2.57	.5065	7.539	3.415	2.208	.4715	.1113
2.58	.5056	7.599	3.426	2.218	.4677	.1105
2.59	.5047	7.659	3.438	2.228	.4639	.1097
2.60	.5039	7.720	3.449	2.238	.4601	.1089
2.61	.5030	7.781	3.460	2.249	.4564	.1081
2.62	.5022	7.842	3.471	2.259	.4526	.1074
2.63	.5013	7.903	3.483	2.269	.4489	.1066
2.64	.5005	7.965	3.494	2.280	.4452	.1058
2.65	.4996	8.026	3.505	2.290	.4416	.1051
2.66	.4988	8.088	3.516	2.301	.4379	.1043
2.67	.4980	8.150	3.527	2.311	.4343	.1036
2.68	.4972	8.213	3.537	2.322	.4307	.1028
2.69	.4964	8.275	3.548	2.332	.4271	.1021
2.70	.4956	8.338	3.559	2.343	.4236	.1014
2.71	.4949	8.401	3.570	2.354	.4201	.1007
2.72	.4941	8.465	3.580	2.364	.4166	.9998 $^{-1}$
2.73	.4933	8.528	3.591	2.375	.4131	.9929 $^{-1}$
2.74	.4926	8.592	3.601	2.386	.4097	.9860 $^{-1}$
2.75	.4918	8.656	3.612	2.397	.4062	.9792 $^{-1}$
2.76	.4911	8.721	3.622	2.407	.4028	.9724 $^{-1}$
2.77	.4903	8.785	3.633	2.418	.3994	.9658 $^{-1}$
2.78	.4896	8.850	3.643	2.429	.3961	.9591 $^{-1}$
2.79	.4889	8.915	3.653	2.440	.3928	.9526 $^{-1}$
2.80	.4882	8.980	3.664	2.451	.3895	.9461 $^{-1}$
2.81	.4875	9.045	3.674	2.462	.3862	.9397 $^{-1}$
2.82	.4868	9.111	3.684	2.473	.3829	.9334 $^{-1}$
2.83	.4861	9.177	3.694	2.484	.3797	.9271 $^{-1}$
2.84	.4854	9.243	3.704	2.496	.3765	.9209 $^{-1}$
2.85	.4847	9.310	3.714	2.507	.3733	.9147 $^{-1}$
2.86	.4840	9.376	3.724	2.518	.3701	.9086 $^{-1}$
2.87	.4833	9.443	3.734	2.529	.3670	.9026 $^{-1}$
2.88	.4827	9.510	3.743	2.540	.3639	.8966 $^{-1}$
2.89	.4820	9.577	3.753	2.552	.3608	.8906 $^{-1}$

Table B.1 (continued) Normal shock tables ($\gamma = 1.4$)

M_1	M_2	p_2/p_1	ρ_2/ρ_1	T_2/T_1	p_{t2}/p_{t1}	p_1/p_{t2}
2.90	.4814	9.645	3.763	2.563	.3577	.8848 $^{-1}$
2.91	.4807	9.713	3.773	2.575	.3547	.8790 $^{-1}$
2.92	.4801	9.781	3.782	2.586	.3517	.8732 $^{-1}$
2.93	.4795	9.849	3.792	2.598	.3487	.8675 $^{-1}$
2.94	.4788	9.918	3.801	2.609	.3457	.8619 $^{-1}$
2.95	.4782	9.986	3.811	2.621	.3428	.8563 $^{-1}$
2.96	.4776	10.06	3.820	2.632	.3398	.8507 $^{-1}$
2.97	.4770	10.12	3.829	2.644	.3369	.8453 $^{-1}$
2.98	.4764	10.19	3.839	2.656	.3340	.8398 $^{-1}$
2.99	.4758	10.26	3.848	2.667	.3312	.8345 $^{-1}$
3.00	.4752	10.33	3.857	2.679	.3283	.8291 $^{-1}$
3.01	.4746	10.40	3.866	2.691	.3255	.8238 $^{-1}$
3.02	.4740	10.47	3.875	2.703	.3227	.8186 $^{-1}$
3.03	.4734	10.54	3.884	2.714	.3200	.8134 $^{-1}$
3.04	.4729	10.62	3.893	2.726	.3172	.8083 $^{-1}$
3.05	.4723	10.69	3.902	2.738	.3145	.8032 $^{-1}$
3.06	.4717	10.76	3.911	2.750	.3118	.7982 $^{-1}$
3.07	.4712	10.83	3.920	2.762	.3091	.7932 $^{-1}$
3.08	.4706	10.90	3.929	2.774	.3065	.7882 $^{-1}$
3.09	.4701	10.97	3.938	2.786	.3038	.7833 $^{-1}$
3.10	.4695	11.05	3.947	2.799	.3012	.7785 $^{-1}$
3.11	.4690	11.12	3.955	2.811	.2986	.7737 $^{-1}$
3.12	.4685	11.19	3.964	2.823	.2960	.7689 $^{-1}$
3.13	.4679	11.26	3.973	2.835	.2935	.7642 $^{-1}$
3.14	.4674	11.34	3.981	2.848	.2910	.7595 $^{-1}$
3.15	.4669	11.41	3.990	2.860	.2885	.7549 $^{-1}$
3.16	.4664	11.48	3.998	2.872	.2860	.7503 $^{-1}$
3.17	.4659	11.56	4.006	2.885	.2835	.7457 $^{-1}$
3.18	.4654	11.63	4.015	2.897	.2811	.7412 $^{-1}$
3.19	.4648	11.71	4.023	2.909	.2786	.7367 $^{-1}$
3.20	.4643	11.78	4.031	2.922	.2762	.7323 $^{-1}$
3.21	.4639	11.85	4.040	2.935	.2738	.7279 $^{-1}$
3.22	.4634	11.93	4.048	2.947	.2715	.7235 $^{-1}$
3.23	.4629	12.01	4.056	2.960	.2691	.7192 $^{-1}$
3.24	.4624	12.08	4.064	2.972	.2668	.7149 $^{-1}$
3.25	.4619	12.16	4.072	2.985	.2645	.7107 $^{-1}$
3.26	.4614	12.23	4.080	2.998	.2622	.7065 $^{-1}$
3.27	.4610	12.31	4.088	3.011	.2600	.7023 $^{-1}$
3.28	.4605	12.38	4.096	3.023	.2577	.6982 $^{-1}$
3.29	.4600	12.46	4.104	3.036	.2555	.6941 $^{-1}$
3.30	.4596	12.54	4.112	3.049	.2533	.6900 $^{-1}$
3.31	.4591	12.62	4.120	3.062	.2511	.6860 $^{-1}$
3.32	.4587	12.69	4.128	3.075	.2489	.6820 $^{-1}$
3.33	.4582	12.77	4.135	3.088	.2468	.6781 $^{-1}$
3.34	.4578	12.85	4.143	3.101	.2446	.6741 $^{-1}$
3.35	.4573	12.93	4.151	3.114	.2425	.6702 $^{-1}$
3.36	.4569	13.00	4.158	3.127	.2404	.6664 $^{-1}$
3.37	.4565	13.08	4.166	3.141	.2383	.6626 $^{-1}$
3.38	.4560	13.16	4.173	3.154	.2363	.6588 $^{-1}$
3.39	.4556	13.24	4.181	3.167	.2342	.6550 $^{-1}$
3.40	.4552	13.32	4.188	3.180	.2322	.6513 $^{-1}$
3.41	.4548	13.40	4.196	3.194	.2302	.6476 $^{-1}$
3.42	.4544	13.48	4.203	3.207	.2282	.6439 $^{-1}$
3.43	.4540	13.56	4.211	3.220	.2263	.6403 $^{-1}$
3.44	.4535	13.64	4.218	3.234	.2243	.6367 $^{-1}$
3.45	.4531	13.72	4.225	3.247	.2224	.6331 $^{-1}$
3.46	.4527	13.80	4.232	3.261	.2205	.6296 $^{-1}$
3.47	.4523	13.88	4.240	3.274	.2186	.6261 $^{-1}$
3.48	.4519	13.96	4.247	3.288	.2167	.6226 $^{-1}$
3.49	.4515	14.04	4.254	3.301	.2148	.6191 $^{-1}$
3.50	.4512	14.13	4.261	3.315	.2129	.6157 $^{-1}$
3.51	.4508	14.21	4.268	3.329	.2111	.6123 $^{-1}$
3.52	.4504	14.29	4.275	3.343	.2093	.6089 $^{-1}$
3.53	.4500	14.37	4.282	3.356	.2075	.6056 $^{-1}$
3.54	.4496	14.45	4.289	3.370	.2057	.6023 $^{-1}$

Table B.1 (continued)   Normal shock tables ($\gamma = 1.4$)

M_1	M_2	p_2/p_1	ρ_2/ρ_1	T_2/T_1	p_{t2}/p_{t1}	p_1/p_{t2}
3.55	.4492	14.54	4.296	3.384	.2039	.5990 $^{-1}$
3.56	.4489	14.62	4.303	3.398	.2022	.5957 $^{-1}$
3.57	.4485	14.70	4.309	3.412	.2004	.5925 $^{-1}$
3.58	.4481	14.79	4.316	3.426	.1987	.5892 $^{-1}$
3.59	.4478	14.87	4.323	3.440	.1970	.5861 $^{-1}$
3.60	.4474	14.95	4.330	3.454	.1953	.5829 $^{-1}$
3.61	.4471	15.04	4.336	3.468	.1936	.5798 $^{-1}$
3.62	.4467	15.12	4.343	3.482	.1920	.5767 $^{-1}$
3.63	.4463	15.21	4.350	3.496	.1903	.5736 $^{-1}$
3.64	.4460	15.29	4.356	3.510	.1887	.5705 $^{-1}$
3.65	.4456	15.38	4.363	3.525	.1871	.5675 $^{-1}$
3.66	.4453	15.46	4.369	3.539	.1855	.5645 $^{-1}$
3.67	.4450	15.55	4.376	3.553	.1839	.5615 $^{-1}$
3.68	.4446	15.63	4.382	3.568	.1823	.5585 $^{-1}$
3.69	.4443	15.72	4.388	3.582	.1807	.5556 $^{-1}$
3.70	.4439	15.81	4.395	3.596	.1792	.5526 $^{-1}$
3.71	.4436	15.89	4.401	3.611	.1777	.5497 $^{-1}$
3.72	.4433	15.98	4.408	3.625	.1761	.5469 $^{-1}$
3.73	.4430	16.07	4.414	3.640	.1746	.5440 $^{-1}$
3.74	.4426	16.15	4.420	3.654	.1731	.5412 $^{-1}$
3.75	.4423	16.24	4.426	3.669	.1717	.5384 $^{-1}$
3.76	.4420	16.33	4.432	3.684	.1702	.5356 $^{-1}$
3.77	.4417	16.42	4.439	3.698	.1687	.5328 $^{-1}$
3.78	.4414	16.50	4.445	3.713	.1673	.5301 $^{-1}$
3.79	.4410	16.59	4.451	3.728	.1659	.5274 $^{-1}$
3.80	.4407	16.68	4.457	3.743	.1645	.5247 $^{-1}$
3.81	.4404	16.77	4.463	3.758	.1631	.5220 $^{-1}$
3.82	.4401	16.86	4.469	3.772	.1617	.5193 $^{-1}$
3.83	.4398	16.95	4.475	3.787	.1603	.5167 $^{-1}$
3.84	.4395	17.04	4.481	3.802	.1589	.5140 $^{-1}$
3.85	.4392	17.13	4.487	3.817	.1576	.5114 $^{-1}$
3.86	.4389	17.22	4.492	3.832	.1563	.5089 $^{-1}$
3.87	.4386	17.31	4.498	3.847	.1549	.5063 $^{-1}$
3.88	.4383	17.40	4.504	3.863	.1536	.5038 $^{-1}$
3.89	.4380	17.49	4.510	3.878	.1523	.5012 $^{-1}$
3.90	.4377	17.58	4.516	3.893	.1510	.4987 $^{-1}$
3.91	.4375	17.67	4.521	3.908	.1497	.4962 $^{-1}$
3.92	.4372	17.76	4.527	3.923	.1485	.4938 $^{-1}$
3.93	.4369	17.85	4.533	3.939	.1472	.4913 $^{-1}$
3.94	.4366	17.94	4.538	3.954	.1460	.4889 $^{-1}$
3.95	.4363	18.04	4.544	3.969	.1448	.4865 $^{-1}$
3.96	.4360	18.13	4.549	3.985	.1435	.4841 $^{-1}$
3.97	.4358	18.22	4.555	4.000	.1423	.4817 $^{-1}$
3.98	.4355	18.31	4.560	4.016	.1411	.4793 $^{-1}$
3.99	.4352	18.41	4.566	4.031	.1399	.4770 $^{-1}$
4.00	.4350	18.50	4.571	4.047	.1388	.4747 $^{-1}$
4.01	.4347	18.59	4.577	4.062	.1376	.4723 $^{-1}$
4.02	.4344	18.69	4.582	4.078	.1364	.4700 $^{-1}$
4.03	.4342	18.78	4.588	4.094	.1353	.4678 $^{-1}$
4.04	.4339	18.88	4.593	4.110	.1342	.4655 $^{-1}$
4.05	.4336	18.97	4.598	4.125	.1330	.4633 $^{-1}$
4.06	.4334	19.06	4.604	4.141	.1319	.4610 $^{-1}$
4.07	.4331	19.16	4.609	4.157	.1308	.4588 $^{-1}$
4.08	.4329	19.25	4.614	4.173	.1297	.4566 $^{-1}$
4.09	.4326	19.35	4.619	4.189	.1286	.4544 $^{-1}$
4.10	.4324	19.45	4.624	4.205	.1276	.4523 $^{-1}$
4.11	.4321	19.54	4.630	4.221	.1265	.4501 $^{-1}$
4.12	.4319	19.64	4.635	4.237	.1254	.4480 $^{-1}$
4.13	.4316	19.73	4.640	4.253	.1244	.4459 $^{-1}$
4.14	.4314	19.83	4.645	4.269	.1234	.4438 $^{-1}$
4.15	.4311	19.93	4.650	4.285	.1223	.4417 $^{-1}$
4.16	.4309	20.02	4.655	4.301	.1213	.4396 $^{-1}$
4.17	.4306	20.12	4.660	4.318	.1203	.4375 $^{-1}$
4.18	.4304	20.22	4.665	4.334	.1193	.4355 $^{-1}$
4.19	.4302	20.32	4.670	4.350	.1183	.4334 $^{-1}$

M_1	M_2	p_2/p_1	ρ_2/ρ_1	T_2/T_1	p_{t2}/p_{t1}	p_1/p_{t2}
4.20	.4299	20.41	4.675	4.367	.1173	.4314 $^{-1}$
4.21	.4297	20.51	4.680	4.383	.1164	.4294 $^{-1}$
4.22	.4295	20.61	4.685	4.399	.1154	.4274 $^{-1}$
4.23	.4292	20.71	4.690	4.416	.1144	.4255 $^{-1}$
4.24	.4290	20.81	4.694	4.432	.1135	.4235 $^{-1}$
4.25	.4288	20.91	4.699	4.449	.1126	.4215 $^{-1}$
4.26	.4286	21.01	4.704	4.466	.1116	.4196 $^{-1}$
4.27	.4283	21.11	4.709	4.482	.1107	.4177 $^{-1}$
4.28	.4281	21.20	4.713	4.499	.1098	.4158 $^{-1}$
4.29	.4279	21.30	4.718	4.516	.1089	.4139 $^{-1}$
4.30	.4277	21.41	4.723	4.532	.1080	.4120 $^{-1}$
4.31	.4275	21.51	4.728	4.549	.1071	.4101 $^{-1}$
4.32	.4272	21.61	4.732	4.566	.1062	.4082 $^{-1}$
4.33	.4270	21.71	4.737	4.583	.1054	.4064 $^{-1}$
4.34	.4268	21.81	4.741	4.600	.1045	.4046 $^{-1}$
4.35	.4266	21.91	4.746	4.617	.1036	.4027 $^{-1}$
4.36	.4264	22.01	4.751	4.633	.1028	.4009 $^{-1}$
4.37	.4262	22.11	4.755	4.651	.1020	.3991 $^{-1}$
4.38	.4260	22.22	4.760	4.668	.1011	.3973 $^{-1}$
4.39	.4258	22.32	4.764	4.685	.1003	.3956 $^{-1}$
4.40	.4255	22.42	4.768	4.702	.9948 $^{-1}$	.3938 $^{-1}$
4.41	.4253	22.52	4.773	4.719	.9867 $^{-1}$	.3921 $^{-1}$
4.42	.4251	22.63	4.777	4.736	.9787 $^{-1}$	.3903 $^{-1}$
4.43	.4249	22.73	4.782	4.753	.9707 $^{-1}$	.3886 $^{-1}$
4.44	.4247	22.83	4.786	4.771	.9628 $^{-1}$	.3869 $^{-1}$
4.45	.4245	22.94	4.790	4.788	.9550 $^{-1}$	.3852 $^{-1}$
4.46	.4243	23.04	4.795	4.805	.9473 $^{-1}$	.3835 $^{-1}$
4.47	.4241	23.14	4.799	4.823	.9596 $^{-1}$	.3818 $^{-1}$
4.48	.4239	23.25	4.803	4.840	.9320 $^{-1}$	.3801 $^{-1}$
4.49	.4237	23.35	4.808	4.858	.9244 $^{-1}$	.3785 $^{-1}$
4.50	.4236	23.46	4.812	4.875	.9170 $^{-1}$	.3768 $^{-1}$
4.51	.4234	23.56	4.816	4.893	.9096 $^{-1}$	.3752 $^{-1}$
4.52	.4232	23.67	4.820	4.910	.9022 $^{-1}$	.3735 $^{-1}$
4.53	.4230	23.77	4.824	4.928	.8950 $^{-1}$	.3719 $^{-1}$
4.54	.4228	23.88	4.829	4.946	.8878 $^{-1}$	.3703 $^{-1}$
4.55	.4226	23.99	4.833	4.963	.8806 $^{-1}$	.3687 $^{-1}$
4.56	.4224	24.09	4.837	4.981	.8735 $^{-1}$	.3671 $^{-1}$
4.57	.4222	24.20	4.841	4.999	.8665 $^{-1}$	.3656 $^{-1}$
4.58	.4220	24.31	4.845	5.017	.8596 $^{-1}$	.3640 $^{-1}$
4.59	.4219	24.41	4.849	5.034	.8527 $^{-1}$	.3624 $^{-1}$
4.60	.4217	24.52	4.853	5.052	.8459 $^{-1}$	.3609 $^{-1}$
4.61	.4215	24.63	4.857	5.070	.8391 $^{-1}$	.3593 $^{-1}$
4.62	.4213	24.74	4.861	5.088	.8324 $^{-1}$	.3578 $^{-1}$
4.63	.4211	24.84	4.865	5.106	.8257 $^{-1}$	.3563 $^{-1}$
4.64	.4210	24.95	4.869	5.124	.8192 $^{-1}$	.3548 $^{-1}$
4.65	.4208	25.06	4.873	5.143	.8126 $^{-1}$	.3533 $^{-1}$
4.66	.4206	25.17	4.877	5.160	.8062 $^{-1}$	.3518 $^{-1}$
4.67	.4204	25.28	4.881	5.179	.7998 $^{-1}$	.3503 $^{-1}$
4.68	.4203	25.39	4.885	5.197	.7934 $^{-1}$	.3488 $^{-1}$
4.69	.4201	25.50	4.889	5.215	.7871 $^{-1}$	.3474 $^{-1}$
4.70	.4199	25.61	4.893	5.233	.7809 $^{-1}$	.3459 $^{-1}$
4.71	.4197	25.71	4.896	5.252	.7747 $^{-1}$	.3445 $^{-1}$
4.72	.4196	25.82	4.900	5.270	.7685 $^{-1}$	.3431 $^{-1}$
4.73	.4194	25.94	4.904	5.289	.7625 $^{-1}$	.3416 $^{-1}$
4.74	.4192	26.05	4.908	5.307	.7564 $^{-1}$	.3402 $^{-1}$
4.75	.4191	26.16	4.912	5.325	.7505 $^{-1}$	.3388 $^{-1}$
4.76	.4189	26.27	4.915	5.344	.7445 $^{-1}$	.3374 $^{-1}$
4.77	.4187	26.38	4.919	5.363	.7387 $^{-1}$	.3360 $^{-1}$
4.78	.4186	26.49	4.923	5.381	.7329 $^{-1}$	.3346 $^{-1}$
4.79	.4184	26.60	4.926	5.400	.7271 $^{-1}$	.3333 $^{-1}$
4.80	.4183	26.71	4.930	5.418	.7214 $^{-1}$	.3319 $^{-1}$
4.81	.4181	26.83	4.934	5.437	.7157 $^{-1}$	.3305 $^{-1}$
4.82	.4179	26.94	4.937	5.456	.7101 $^{-1}$	.3292 $^{-1}$
4.83	.4178	27.05	4.941	5.475	.7046 $^{-1}$	.3278 $^{-1}$
4.84	.4176	27.16	4.945	5.494	.6991 $^{-1}$	.3265 $^{-1}$

Table B.1 (continued)   Normal shock tables ($\gamma = 1.4$)

M_1	M_2	p_2/p_1	ρ_2/ρ_1	T_2/T_1	p_{t2}/p_{t1}	p_1/p_{t2}
4.85	.4175	27.28	4.948	5.512	$.6936^{-1}$	$.3252^{-1}$
4.86	.4173	27.39	4.952	5.531	$.6882^{-1}$	$.3239^{-1}$
4.87	.4172	27.50	4.955	5.550	$.6828^{-1}$	$.3226^{-1}$
4.88	.4170	27.62	4.959	5.569	$.6775^{-1}$	$.3213^{-1}$
4.89	.4169	27.73	4.962	5.588	$.6722^{-1}$	$.3200^{-1}$
4.90	.4167	27.85	4.966	5.607	$.6670^{-1}$	$.3187^{-1}$
4.91	.4165	27.96	4.969	5.626	$.6618^{-1}$	$.3174^{-1}$
4.92	.4164	28.07	4.973	5.646	$.6567^{-1}$	$.3161^{-1}$
4.93	.4163	28.19	4.976	5.665	$.6516^{-1}$	$.3149^{-1}$
4.94	.4161	28.30	4.980	5.684	$.6465^{-1}$	$.3136^{-1}$
4.95	.4160	28.42	4.983	5.703	$.6415^{-1}$	$.3124^{-1}$
4.96	.4158	28.54	4.987	5.723	$.6366^{-1}$	$.3111^{-1}$
4.97	.4157	28.65	4.990	5.742	$.6317^{-1}$	$.3099^{-1}$
4.98	.4155	28.77	4.993	5.761	$.6268^{-1}$	$.3087^{-1}$
4.99	.4154	28.88	4.997	5.781	$.6220^{-1}$	$.3075^{-1}$
5.00	.4152	29.00	5.000	5.800	$.6172^{-1}$	$.3062^{-1}$
6.00	.4042	41.83	5.268	7.941	$.2965^{-1}$	$.2136^{-1}$
7.00	.3974	57.00	5.444	10.47	$.1535^{-1}$	$.1574^{-1}$
8.00	.3929	74.50	5.565	13.39	$.8488^{-2}$	$.1207^{-1}$
9.00	.3898	94.33	5.651	16.69	$.4964^{-2}$	$.9546^{-2}$
10.00	.3876	116.5	5.714	20.39	$.3045^{-2}$	$.7739^{-2}$

Material in Table B.1 has been adapted from NACA Report 1135, "Equations, Tables, and Charts for Compressible Flow," Ames Research Staff, 1953.

Table B.2 Normal shock tables ($\gamma = 1.3$)

M_1	M_2	p_2/p_1	ρ_2/ρ_1	T_2/T_1	p_{t_2}/p_{t_1}	p_1/p_{t_2}
1.00	1.0000	1.000	1.000	1.000	1.0000	0.5457
1.05	.9530	1.116	1.088	1.026	.9998	.5152
1.10	.9112	1.237	1.178	1.051	.9989	.4859
1.15	.8739	1.364	1.269	1.075	.9966	.4581
1.20	.8403	1.498	1.362	1.100	.9925	.4318
1.25	.8100	1.636	1.456	1.124	.9866	.4070
1.30	.7825	1.780	1.551	1.148	.9786	.3839
1.35	.7575	1.930	1.646	1.172	.9684	.3623
1.40	.7346	2.085	1.742	1.197	.9562	.3422
1.45	.7136	2.246	1.838	1.222	.9421	.3236
1.50	.6942	2.413	1.935	1.247	.9261	.3063
1.55	.6764	2.585	2.031	1.273	.9084	.2901
1.60	.6599	2.763	2.127	1.299	.8891	.2751
1.65	.6446	2.947	2.223	1.326	.8684	.2611
1.70	.6304	3.137	2.318	1.353	.8466	.2481
1.75	.6172	3.332	2.413	1.380	.8238	.2360
1.80	.6048	3.532	2.507	1.409	.8001	.2246
1.85	.5933	3.738	2.601	1.437	.7758	.2140
1.90	.5825	3.950	2.694	1.467	.7510	.2042
1.95	.5724	4.168	2.785	1.497	.7259	.1949
2.00	.5629	4.391	2.875	1.527	.7006	.1862
2.05	.5539	4.621	2.964	1.559	.6752	.1781
2.10	.5455	4.855	3.052	1.590	.6499	.1705
2.15	.5376	5.095	3.139	1.623	.6248	.1633
2.20	.5301	5.341	3.225	1.656	.6000	.1566
2.25	.5230	5.592	3.309	1.690	.5755	.1502
2.30	.5163	5.849	3.392	1.725	.5515	.1442
2.35	.5100	6.112	3.474	1.760	.5280	.1386
2.40	.5040	6.381	3.554	1.796	.5050	.1333
2.45	.4983	6.655	3.633	1.832	.4827	.1282
2.50	.4929	6.935	3.710	1.869	.4610	.1235
2.55	.4878	7.220	3.786	1.907	.4400	.1190
2.60	.4829	7.511	3.860	1.946	.4196	.1147
2.65	.4782	7.808	3.933	1.985	.3999	.1107
2.70	.4738	8.110	4.005	2.025	.3810	.1068
2.75	.4696	8.418	4.075	2.066	.3628	.1032
2.80	.4655	8.732	4.144	2.108	.3452	$.9970^{-1}$
2.85	.4616	9.052	4.211	2.150	.3284	$.9643^{-1}$
2.90	.4579	9.377	4.277	2.193	.3123	$.9328^{-1}$
2.95	.4544	9.708	4.341	2.236	.2969	$.9025^{-1}$

Table B.2 (continued)   Normal shock tables ($\gamma = 1.3$)

M_1	M_2	p_2/p_1	ρ_2/ρ_1	T_2/T_1	p_{t_2}/p_{t_1}	p_1/p_{t_2}
3.00	.4511	10.04	4.404	2.280	.2822	$.8741^{-1}$
3.50	.4241	13.72	4.964	2.763	.1677	$.6498^{-1}$
4.00	.4058	17.96	5.412	3.318	$.9932^{-1}$	$.5010^{-1}$
4.50	.3927	22.76	5.768	3.946	$.5941^{-1}$	$.3971^{-1}$
5.00	.3832	28.13	6.053	4.648	$.3612^{-1}$	$.3236^{-1}$
6.00	.3704	40.57	6.469	6.271	$.1422^{-1}$	$.2257^{-1}$
7.00	.3625	55.26	6.749	8.189	$.6098^{-2}$	$.1663^{-1}$
8.00	.3573	72.22	6.943	10.401	$.2826^{-2}$	$.1276^{-1}$
9.00	.3536	91.43	7.084	12.908	$.1404^{-2}$	$.1009^{-1}$
10.00	.3510	112.91	7.188	15.710	$.7408^{-3}$	$.8174^{-2}$

Table B.3   Normal shock tables ($\gamma = 5/3$)

M_1	M_2	p_2/p_1	ρ_2/ρ_1	T_2/T_1	p_{t_2}/p_{t_1}	p_1/p_{t_2}
1.00	1.0000	1.000	1.000	1.000	1.0000	0.4871
1.05	.9535	1.128	1.075	1.049	.9999	.4573
1.10	.9131	1.263	1.150	1.098	.9990	.4291
1.15	.8776	1.403	1.224	1.147	.9969	.4025
1.20	.8462	1.550	1.297	1.195	.9933	.3778
1.25	.8184	1.703	1.370	1.243	.9882	.3548
1.30	.7934	1.863	1.441	1.292	.9813	.3335
1.35	.7710	2.028	1.512	1.342	.9727	.3138
1.40	.7508	2.200	1.581	1.392	.9626	.2956
1.45	.7324	2.378	1.648	1.443	.9510	.2787
1.50	.7157	2.562	1.714	1.495	.9380	.2631
1.55	.7004	2.753	1.779	1.548	.9238	.2487
1.60	.6864	2.950	1.842	1.602	.9085	.2354
1.65	.6736	3.153	1.903	1.657	.8923	.2230
1.70	.6618	3.363	1.963	1.713	.8752	.2115
1.75	.6508	3.578	2.021	1.771	.8575	.2009
1.80	.6407	3.800	2.077	1.830	.8392	.1910
1.85	.6314	4.028	2.132	1.890	.8205	.1817
1.90	.6226	4.263	2.185	1.951	.8015	.1731
1.95	.6145	4.503	2.236	2.014	.7823	.1651
2.00	.6070	4.750	2.286	2.078	.7630	.1575
2.05	.5999	5.003	2.334	2.144	.7436	.1506
2.10	.5933	5.263	2.381	2.211	.7243	.1140
2.15	.5871	5.528	2.426	2.279	.7051	.1378
2.20	.5813	5.800	2.469	2.349	.6860	.1320
2.25	.5759	6.078	2.512	2.420	.6672	.1266
2.30	.5707	6.363	2.552	2.493	.6486	.1215
2.35	.5659	6.653	2.592	2.567	.6302	.1166
2.40	.5613	6.950	2.630	2.642	.6123	.1121
2.45	.5570	7.253	2.667	2.720	.5947	.1078
2.50	.5530	7.563	2.703	2.798	.5774	.1037
2.55	.5491	7.878	2.737	2.878	.5606	$.9990^{-1}$
2.60	.5455	8.200	2.770	2.960	.5441	$.9627^{-1}$
2.65	.5420	8.528	2.803	3.043	.5280	$.9283^{-1}$
2.70	.5398	8.863	2.834	3.127	.5124	$.8957^{-1}$
2.75	.5357	9.203	2.864	3.213	.4971	$.8648^{-1}$
2.80	.5327	9.550	2.893	3.301	.4823	$.8353^{-1}$
2.85	.5299	9.903	2.921	3.390	.4680	$.8074^{-1}$
2.90	.5272	10.26	2.948	3.491	.4540	$.7808^{-1}$
2.95	.5247	10.63	2.975	3.573	.4404	$.7555^{-1}$
3.00	0.5222	11.00	3.000	3.666	0.4273	0.7314^{-1}
3.50	.5031	15.06	3.213	4.688	.3166	$.5122^{-1}$
4.00	.4904	19.75	3.368	5.863	.2373	$.4175^{-1}$
4.50	.4816	25.06	3.484	7.194	.1806	$.3312^{-1}$
5.00	.4752	31.00	3.571	8.680	.1397	$.2691^{-1}$
6.00	.4667	44.75	3.692	12.120	$.8751^{-1}$	$.1875^{-1}$
7.00	.4616	61.00	3.769	16.184	$.5789^{-1}$	$.1381^{-1}$
8.00	.4583	79.75	3.821	20.872	$.4007^{-1}$	$.1059^{-1}$
9.00	.4560	101.00	3.857	26.185	$.2879^{-1}$	$.8374^{-2}$
10.00	.4543	124.75	3.883	32.123	$.2133^{-1}$	$.6789^{-2}$

Oblique shock charts ($\gamma = 1.4$)

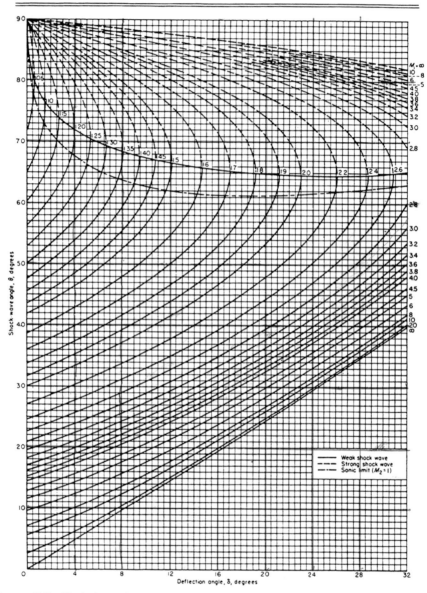

Figure C.1 Variation of shock-wave angle with flow-deflection angle for various upstream Mach numbers

Material in Appendix C is from NACA Report 1135, "Equations, Tables, and Charts for Compressible Flow," Ames Research Staff, 1953.

Oblique shock charts ($\gamma = 1.4$)

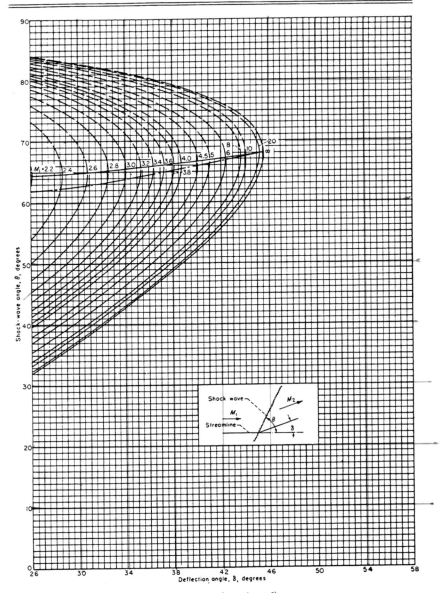

Figure C.1 (continued)

Oblique shock charts ($\gamma = 1.4$)

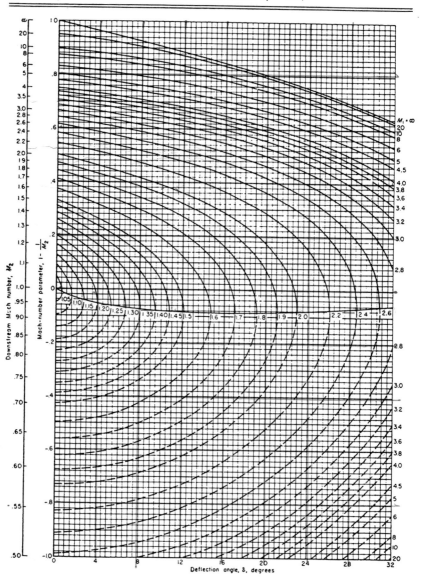

Figure C.2 Variation of Mach number downstream of a shock wave with flow-deflection angle for various upstream Mach numbers

Oblique shock charts ($\gamma = 1.4$)

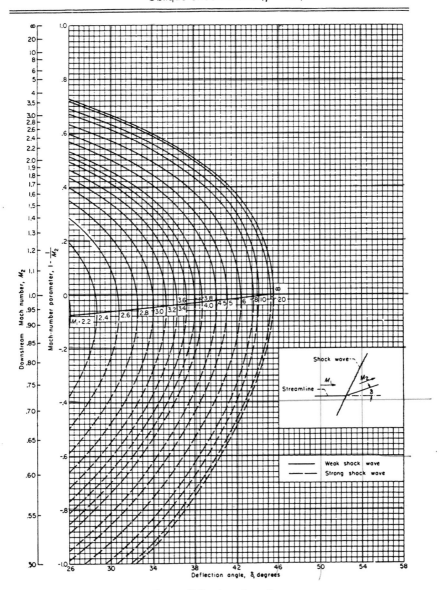

Figure C.2 (continued)

Conical shock charts ($\gamma = 1.405$)

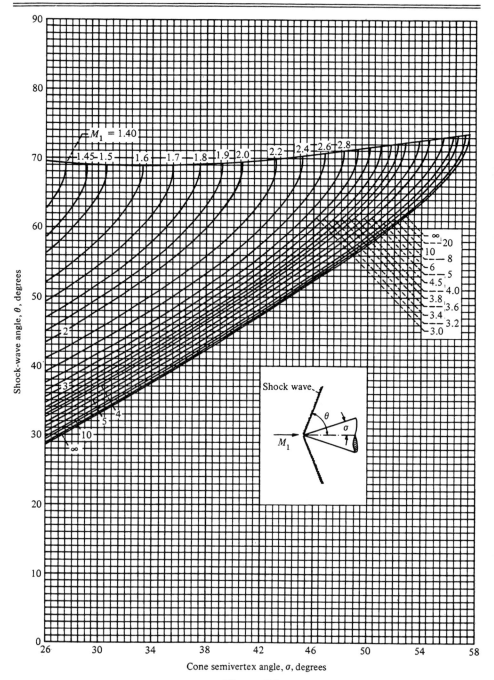

Figure C.3

Conical shock charts ($\gamma = 1.405$)

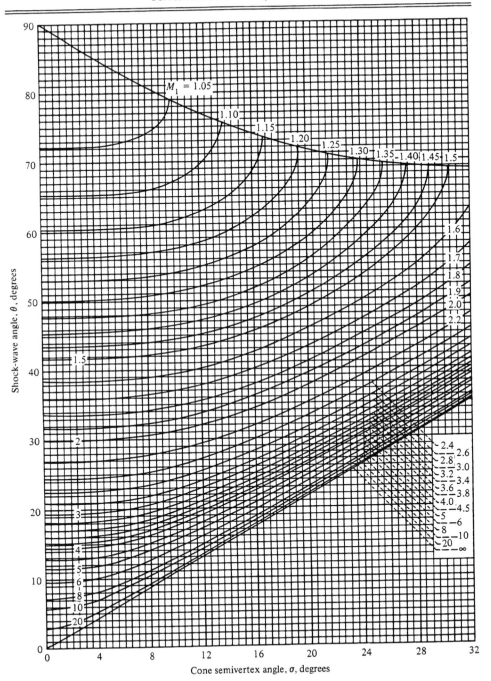

Figure C.3 (continued)

Conical shock charts ($\gamma = 1.405$)

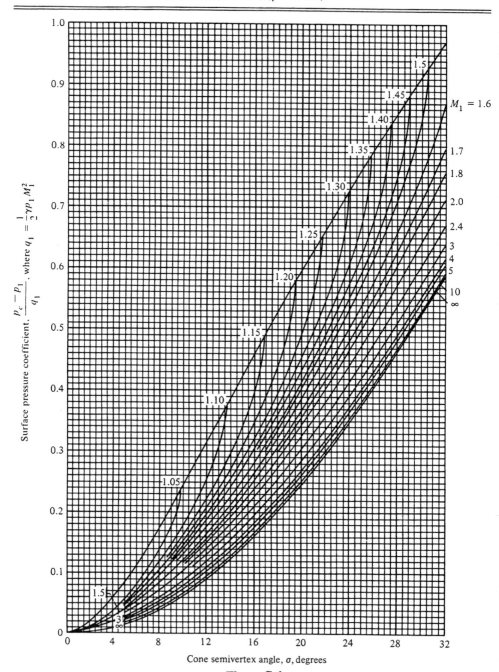

Cone semivertex angle, σ, degrees

Figure C.4

Conical shock charts ($\gamma = 1.405$)

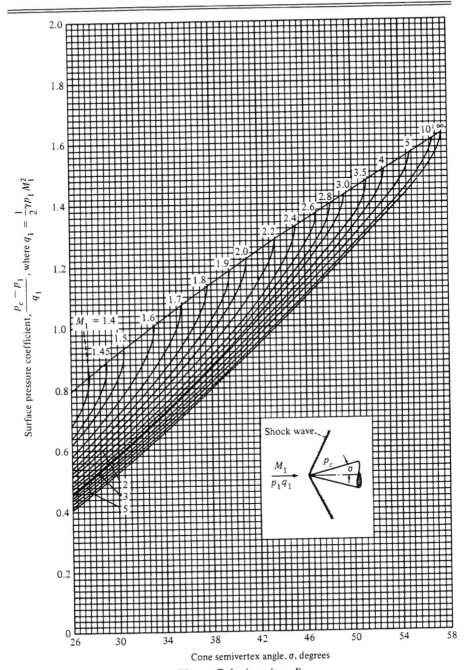

Figure C.4 (continued)

Conical shock charts ($\gamma = 1.405$)

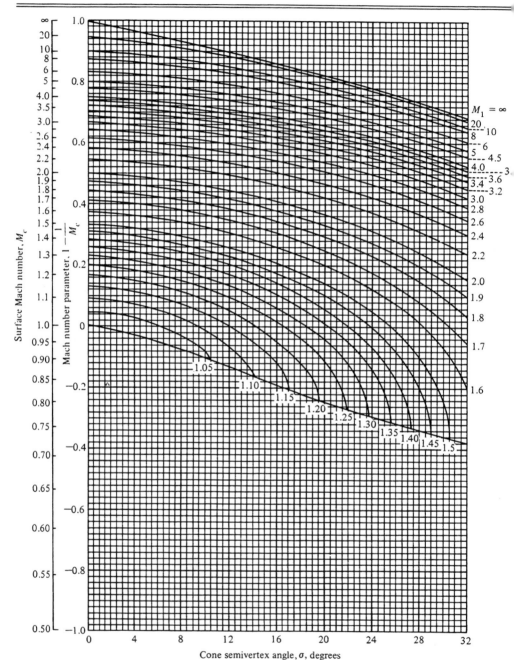

Figure C.5

Conical shock charts ($\gamma = 1.405$)

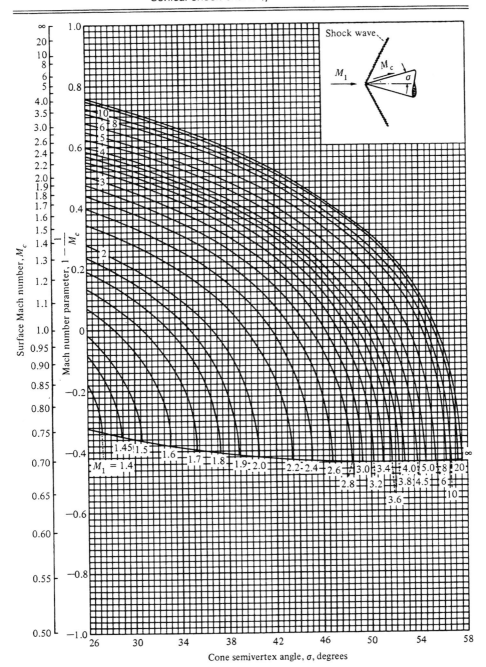

Figure C.5 (continued)

APPENDIX D

Prandtl Meyer functions ($\gamma = 1.4$)

M	ν	μ	M	ν	μ
1.00	0	90.00	1.60	14.861	38.68
1.01	.04473	81.93	1.61	15.156	38.40
1.02	.1257	78.64	1.62	15.452	38.12
1.03	.2294	76.14	1.63	15.747	37.84
1.04	.3510	74.06	1.64	16.043	37.57
1.05	.4874	72.25	1.65	16.338	37.31
1.06	.6367	70.63	1.66	16.633	37.04
1.07	.7973	69.16	1.67	16.928	36.78
1.08	.9680	67.81	1.68	17.222	36.53
1.09	1.148	66.55	1.69	17.516	36.28
1.10	1.336	65.38	1.70	17.810	36.03
1.11	1.532	64.28	1.71	18.103	35.79
1.12	1.735	63.23	1.72	18.397	35.55
1.13	1.944	62.25	1.73	18.689	35.31
1.14	2.160	61.31	1.74	18.981	35.08
1.15	2.381	60.41	1.75	19.273	34.85
1.16	2.607	59.55	1.76	19.565	34.62
1.17	2.839	58.73	1.77	19.855	34.40
1.18	3.074	57.94	1.78	20.146	34.18
1.19	3.314	57.18	1.79	20.436	33.96
1.20	3.558	56.44	1.80	20.725	33.75
1.21	3.806	55.74	1.81	21.014	33.54
1.22	4.057	55.05	1.82	21.302	33.33
1.23	4.312	54.39	1.83	21.590	33.12
1.24	4.569	53.75	1.84	21.877	32.92
1.25	4.830	53.13	1.85	22.163	32.72
1.26	5.093	52.53	1.86	22.449	32.52
1.27	5.359	51.94	1.87	22.735	32.33
1.28	5.627	51.38	1.88	23.019	32.13
1.29	5.898	50.82	1.89	23.303	31.94
1.30	6.170	50.28	1.90	23.586	31.76
1.31	6.445	49.76	1.91	23.869	31.57
1.32	6.721	49.25	1.92	24.151	31.39
1.33	7.000	48.75	1.93	24.432	31.21
1.34	7.280	48.27	1.94	24.712	31.03
1.35	7.561	47.79	1.95	24.992	30.85
1.36	7.844	47.33	1.96	25.271	30.68
1.37	8.128	46.88	1.97	25.549	30.51
1.38	8.413	46.44	1.98	25.827	30.33
1.39	8.699	46.01	1.99	26.104	30.17
1.40	8.987	45.58	2.00	26.380	30.00
1.41	9.276	45.17	2.01	26.655	29.84
1.42	9.565	44.77	2.02	26.929	29.67
1.43	9.855	44.37	2.03	27.203	29.51
1.44	10.146	43.98	2.04	27.476	29.35
1.45	10.438	43.60	2.05	27.748	29.20
1.46	10.731	43.23	2.06	28.020	29.04
1.47	11.023	42.86	2.07	28.290	28.89
1.48	11.317	42.51	2.08	28.560	28.74
1.49	11.611	42.16	2.09	28.829	28.59
1.50	11.905	41.81	2.10	29.097	28.44
1.51	12.200	41.47	2.11	29.364	28.29
1.52	12.495	41.14	2.12	29.631	28.14
1.53	12.790	40.81	2.13	29.897	28.00
1.54	13.086	40.49	2.14	30.161	27.86
1.55	13.381	40.18	2.15	30.425	27.72
1.56	13.677	39.87	2.16	30.689	27.58
1.57	13.973	39.56	2.17	30.951	27.44
1.58	14.269	39.27	2.18	31.212	27.30
1.59	14.564	38.97	2.19	31.473	27.17

Prandtl Meyer functions ($\gamma = 1.4$)

M	ν	μ	M	ν	μ
2.20	31.732	27.04	2.85	46.778	20.54
2.21	31.991	26.90	2.86	46.982	20.47
2.22	32.250	26.77	2.87	47.185	20.39
2.23	32.507	26.64	2.88	47.388	20.32
2.24	32.763	26.51	2.89	47.589	20.24
2.25	33.018	26.39	2.90	47.790	20.17
2.26	33.273	26.26	2.91	47.990	20.10
2.27	33.527	26.14	2.92	48.190	20.03
2.28	33.780	26.01	2.93	48.388	19.96
2.29	34.032	25.89	2.94	48.586	19.89
2.30	34.283	25.77	2.95	48.783	19.81
2.31	34.533	25.65	2.96	48.980	19.75
2.32	34.783	25.53	2.97	49.175	19.68
2.33	35.031	25.42	2.98	49.370	19.61
2.34	35.279	25.30	2.99	49.564	19.54
2.35	35.526	25.18	3.00	49.757	19.47
2.36	35.771	25.07	3.01	49.950	19.40
2.37	36.017	24.96	3.02	50.142	19.34
2.38	36.261	24.85	3.03	50.333	19.27
2.39	36.504	24.73	3.04	50.523	19.20
2.40	36.746	24.62	3.05	50.713	19.14
2.41	36.988	24.52	3.06	50.902	19.07
2.42	37.229	24.41	3.07	51.090	19.01
2.43	37.469	24.30	3.08	51.277	18.95
2.44	37.708	24.19	3.09	51.464	18.88
2.45	37.946	24.09	3.10	51.650	18.82
2.46	38.183	23.99	3.11	51.835	18.76
2.47	38.420	23.88	3.12	52.020	18.69
2.48	38.655	23.78	3.13	52.203	18.63
2.49	38.890	23.68	3.14	52.386	18.57
2.50	39.124	23.58	3.15	52.569	18.51
2.51	39.357	23.48	3.16	52.751	18.45
2.52	39.589	23.38	3.17	52.931	18.39
2.53	39.820	23.28	3.18	53.112	18.33
2.54	40.050	23.18	3.19	53.292	18.27
2.55	40.280	23.09	3.20	53.470	18.21
2.56	40.509	22.99	3.21	53.648	18.15
2.57	40.736	22.91	3.22	53.826	18.09
2.58	40.963	22.81	3.23	54.003	18.03
2.59	41.189	22.71	3.24	54.179	17.98
2.60	41.415	22.62	3.25	54.355	17.92
2.61	41.639	22.53	3.26	54.529	17.86
2.62	41.863	22.44	3.27	54.703	17.81
2.63	42.086	22.35	3.28	54.877	17.75
2.64	42.307	22.26	3.29	55.050	17.70
2.65	42.529	22.17	3.30	55.222	17.64
2.66	42.749	22.08	3.31	55.393	17.58
2.67	42.968	22.00	3.32	55.564	17.53
2.68	43.187	21.91	3.33	55.734	17.48
2.69	43.405	21.82	3.34	55.904	17.42
2.70	43.621	21.74	3.35	56.073	17.37
2.71	43.838	21.65	3.36	56.241	17.31
2.72	44.053	21.57	3.37	56.409	17.26
2.73	44.267	21.49	3.38	56.576	17.21
2.74	44.481	21.41	3.39	56.742	17.16
2.75	44.694	21.32	3.40	56.907	17.10
2.76	44.906	21.24	3.41	57.073	17.05
2.77	45.117	21.16	3.42	57.237	17.00
2.78	45.327	21.08	3.43	57.401	16.95
2.79	45.537	21.00	3.44	57.564	16.90
2.80	45.746	20.92	3.45	57.726	16.85
2.81	45.954	20.85	3.46	57.888	16.80
2.82	46.161	20.77	3.47	58.050	16.75
2.83	46.368	20.69	3.48	58.210	16.70
2.84	46.573	20.62	3.49	58.370	16.65

Prandtl Meyer functions ($\gamma = 1.4$)

M	ν	μ	M	ν	μ
3.50	58.530	16.60	4.15	67.713	13.94
3.51	58.689	16.55	4.16	67.838	13.91
3.52	58.847	16.51	4.17	67.963	13.88
3.53	59.004	16.46	4.18	68.087	13.84
3.54	59.162	16.41	4.19	68.210	13.81
3.55	59.318	16.36	4.20	68.333	13.77
3.56	59.474	16.31	4.21	68.456	13.74
3.57	59.629	16.27	4.22	68.578	13.71
3.58	59.754	16.22	4.23	68.700	13.67
3.59	59.938	16.17	4.24	68.821	13.64
3.60	60.091	16.13	4.25	68.942	13.61
3.61	60.244	16.08	4.26	69.053	13.58
3.62	60.397	16.04	4.27	69.183	13.54
3.63	60.549	15.99	4.28	69.302	13.51
3.64	60.700	15.95	4.29	69.422	13.48
3.65	60.851	15.90	4.30	69.541	13.45
3.66	61.000	15.86	4.31	69.659	13.42
3.67	61.150	15.81	4.32	69.777	13.38
3.68	61.299	15.77	4.33	69.895	13.35
3.69	61.447	15.72	4.34	70.012	13.32
3.70	61.595	15.68	4.35	70.128	13.29
3.71	61.743	15.64	4.36	70.245	13.26
3.72	61.889	15.59	4.37	70.361	13.23
3.73	62.036	15.55	4.38	70.476	13.20
3.74	62.181	15.51	4.39	70.591	13.17
3.75	62.326	15.47	4.40	70.706	13.14
3.76	62.471	15.42	4.41	70.820	13.11
3.77	62.615	15.38	4.42	70.934	13.08
3.78	62.758	15.34	4.43	71.048	13.05
3.79	62.901	15.30	4.44	71.161	13.02
3.80	63.044	15.26	4.45	71.274	12.99
3.81	63.186	15.22	4.46	71.386	12.96
3.82	63.327	15.18	4.47	71.498	12.93
3.83	63.468	15.14	4.48	71.610	12.90
3.84	63.608	15.10	4.49	71.721	12.87
3.85	63.748	15.06	4.50	71.832	12.84
3.86	63.887	15.02	4.51	71.942	12.81
3.87	64.026	14.98	4.52	72.052	12.78
3.88	64.164	14.94	4.53	72.162	12.75
3.89	64.302	14.90	4.54	72.271	12.73
3.90	64.440	14.86	4.55	72.380	12.70
3.91	64.576	14.82	4.56	72.489	12.67
3.92	64.713	14.78	4.57	72.597	12.64
3.93	64.848	14.74	4.58	72.705	12.61
3.94	64.983	14.70	4.59	72.812	12.58
3.95	65.118	14.67	4.60	72.919	12.56
3.96	65.253	14.63	4.61	73.026	12.53
3.97	65.386	14.59	4.62	73.132	12.50
3.98	65.520	14.55	4.63	73.238	12.47
3.99	65.652	14.52	4.64	73.344	12.45
4.00	65.785	14.48	4.65	73.449	12.42
4.01	65.917	14.44	4.66	73.554	12.39
4.02	66.048	14.40	4.67	73.659	12.37
4.03	66.179	14.37	4.68	73.763	12.34
4.04	66.309	14.33	4.69	73.867	12.31
4.05	66.439	14.30	4.70	73.970	12.28
4.06	66.569	14.26	4.71	74.073	12.26
4.07	66.698	14.22	4.72	74.176	12.23
4.08	66.826	14.19	4.73	74.279	12.21
4.09	66.954	14.15	4.74	74.381	12.18
4.10	67.082	14.12	4.75	74.483	12.15
4.11	67.209	14.08	4.76	74.584	12.13
4.12	67.336	14.05	4.77	74.685	12.10
4.13	67.462	14.01	4.78	74.786	12.08
4.14	67.588	13.98	4.79	74.886	12.05

APPENDIX D (continued)

Prandtl Meyer functions $(\gamma = 1.4)$

M	ν	μ	M	ν	μ
4.80	74.986	12.03	4.90	75.969	11.78
4.81	75.086	12.00	4.91	76.066	11.75
4.82	75.186	11.97	4.92	76.162	11.73
4.83	75.285	11.95	4.93	76.258	11.70
4.84	75.383	11.92	4.94	76.353	11.68
4.85	75.482	11.90	4.95	76.449	11.66
4.86	75.580	11.87	4.96	76.544	11.63
4.87	75.678	11.85	4.97	76.638	11.61
4.88	75.775	11.83	4.98	76.732	11.58
4.89	75.872	11.80	4.99	76.826	11.56
			5.00	76.920	11.54

Material in Appendix D is from NACA Report 1135, "Equations, Tables, and Charts for Compressible Flow," Ames Research Staff, 1953.

APPENDIX E

Fanno line flow $(\gamma = 1.4)$

M	T/T^*	p/p^*	p_t/p_t^*	V/V^*	fl_{max}/D
0	1.2000	∞	∞	0	∞
0.01	1.2000	109.544	57.874	.01095	7134.40
.02	1.1999	54.770	28.942	.02191	1778.45
.03	1.1998	36.511	19.300	.03286	787.08
.04	1.1996	27.382	14.482	.04381	440.35
.05	1.1994	21.903	11.5914	.05476	280.02
.06	1.1991	18.251	9.6659	.06570	193.03
.07	1.1988	15.642	8.2915	.07664	140.66
.08	1.1985	13.684	7.2616	.08758	106.72
.09	1.1981	12.162	6.4614	.09851	83.496
.10	1.1976	10.9435	5.8218	.10943	66.922
.11	1.1971	9.9465	5.2992	.12035	54.688
.12	1.1966	9.1156	4.8643	.13126	45.408
.13	1.1960	8.4123	4.4968	.14216	38.207
.14	1.1953	7.8093	4.1824	.15306	32.511
.15	1.1946	7.2866	3.9103	.16395	27.932
.16	1.1939	6.8291	3.6727	.17482	24.198
.17	1.1931	6.4252	3.4635	.18568	21.115
.18	1.1923	6.0662	3.2779	.19654	18.543
.19	1.1914	5.7448	3.1123	.20739	16.375
.20	1.1905	5.4555	2.9635	.21822	14.533
.21	1.1895	5.1936	2.8293	.22904	12.956
.22	1.1885	4.9554	2.7076	.23984	11.596
.23	1.1874	4.7378	2.5968	.25063	10.416
.24	1.1863	4.5383	2.4956	.26141	9.3865
.25	1.1852	4.3546	2.4027	.27217	8.4834
.26	1.1840	4.1850	2.3173	.28291	7.6876
.27	1.1828	4.0280	2.2385	.29364	6.9832
.28	1.1815	3.8820	2.1656	.30435	6.3572
.29	1.1802	3.7460	2.0979	.31504	5.7989

401

Fanno line flow ($\gamma = 1.4$)

M	T/T^*	p/p^*	p_t/p_t^*	V/V^*	fl_{max}/D
.30	1.1788	3.6190	2.0351	.32572	5.2992
.31	1.1774	3.5002	1.9765	.33637	4.8507
.32	1.1759	3.3888	1.9219	.34700	4.4468
.33	1.1744	3.2840	1.8708	.35762	4.0821
.34	1.1729	3.1853	1.8229	.36822	3.7520
.35	1.1713	3.0922	1.7780	.37880	3.4525
.36	1.1697	3.0042	1.7358	.38935	3.1801
.37	1.1680	2.9209	1.6961	.39988	2.9320
.38	1.1663	2.8420	1.6587	.41039	2.7055
.39	1.1646	2.7671	1.6234	.42087	2.4983
.40	1.1628	2.6958	1.5901	.43133	2.3085
.41	1.1610	2.6280	1.5587	.44177	2.1344
.42	1.1591	2.5634	1.5289	.45218	1.9744
.43	1.1572	2.5017	1.5007	.46257	1.8272
.44	1.1553	2.4428	1.4739	.47293	1.6915
.45	1.1533	2.3865	1.4486	.48326	1.5664
.46	1.1513	2.3326	1.4246	.49357	1.4509
.47	1.1492	2.2809	1.4018	.50385	1.3442
.48	1.1471	2.2314	1.3801	.51410	1.2453
.49	1.1450	2.1838	1.3595	.52433	1.1539
.50	1.1429	2.1381	1.3399	.53453	1.06908
.51	1.1407	2.0942	1.3212	.54469	.99042
.52	1.1384	2.0519	1.3034	.55482	.91741
.53	1.1362	2.0112	1.2864	.56493	.84963
.54	1.1339	1.9719	1.2702	.57501	.78662
.55	1.1315	1.9341	1.2549	.58506	.72805
.56	1.1292	1.8976	1.2403	.59507	.67357
.57	1.1268	1.8623	1.2263	.60505	.62286
.58	1.1244	1.8282	1.2130	.61500	.57568
.59	1.1219	1.7952	1.2003	.62492	.53174
.60	1.1194	1.7634	1.1882	.63481	.49081
.61	1.1169	1.7325	1.1766	.64467	.45270
.62	1.1144	1.7026	1.1656	.65449	.41720
.63	1.1118	1.6737	1.1551	.66427	.38411
.64	1.1091	1.6456	1.1451	.67402	.35330
.65	1.10650	1.6183	1.1356	.68374	.32460
.66	1.10383	1.5919	1.1265	.69342	.29785
.67	1.10114	1.5662	1.1179	.70306	.27295
.68	1.09842	1.5413	1.1097	.71267	.24978
.69	1.09567	1.5170	1.1018	.72225	.22821
.70	1.09290	1.4934	1.09436	.73179	.20814
.71	1.09010	1.4705	1.08729	.74129	.18949
.72	1.08727	1.4482	1.08057	.75076	.17215
.73	1.08442	1.4265	1.07419	.76019	.15606
.74	1.08155	1.4054	1.06815	.76958	.14113

Fanno line flow ($\gamma = 1.4$)

M	T/T^*	p/p^*	p_t/p_t^*	V/V^*	fl_{max}/D
.75	1.07865	1.3848	1.06242	.77893	.12728
.76	1.07573	1.3647	1.05700	.78825	.11446
.77	1.07279	1.3451	1.05188	.79753	.10262
.78	1.06982	1.3260	1.04705	.80677	.09167
.79	1.06684	1.3074	1.04250	.81598	.08159
.80	1.06383	1.2892	1.03823	.82514	.07229
.81	1.06080	1.2715	1.03422	.83426	.06375
.82	1.05775	1.2542	1.03047	.84334	.05593
.83	1.05468	1.2373	1.02696	.85239	.04878
.84	1.05160	1.2208	1.02370	.86140	.04226
.85	1.04849	1.2047	1.02067	.87037	.03632
.86	1.04537	1.1889	1.01787	.87929	.03097
.87	1.04223	1.1735	1.01529	.88818	.02613
.88	1.03907	1.1584	1.01294	.89703	.02180
.89	1.03589	1.1436	1.01080	.90583	.01793
.90	1.03270	1.12913	1.00887	.91459	.014513
.91	1.02950	1.11500	1.00714	.92332	.011519
.92	1.02627	1.10114	1.00560	.93201	.008916
.93	1.02304	1.08758	1.00426	.94065	.006694
.94	1.01978	1.07430	1.00311	.94925	.004815
.95	1.01652	1.06129	1.00215	.95782	.003280
.96	1.01324	1.04854	1.00137	.96634	.002056
.97	1.00995	1.03605	1.00076	.97481	.001135
.98	1.00664	1.02379	1.00033	.98324	.000493
.99	1.00333	1.01178	1.00008	.99164	.000120
1.00	1.00000	1.00000	1.00000	1.00000	0
1.01	.99666	.98844	1.00008	1.00831	.000114
1.02	.99331	.97711	1.00033	1.01658	.000458
1.03	.98995	.96598	1.00073	1.02481	.001013
1.04	.98658	.95506	1.00130	1.03300	.001771
1.05	.98320	.94435	1.00203	1.04115	.002712
1.06	.97982	.93383	1.00291	1.04925	.003837
1.07	.97642	.92350	1.00394	1.05731	.005129
1.08	.97302	.91335	1.00512	1.06533	.006582
1.09	.96960	.90338	1.00645	1.07331	.008185
1.10	.96618	.89359	1.00793	1.08124	.009933
1.11	.96276	.88397	1.00955	1.08913	.011813
1.12	.95933	.87451	1.01131	1.09698	.013824
1.13	.95589	.86522	1.01322	1.10479	.015949
1.14	.95244	.85608	1.01527	1.11256	.018187
1.15	.94899	.84710	1.01746	1.1203	.02053
1.16	.94554	.83827	1.01978	1.1280	.02298
1.17	.94208	.82958	1.02224	1.1356	.02552
1.18	.93862	.82104	1.02484	1.1432	.02814
1.19	.93515	.81263	1.02757	1.1508	.03085

Fanno line flow ($\gamma = 1.4$)

M	T/T^*	p/p^*	p_t/p_t^*	V/V^*	fl_{max}/D
1.20	.93168	.80436	1.03044	1.1583	.03364
1.21	.92820	.79623	1.03344	1.1658	.03650
1.22	.92473	.78822	1.03657	1.1732	.03942
1.23	.92125	.78034	1.03983	1.1806	.04241
1.24	.91777	.77258	1.04323	1.1879	.04547
1.25	.91429	.76495	1.04676	1.1952	.04858
1.26	.91080	.75743	1.05041	1.2025	.05174
1.27	.90732	.75003	1.05419	1.2097	.05494
1.28	.90383	.74274	1.05809	1.2169	.05820
1.29	.90035	.73556	1.06213	1.2240	.06150
1.30	.89686	.72848	1.06630	1.2311	.06483
1.31	.89338	.72152	1.07060	1.2382	.06820
1.32	.88989	.71465	1.07502	1.2452	.07161
1.33	.88641	.70789	1.07957	1.2522	.07504
1.34	.88292	.70123	1.08424	1.2591	.07850
1.35	.87944	.69466	1.08904	1.2660	.08199
1.36	.87596	.68818	1.09397	1.2729	.08550
1.37	.87249	.68180	1.09902	1.2797	.08904
1.38	.86901	.67551	1.10419	1.2864	.09259
1.39	.86554	.66931	1.10948	1.2932	.09616
1.40	.86207	.66320	1.1149	1.2999	.09974
1.41	.85860	.65717	1.1205	1.3065	.10333
1.42	.85514	.65122	1.1262	1.3131	.10694
1.43	.85168	.64536	1.1320	1.3197	.11056
1.44	.84822	.63958	1.1379	1.3262	.11419
1.45	.84477	63387	1.1440	1.3327	.11782
1.46	.84133	.62824	1.1502	1.3392	.12146
1.47	.83788	.62269	1.1565	1.3456	.12510
1.48	.83445	.61722	1.1629	1.3520	.12875
1.49	.83101	.61181	1.1695	1.3583	.13240
1.50	.82759	.60648	1.1762	1.3646	.13605
1.51	.82416	.60122	1.1830	1.3708	.13970
1.52	.82075	.59602	1.1899	1.3770	.14335
1.53	.81734	.59089	1.1970	1.3832	.14699
1.54	.81394	.58583	1.2043	1.3894	.15063
1.55	.81054	.58084	1.2116	1.3955	.15427
1.56	.80715	.57591	1.2190	1.4015	.15790
1.57	.80376	.57104	1.2266	1.4075	.16152
1.58	.80038	.56623	1.2343	1.4135	.16514
1.59	.79701	.56148	1.2422	1.4195	.16876
1.60	.79365	.55679	1.2502	1.4254	.17236
1.61	.79030	.55216	1.2583	1.4313	.17595
1.62	.78695	.54759	1.2666	1.4371	.17953
1.63	.78361	.54308	1.2750	1.4429	.18311
1.64	.78028	.53862	1.2835	1.4487	.18667

Fanno line flow ($\gamma = 1.4$)

M	T/T^*	p/p_t^*	p_t/p_t^*	V/V^*	fl_{max}/D
1.65	.77695	.53421	1.2922	1.4544	.19022
1.66	.77363	.52986	1.3010	1.4601	.19376
1.67	.77033	.52556	1.3099	1.4657	.19729
1.68	.76703	.52131	1.3190	1.4713	.20081
1.69	.76374	.51711	1.3282	1.4769	.20431
1.70	.76046	.51297	1.3376	1.4825	.20780
1.71	.75718	.50887	1.3471	1.4880	.21128
1.72	.75392	.50482	1.3567	1.4935	.21474
1.73	.75067	.50082	1.3665	1.4989	.21819
1.74	.74742	.49686	1.3764	1.5043	.22162
1.75	.74419	.49295	1.3865	1.5097	.22504
1.76	.74096	.48909	1.3967	1.5150	.22844
1.77	.73774	.48527	1.4070	1.5203	.23183
1.78	.73453	.48149	1.4175	1.5256	.23520
1.79	.73134	.47776	1.4282	1.5308	.23855
1.80	.72816	.47407	1.4390	1.5360	.24189
1.81	.72498	.47042	1.4499	1.5412	.24521
1.82	.72181	.46681	1.4610	1.5463	.24851
1.83	.71865	.46324	1.4723	1.5514	.25180
1.84	.71551	.45972	1.4837	1.5564	.25507
1.85	.71238	.45623	1.4952	1.5614	.25832
1.86	.70925	.45278	1.5069	1.5664	.26156
1.87	.70614	.44937	1.5188	1.5714	.26478
1.88	.70304	.44600	1.5308	1.5763	.26798
1.89	.69995	.44266	1.5429	1.5812	.27116
1.90	.69686	.43936	1.5552	1.5861	.27433
1.91	.69379	.43610	1.5677	1.5909	.27748
1.92	.69074	.43287	1.5804	1.5957	.28061
1.93	.68769	.42967	1.5932	1.6005	.28372
1.94	.68465	.42651	1.6062	1.6052	.28681
1.95	.68162	.42339	1.6193	1.6099	.28989
1.96	.67861	.42030	1.6326	1.6146	.29295
1.97	.67561	.41724	1.6461	1.6193	.29599
1.98	.67262	.41421	1.6597	1.6239	.29901
1.99	.66964	.41121	1.6735	1.6284	.30201
2.00	.66667	.40825	1.6875	1.6330	.30499
2.01	.66371	.40532	1.7017	1.6375	.30796
2.02	.66076	.40241	1.7160	1.6420	.31091
2.03	.65783	.39954	1.7305	1.6465	.31384
2.04	.65491	.39670	1.7452	1.6509	.31675
2.05	.65200	.39389	1.7600	1.6553	.31965
2.06	.64910	.39110	1.7750	1.6597	.32253
2.07	.64621	.38834	1.7902	1.6640	.32538
2.08	.64333	.38562	1.8056	1.6683	.32822
2.09	.64047	.38292	1.8212	1.6726	.33104

APPENDIX E (continued)

Fanno line flow ($\gamma = 1.4$)

M	T/T^*	p/p^*	p_t/p_t^*	V/V^*	fl_{max}/D
2.10	.63762	.38024	1.8369	1.6769	.33385
2.11	.63478	.37760	1.8528	1.6811	.33664
2.12	.63195	.37498	1.8690	1.6853	.33940
2.13	.62914	.37239	1.8853	1.6895	.34215
2.14	.62633	.36982	1.9018	1.6936	.34488
2.15	.62354	.36728	1.9185	1.6977	.34760
2.16	.62076	.36476	1.9354	1.7018	.35030
2.17	.61799	.36227	1.9525	1.7059	.35298
2.18	.61523	.35980	1.9698	1.7099	.35564
2.19	.61249	.35736	1.9873	1.7139	.35828
2.20	.60976	.35494	2.0050	1.7179	.36091
2.21	.60704	.35254	2.0228	1.7219	.36352
2.22	.60433	.35017	2.0409	1.7258	.36611
2.23	.60163	.34782	2.0592	1.7297	.36868
2.24	.59895	.34550	2.0777	1.7336	.37124
2.25	.59627	.34319	2.0964	1.7374	.37378
2.26	.59361	.34091	2.1154	1.7412	.37630
2.27	.59096	.33865	2.1345	1.7450	.37881
2.28	.58833	.33641	2.1538	1.7488	.38130
2.29	.58570	.33420	2.1733	1.7526	.38377
2.30	.58309	.33200	2.1931	1.7563	.38623
2.31	.58049	.32983	2.2131	1.7600	.38867
2.32	.57790	.32767	2.2333	1.7637	.39109
2.33	.57532	.32554	2.2537	1.7673	.39350
2.34	.57276	.32342	2.2744	1.7709	.39589
2.35	.57021	.32133	2.2953	1.7745	.39826
2.36	.56767	.31925	2.3164	1.7781	.40062
2.37	.56514	.31720	2.3377	1.7817	.40296
2.38	.56262	.31516	2.3593	1.7852	.40528
2.39	.56011	.31314	2.3811	1.7887	.40760
2.40	.55762	.31114	2.4031	1.7922	.40989
2.41	.55514	.30916	2.4254	1.7956	.41216
2.42	.55267	.30720	2.4479	1.7991	.41442
2.43	.55021	.30525	2.4706	1.8025	.41667
2.44	.54776	.30332	2.4936	1.8059	.41891
2.45	.54533	.30141	2.5168	1.8092	.42113
2.46	.54291	.29952	2.5403	1.8126	.42333
2.47	.54050	.29765	2.5640	1.8159	.42551
2.48	.53810	.29579	2.5880	1.8192	.42768
2.49	.53571	.29395	2.6122	1.8225	.42983
2.50	.53333	.29212	2.6367	1.8257	.43197
2.51	.53097	.29031	2.6615	1.8290	.43410
2.52	.52862	.28852	2.6865	1.8322	.43621
2.53	.52627	.28674	2.7117	1.8354	.43831
2.54	.52394	.28498	2.7372	1.8386	.44040

Fanno line flow ($\gamma = 1.4$)

M	T/T^*	p/p^*	p_t/p_t^*	V/V^*	fl_{max}/D
2.55	.52163	.28323	2.7630	1.8417	.44247
2.56	.51932	.28150	2.7891	1.8448	.44452
2.57	.51702	.27978	2.8154	1.8479	.44655
2.58	.51474	.27808	2.8420	1.8510	.44857
2.59	.51247	.27640	2.8689	1.8541	.45059
2.60	.51020	.27473	2.8960	1.8571	.45259
2.61	.50795	.27307	2.9234	1.8602	.45457
2.62	.50571	.27143	2.9511	1.8632	.45654
2.63	.50349	.26980	2.9791	1.8662	.45850
2.64	.50127	.26818	3.0074	1.8691	.46044
2.65	.49906	.26658	3.0359	1.8721	.46237
2.66	.49687	.26499	3.0647	1.8750	.46429
2.67	.49469	.26342	3.0938	1.8779	.46619
2.68	.49251	.26186	3.1234	1.8808	.46807
2.69	.49035	.26032	3.1530	1.8837	.46996
2.70	.48820	.25878	3.1830	1.8865	.47182
2.71	.48606	.25726	3.2133	1.8894	.47367
2.72	.48393	.25575	3.2440	1.8922	.47551
2.73	.48182	.25426	3.2749	1.8950	.47734
2.74	.47971	.25278	3.3061	1.8978	.47915
2.75	.47761	.25131	3.3376	1.9005	.48095
2.76	.47553	.24985	3.3695	1.9032	.48274
2.77	.47346	.24840	3.4017	1.9060	.48452
2.78	.47139	.24697	3.4342	1.9087	.48628
2.79	.46933	.24555	3.4670	1.9114	.48803
2.80	.46729	.24414	3.5001	1.9140	.48976
2.81	.46526	.24274	3.5336	1.9167	.49148
2.82	.46324	.24135	3.5674	1.9193	.49321
2.83	.46122	.23997	3.6015	1.9220	.49491
2.84	.45922	.23861	3.6359	1.9246	.49660
2.85	.45723	.23726	3.6707	1.9271	.49828
2.86	.45525	.23592	3.7058	1.9297	.49995
2.87	.45328	.23458	3.7413	1.9322	.50161
2.88	.45132	.23326	3.7771	1.9348	.50326
2.89	.44937	.23196	3.8133	1.9373	.50489
2.90	.44743	.23066	3.8498	1.9398	.50651
2.91	.44550	.22937	3.8866	1.9423	.50812
2.92	.44358	.22809	3.9238	1.9448	.50973
2.93	.44167	.22682	3.9614	1.9472	.51133
2.94	.43977	.22556	3.9993	1.9497	.51291
2.95	.43788	.22431	4.0376	1.9521	.51447
2.96	.43600	.22307	4.0763	1.9545	.51603
2.97	.43413	.22185	4.1153	1.9569	.51758
2.98	.43226	.22063	4.1547	1.9592	.51912
2.99	.43041	.21942	4.1944	1.9616	.52064

APPENDIX E (continued)

Fanno line flow ($\gamma = 1.4$)

M	T/T^*	p/p^*	p_t/p_t^*	V/V^*	fl_{max}/D
3.0	.42857	.21822	4.2346	1.9640	.52216
3.5	.34783	.16850	6.7896	2.0642	.58643
4.0	.28571	.13363	10.719	2.1381	.63306
4.5	.23762	.10833	16.562	2.1936	.66764
5.0	.20000	.08944	25.000	2.2361	.69381
6.0	.14634	.06376	53.180	2.2953	.72987
7.0	.11111	.04762	104.14	2.3333	.75281
8.0	.08696	.03686	190.11	2.3591	.76820
9.0	.06977	.02935	327.19	2.3772	.77898
10.0	.05714	.02390	535.94	2.3905	.78683
∞	0	0	∞	2.4495	.82153

This table has been adapted from J. H. Keenan and J. Kaye, *Gas Tables*, John Wiley and Sons, Inc., New York, 1948, and is reprinted by permission of the publisher.

APPENDIX F

Rayleigh line flow ($\gamma = 1.4$)

M	T_t/T_t^*	T/T^*	p/p^*	p_t/p_t^*	V/V^*
0	0	0	2.4000	1.2679	0
0.01	.000480	.000576	2.3997	1.2678	.000240
.02	.00192	.00230	2.3987	1.2675	.000959
.03	.00431	.00516	2.3970	1.2671	.00216
.04	.00765	.00917	2.3946	1.2665	.00383
.05	.01192	.01430	2.3916	1.2657	.00598
.06	.01712	.02053	2.3880	1.2647	.00860
.07	.02322	.02784	2.3837	1.2636	.01168
.08	.03021	.03621	2.3787	1.2623	.01522
.09	.03807	.04562	2.3731	1.2608	.01922
.10	.04678	.05602	2.3669	1.2591	.02367
.11	.05630	.06739	2.3600	1.2573	.02856
.12	.06661	.07970	2.3526	1.2554	.03388
.13	.07768	.09290	2.3445	1.2533	.03962
.14	.08947	.10695	2.3359	1.2510	.04578
.15	.10196	.12181	2.3267	1.2486	.05235
.16	.11511	.13743	2.3170	1.2461	.05931
.17	.12888	.15377	2.3067	1.2434	.06666
.18	.14324	.17078	2.2959	1.2406	.07438
.19	.15814	.18841	2.2845	1.2377	.08247
.20	.17355	.20661	2.2727	1.2346	.09091
.21	.18943	.22533	2.2604	1.2314	.09969
.22	.20574	.24452	2.2477	1.2281	.10879
.23	.22244	.26413	2.2345	1.2248	.11820
.24	.23948	.28411	2.2209	1.2213	.12792

Rayleigh line flow ($\gamma = 1.4$)

M	T_t/T_t^*	T/T^*	p/p^*	p_t/p_t^*	V/V^*
.25	.25684	.30440	2.2069	1.2177	.13793
.26	.27446	.32496	2.1925	1.2140	.14821
.27	.29231	.34573	2.1777	1.2102	.15876
.28	.31035	.36667	2.1626	1.2064	.16955
.29	.32855	.38773	2.1472	1.2025	.18058
.30	.34686	.40887	2.1314	1.1985	.19183
.31	.36525	.43004	2.1154	1.1945	.20329
.32	.38369	.45119	2.0991	1.1904	.21494
.33	.40214	.47228	2.0825	1.1863	.22678
.34	.42057	.49327	2.0657	1.1821	.23879
.35	.43894	.51413	2.0487	1.1779	.25096
.36	.45723	.53482	2.0314	1.1737	.26327
.37	.47541	.55530	2.0140	1.1695	.27572
.38	.49346	.57553	1.9964	1.1652	.28828
.39	.51134	.59549	1.9787	1.1609	.30095
.40	.52903	.61515	1.9608	1.1566	.31372
.41	.54651	.63448	1.9428	1.1523	.32658
.42	.56376	.65345	1.9247	1.1480	.33951
.43	.58075	.67205	1.9065	1.1437	.35251
.44	.59748	.69025	1.8882	1.1394	.36556
.45	.61393	.70803	1.8699	1.1351	.37865
.46	.63007	.72538	1.8515	1.1308	.39178
.47	.64589	.74228	1.8331	1.1266	.40493
.48	.66139	.75871	1.8147	1.1224	.41810
.49	.67655	.77466	1.7962	1.1182	.43127
.50	.69136	.79012	1.7778	1.1140	.44445
.51	.70581	.80509	1.7594	1.1099	.45761
.52	.71990	.81955	1.7410	1.1059	.47075
.53	.73361	.83351	1.7226	1.1019	.48387
.54	.74695	.84695	1.7043	1.0979	.49696
.55	.75991	.85987	1.6860	1.09397	.51001
.56	.77248	.87227	1.6678	1.09010	.52302
.57	.78467	.88415	1.6496	1.08630	.53597
.58	.79647	.89552	1.6316	1.08255	.54887
.59	.80789	.90637	1.6136	1.07887	.56170
.60	.81892	.91670	1.5957	1.07525	.57447
.61	.82956	.92653	1.5780	1.07170	.58716
.62	.83982	.93585	1.5603	1.06821	.59978
.63	.84970	.94466	1.5427	1.06480	.61232
.64	.85920	.95298	1.5253	1.06146	.62477
.65	.86833	.96081	1.5080	1.05820	.63713
.66	.87709	.96816	1.4908	1.05502	.64941
.67	.88548	.97503	1.4738	1.05192	.66159
.68	.89350	.98144	1.4569	1.04890	.67367
.69	.90117	.98739	1.4401	1.04596	.68564

Rayleigh line flow ($\gamma = 1.4$)

M	T_t/T_t^*	T/T^*	p/p^*	p_t/p_t^*	V/V^*
.70	.90850	.99289	1.4235	1.04310	.69751
.71	.91548	.99796	1.4070	1.04033	.70927
.72	.92212	1.00260	1.3907	1.03764	.72093
.73	.92843	1.00682	1.3745	1.03504	.73248
.74	.93442	1.01062	1.3585	1.03253	.74392
.75	.94009	1.01403	1.3427	1.03010	.75525
.76	.94546	1.01706	1.3270	1.02776	.76646
.77	.95052	1.01971	1.3115	1.02552	.77755
.78	.95528	1.02198	1.2961	1.02337	.78852
.79	.95975	1.02390	1.2809	1.02131	.79938
.80	.96394	1.02548	1.2658	1.01934	.81012
.81	.96786	1.02672	1.2509	1.01746	.82075
.82	.97152	1.02763	1.2362	1.01569	.83126
.83	.97492	1.02823	1.2217	1.01399	.84164
.84	.97807	1.02853	1.2073	1.01240	.85190
.85	.98097	1.02854	1.1931	1.01091	.86204
.86	.98363	1.02826	1.1791	1.00951	.87206
.87	.98607	1.02771	1.1652	1.00819	.88196
.88	.98828	1.02690	1.1515	1.00698	.89175
.89	.99028	1.02583	1.1380	1.00587	.90142
.90	.99207	1.02451	1.1246	1.04485	.91097
.91	.99366	1.02297	1.1114	1.00393	.92039
.92	.99506	1.02120	1.09842	1.00310	.92970
.93	.99627	1.01921	1.08555	1.00237	.93889
.94	.99729	1.01702	1.07285	1.00174	.94796
.95	.99814	1.01463	1.06030	1.00121	.95692
.96	.99883	1.01205	1.04792	1.00077	.96576
.97	.99935	1.00929	1.03570	1.00043	.97449
.98	.99972	1.00636	1.02364	1.00019	.98311
.99	.99993	1.00326	1.01174	1.00004	.99161
1.00	1.00000	1.00000	1.00000	1.00000	1.00000
1.01	.99993	.99659	.98841	1.00004	1.00828
1.02	.99973	.99304	.97697	1.00019	1.01644
1.03	.99940	.98936	.96569	1.00043	1.02450
1.04	.99895	.98553	.95456	1.00077	1.03246
1.05	.99838	.98161	.94358	1.00121	1.04030
1.06	.99769	.97755	.93275	1.00175	1.04804
1.07	.99690	.97339	.92206	1.00238	1.05567
1.08	.99600	.96913	.91152	1.00311	1.06320
1.09	.99501	.96477	.90112	1.00394	1.07062
1.10	.99392	.96031	.89086	1.00486	1.07795
1.11	.99274	.95577	.88075	1.00588	1.08518
1.12	.99148	.95115	.87078	1.00699	1.09230
1.13	.99013	.94646	.86094	1.00820	1.09933
1.14	.98871	.94169	.85123	1.00951	1.10626

Rayleigh line flow ($\gamma = 1.4$)

M	T_t/T_t^*	T/T^*	p/p^*	p_t/p_t^*	V/V^*
1.15	.98721	.93685	.84166	1.01092	1.1131
1.16	.98564	.93195	.83222	1.01243	1.1198
1.17	.98400	.92700	.82292	1.01403	1.1264
1.18	.98230	.92200	.81374	1.01572	1.1330
1.19	.98054	.91695	.80468	1.01752	1.1395
1.20	.97872	.91185	.79576	1.01941	1.1459
1.21	.97685	.90671	.78695	1.02140	1.1522
1.22	.97492	.90153	.77827	1.02348	1.1584
1.23	.97294	.89632	.76971	1.02566	1.1645
1.24	.97092	.89108	.76127	1.02794	1.1705
1.25	.96886	.88581	.75294	1.03032	1.1764
1.26	.96675	.88052	.74473	1.03280	1.1823
1.27	.96461	.87521	.73663	1.03536	1.1881
1.28	.96243	.86988	.72865	1.03803	1.1938
1.29	.96022	.86453	.72078	1.04080	1.1994
1.30	.95798	.85917	.71301	1.04365	1.2050
1.31	.95571	.85380	.70535	1.04661	1.2105
1.32	.95341	.84843	.69780	1.04967	1.2159
1.33	.95108	.84305	.69035	1.05283	1.2212
1.34	.94873	.83766	.68301	1.05608	1.2264
1.35	.94636	.83227	.67577	1.05943	1.2316
1.36	.94397	.82698	.66863	1.06288	1.2367
1.37	.94157	.82151	.66159	1.06642	1.2417
1.38	.93915	.81613	.65464	1.07006	1.2467
1.39	.93671	.81076	.64778	1.07380	1.2516
1.40	.93425	.80540	.64102	1.07765	1.2564
1.41	.93178	.80004	.63436	1.08159	1.2612
1.42	.92931	.79469	.62779	1.08563	1.2659
1.43	.92683	.78936	.62131	1.08977	1.2705
1.44	.92434	.78405	.61491	1.09400	1.2751
1.45	.92184	.77875	.60860	1.0983	1.2796
1.46	.91933	.77346	.60237	1.1028	1.2840
1.47	.91682	.76819	.59623	1.1073	1.2884
1.48	.91431	.76294	.59018	1.1120	1.2927
1.49	.91179	.75771	.58421	1.1167	1.2970
1.50	.90928	.75250	.57831	1.1215	1.3012
1.51	.90676	.74731	.57250	1.1264	1.3054
1.52	.90424	.74215	.56677	1.1315	1.3095
1.53	.90172	.73701	.56111	1.1367	1.3135
1.54	.89920	.73189	.55553	1.1420	1.3175
1.55	.89669	.72680	.55002	1.1473	1.3214
1.56	.89418	.72173	.54458	1.1527	1.3253
1.57	.89167	.71669	.53922	1.1582	1.3291
1.58	.88917	.71168	.53393	1.1639	1.3329
1.59	.88668	.70669	.52871	1.1697	1.3366

Rayleigh line flow ($\gamma = 1.4$)

M	T_t/T_t^*	T/T^*	p/p^*	p_t/p_t^*	V/V^*
1.60	.88419	.70173	.52356	1.1756	1.3403
1.61	.88170	.69680	.51848	1.1816	1.3439
1.62	.87922	.69190	.51346	1.1877	1.3475
1.63	.87675	.68703	.50851	1.1939	1.3511
1.64	.87429	.68219	.50363	1.2002	1.3546
1.65	.87184	.67738	.49881	1.2066	1.3580
1.66	.86940	.67259	.49405	1.2131	1.3614
1.67	.86696	.66784	.48935	1.2197	1.3648
1.68	.86453	.66312	.48471	1.2264	1.3681
1.69	.86211	.65843	.48014	1.2332	1.3713
1.70	.85970	.65377	.47563	1.2402	1.3745
1.71	.85731	.64914	.47117	1.2473	1.3777
1.72	.85493	.64455	.46677	1.2545	1.3809
1.73	.85256	.63999	.46242	1.2618	1.3840
1.74	.85020	.63546	.45813	1.2692	1.3871
1.75	.84785	.63096	.45390	1.2767	1.3901
1.76	.84551	.62649	.44972	1.2843	1.3931
1.77	.84318	.62205	.44559	1.2920	1.3960
1.78	.84087	.61765	.44152	1.2998	1.3989
1.79	.83857	.61328	.43750	1.3078	1.4018
1.80	.83628	.60894	.43353	1.3159	1.4046
1.81	.83400	.60463	.42960	1.3241	1.4074
1.82	.83174	.60036	.42573	1.3324	1.4102
1.83	.82949	.59612	.42191	1.3408	1.4129
1.84	.82726	.59191	.41813	1.3494	1.4156
1.85	.82504	.58773	.41440	1.3581	1.4183
1.86	.82283	.58359	.41072	1.3669	1.4209
1.87	.82064	.57948	.40708	1.3758	1.4235
1.88	.81846	.57540	.40349	1.3848	1.4261
1.89	.81629	.57135	.39994	1.3940	1.4286
1.90	.81414	.56734	.39643	1.4033	1.4311
1.91	.81200	.56336	.39297	1.4127	1.4336
1.92	.80987	.55941	.38955	1.4222	1.4360
1.93	.80776	.55549	.38617	1.4319	1.4384
1.94	.80567	.55160	.38283	1.4417	1.4408
1.95	.80359	.54774	.37954	1.4516	1.4432
1.96	.80152	.54391	.37628	1.4616	1.4455
1.97	.79946	.54012	.37306	1.4718	1.4478
1.98	.79742	.53636	.36988	1.4821	1.4501
1.99	.79540	.53263	.36674	1.4925	1.4523
2.00	.79339	.52893	.36364	1.5031	1.4545
2.01	.79139	.52526	.36057	1.5138	1.4567
2.02	.78941	.52161	.35754	1.5246	1.4589
2.03	.78744	.51800	.35454	1.5356	1.4610
2.04	.78549	.51442	.35158	1.5467	1.4631

Rayleigh line flow ($\gamma = 1.4$)

M	T_t/T_t^*	T/T^*	p/p^*	p_t/p_t^*	V/V^*
2.05	.78355	.51087	.34866	1.5579	1.4652
2.06	.78162	.50735	.34577	1.5693	1.4673
2.07	.77971	.50386	.34291	1.5808	1.4694
2.08	.77781	.50040	.34009	1.5924	1.4714
2.09	.77593	.49697	.33730	1.6042	1.4734
2.10	.77406	.49356	.33454	1.6161	1.4753
2.11	.77221	.49018	.33181	1.6282	1.4773
2.12	.77037	.48683	.32912	1.6404	1.4792
2.13	.76854	.48351	.32646	1.6528	1.4811
2.14	.76673	.48022	.32383	1.6653	1.4830
2.15	.76493	.47696	.32122	1.6780	1.4849
2.16	.76314	.47373	.31864	1.6908	1.4867
2.17	.76137	.47052	.31610	1.7037	1.4885
2.18	.75961	.46734	.31359	1.7168	1.4903
2.19	.75787	.46419	.31110	1.7300	1.4921
2.20	.75614	.46106	.30864	1.7434	1.4939
2.21	.75442	.45796	.30621	1.7570	1.4956
2.22	.75271	.45489	.30381	1.7707	1.4973
2.23	.75102	.45184	.30143	1.7846	1.4990
2.24	.74934	.44882	.29908	1.7986	1.5007
2.25	.74767	.44582	.29675	1.8128	1.5024
2.26	.74602	.44285	.29445	1.8271	1.5040
2.27	.74438	.43990	.29218	1.8416	1.5056
2.28	.74275	.43698	.28993	1.8562	1.5072
2.29	.74114	.43409	.28771	1.8710	1.5088
2.30	.73954	.43122	.28551	1.8860	1.5104
2.31	.73795	.42837	.28333	1.9012	1.5119
2.32	.73638	.42555	.28118	1.9165	1.5134
2.33	.73482	.42276	.27905	1.9320	1.5150
2.34	.73327	.41999	.27695	1.9476	1.5165
2.35	.73173	.41724	.27487	1.9634	1.5180
2.36	.73020	.41451	.27281	1.9794	1.5195
2.37	.72868	.41181	.27077	1.9955	1.5209
2.38	.72718	.40913	.26875	2.0118	1.5223
2.39	.72569	.40647	.26675	2.0283	1.5237
2.40	.72421	.40383	.26478	2.0450	1.5252
2.41	.72274	.40122	.26283	2.0619	1.5266
2.42	.72129	.39863	.26090	2.0789	1.5279
2.43	.71985	.39606	.25899	2.0961	1.5293
2.44	.71842	.39352	.25710	2.1135	1.5306
2.45	.71700	.39100	.25523	2.1311	1.5320
2.46	.71559	.38850	.25337	2.1489	1.5333
2.47	.71419	.38602	.25153	2.1669	1.5346
2.48	.71280	.38356	.24972	2.1850	1.5359
2.49	.71142	.38112	.24793	2.2033	1.5372

Rayleigh line flow ($\gamma = 1.4$)

M	T_t/T_t^*	T/T^*	p/p^*	p_t/p_t^*	V/V^*
2.50	.71005	.37870	.24616	2.2218	1.5385
2.51	.70870	.37630	.24440	2.2405	1.5398
2.52	.70736	.37392	.24266	2.2594	1.5410
2.53	.70603	.37157	.24094	2.2785	1.5422
2.54	.70471	.36923	.23923	2.2978	1.5434
2.55	.70340	.36691	.23754	2.3173	1.5446
2.56	.70210	.36461	.23587	2.3370	1.5458
2.57	.70081	.36233	.23422	2.3569	1.5470
2.58	.69953	.36007	.23258	2.3770	1.5482
2.59	.69825	.35783	.23096	2.3972	1.5494
2.60	.69699	.35561	.22936	2.4177	1.5505
2.61	.69574	.35341	.22777	2.4384	1.5516
2.62	.69450	.35123	.22620	2.4593	1.5527
2.63	.69327	.34906	.22464	2.4804	1.5538
2.64	.69205	.34691	.22310	2.5017	1.5549
2.65	.69084	.34478	.22158	2.5233	1.5560
2.66	.68964	.34267	.22007	2.5451	1.5571
2.67	.68845	.34057	.21857	2.5671	1.5582
2.68	.68727	.33849	.21709	2.5892	1.5593
2.69	.68610	.33643	.21562	2.6116	1.5603
2.70	.68494	.33439	.21417	2.6342	1.5613
2.71	.68378	.33236	.21273	2.6571	1.5623
2.72	.68263	.33035	.21131	2.6802	1.5633
2.73	.68150	.32836	.20990	2.7035	1.5644
2.74	.68038	.32638	.20850	2.7270	1.5654
2.75	.67926	.32442	.20712	2.7508	1.5663
2.76	.67815	.32248	.20575	2.7748	1.5673
2.77	.67704	.32055	.20439	2.7990	1.5683
2.78	.67595	.31864	.20305	2.8235	1.5692
2.79	.67487	.31674	.20172	2.8482	1.5702
2.80	.67380	.31486	.20040	2.8731	1.5711
2.81	.67273	.31299	.19909	2.8982	1.5721
2.82	.67167	.31114	.19780	2.9236	1.5730
2.83	.67062	.30931	.19652	2.9493	1.5739
2.84	.66958	.30749	.19525	2.9752	1.5748
2.85	.66855	.30568	.19399	3.0013	1.5757
2.86	.66752	.30389	.19274	3.0277	1.5766
2.87	.66650	.30211	.19151	3.0544	1.5775
2.88	.66549	.30035	.19029	3.0813	1.5784
2.89	.66449	.29860	.18908	3.1084	1.5792
2.90	.66350	.29687	.18788	3.1358	1.5801
2.91	.66252	.29515	.18669	3.1635	1.5809
2.92	.66154	.29344	.18551	3.1914	1.5818
2.93	.66057	.29175	.18435	3.2196	1.5826
2.94	.65961	.29007	.18320	3.2481	1.5834

Rayleigh line flow ($\gamma = 1.4$)

M	T_t/T_t^*	T/T^*	p/p^*	p_t/p_t^*	V/V^*
2.95	.65865	.28841	.18205	3.2768	1.5843
2.96	.65770	.28676	.18091	3.3058	1.5851
2.97	.65676	.28512	.17978	3.3351	1.5859
2.98	.65583	.28349	.17867	3.3646	1.5867
2.99	.65490	.28188	.17757	3.3944	1.5875
3.00	.65398	.28028	.17647	3.4244	1.5882
3.50	.61580	.21419	.13223	5.3280	1.6198
4.00	.58909	.16831	.10256	8.2268	1.6410
4.50	.56983	.13540	.08177	12.502	1.6559
5.00	.55555	.11111	.06667	18.634	1.6667
6.00	.53633	.07849	.04669	38.946	1.6809
7.00	.52437	.05826	.03448	75.414	1.6896
8.00	.51646	.04491	.02649	136.62	1.6954
9.00	.51098	.03565	.02098	233.88	1.6993
10.00	.50702	.02897	.01702	381.62	1.7021
∞	.48980	0	0	∞	1.7143

This table has been adapted from J. H. Keenan and J. Kaye, *Gas Tables*, John Wiley and Sons, Inc., New York, 1948, and is reprinted by permission of the publisher.

APPENDIX G

Table G.1 Ideal gas properties of various gases

Gas	Chemical Formula	Molecular Mass	Gas Constant, R $(kJ/kg \cdot K)$	c_p $(kJ/kg \cdot K)$	$\gamma = c_p/c_v$
Acetylene	C_2H_2	26.038	0.3193	1.711	1.23
Air		28.967	0.2870	1.004	1.40
Ammonia	NH_3	17.03	0.4882	2.092	1.30
Argon	Ar	39.948	0.2081	0.519	1.67
Butane	C_4H_{10}	58.124	0.1430	1.674	1.09
Carbon dioxide	CO_2	44.009	0.1889	0.845	1.30
Carbon monoxide	CO	28.010	0.2968	1.042	1.40
Ethane	C_2H_6	30.070	0.2765	1.766	1.18
Ethylene	C_2H_4	28.054	0.2964	1.720	1.21
Helium	He	4.003	2.077	5.200	1.67
Hydrogen	H_2	2.016	4.124	14.35	1.40
Methane	CH_4	16.043	0.5182	2.223	1.32
Nitrogen	N_2	28.014	0.2968	1.038	1.40
Oxygen	O_2	31.998	0.2598	0.916	1.40
Propane	C_3H_8	44.097	0.1885	1.690	1.12
Water vapor	H_2O	18.015	0.4615	1.866	1.33

Note: c_p and c_v are taken at 25°C.

This table has been adapted from W.L. Haberman and J.E.A. John, *Engineering Thermodynamics*, Boston, Allyn and Bacon, Inc., 1980, p. 443.

Table G.2 Ideal gas specific heat ratio γ of several gases as function of temperature

$T(K)$	Nitrogen (N_2)	Oxygen (O_2)	Hydrogen (H_2)	Water Vapor
200	1.400	1.399	1.439	1.332
400	1.397	1.382	1.398	1.321
600	1.381	1.350	1.396	1.297
800	1.360	1.327	1.390	1.274
1000	1.341	1.313	1.380	1.253

This table is taken from Hilsenrath et al., *Tables of Thermodynamic and Transport Properties*, Elmsford, N.Y., Pergamon Press, 1960. Originally published as National Bureau of Standards Circular 564.

Table G.3 Thermodynamic properties of air at 1 atmosphere (R for air $= 0.2870$ kJ/kg·K)

$T(K)$	c_p/R	γ	$(h - h_{ref})kJ/kg$	$\dfrac{s - s_{ref}}{R}$
200	3.506	1.406	199.7	22.497
300	3.506	1.402	300.2	23.917
400	3.533	1.396	401.2	24.929
500	3.588	1.387	503.3	25.723
600	3.663	1.376	607.4	26.383
700	3.746	1.365	713.7	26.954
800	3.828	1.354	822.4	27.460
900	3.906	1.345	933.4	27.915
1000	3.979	1.336	1046.6	28.330
1200	4.109	1.322	1278.8	29.068
1400	4.230	1.310	1518.2	29.711
1600	4.352	1.299	1764.5	30.284
1800	4.487	1.288	2018.2	30.804
2000	4.662	1.274	2280.6	31.284

This table is taken from Hilsenrath et al., *Tables of Thermodynamic and Transport Properties*, Elmsford, N.Y., Pergamon Press, 1960. Originally published as National Bureau of Standards Circular 564.

Table G.4 Thermodynamic properties of air at 10 atmospheres (R for air $= 0.2870$ kJ/kg·K)

$T(K)$	c_p/R	γ	$(h - h_{ref})kJ/kg$	$\dfrac{s - s_{ref}}{R}$
200	3.643	1.449	195.3	20.139
300	3.555	1.418	298.2	21.594
400	3.558	1.404	400.2	22.616
500	3.603	1.392	502.9	23.414
600	3.673	1.379	607.3	24.077
700	3.753	1.366	713.8	24.649
800	3.834	1.355	822.7	25.155
900	3.910	1.345	933.8	25.610
1000	3.983	1.336	1047.1	26.025
1200	4.111	1.322	1279.5	26.763
1400	4.232	1.310	1519.0	27.404
1600	4.351	1.299	1765.3	27.977
1800	4.478	1.288	2018.7	28.740
2000	4.613	1.278	2279.5	28.974

This table is taken from Hilsenrath et al., *Tables of Thermodynamic and Transport Properties*, Elmsford, N.Y., Pergamon Press, 1960. Originally published as National Bureau of Standards Circular 564.

Table G.5 Thermodynamic properties of nitrogen at 1 atmosphere (R for nitrogen = 0.2968 kJ/kg · K)

$T\ (K)$	$h - h_{ref}$ kJ/kg	$\dfrac{s - s_{ref}}{R}$
200	207.0	21.6249
250	259.1	22.4085
300	311.2	23.0482
350	363.3	23.5892
400	415.4	24.0586
450	467.8	24.4741
500	520.4	24.8479
550	573.4	25.1884
600	626.9	25.5020
650	681.0	25.7935
700	735.6	26.0662
750	790.8	26.3228
800	846.6	26.5656
900	960.8	27.0154
1000	1075.6	27.4260
1100	1193.3	27.8039
1200	1312.9	28.1543
1300	1434.0	28.4811
1400	1556.6	28.7872
1500	1680.4	29.0751
1600	1805.3	29.3467
1700	1931.2	29.6037
1800	2057.9	29.8477
1900	2185.3	30.0799
2000	2313.4	30.3013

This table is taken from Hilsenrath et al., *Tables of Thermodynamic and Transport Properties*, Elmsford, N.Y., Pergamon Press, 1960. Originally published as National Bureau of Standards Circular 564.

Table G.6 Thermodynamic properties of nitrogen at 10 atmospheres (R for nitrogen = 0.2968 kJ/kg·K)

$T\ (K)$	$h - h_{ref}$ kJ/kg	$\dfrac{s - s_{ref}}{R}$
200	202.7	19.2682
250	256.2	20.0741
300	309.2	20.7248
350	361.9	21.2721
400	414.5	21.7454
450	467.1	22.1634
500	520.0	22.5390
550	573.2	22.8807
600	626.9	23.1953
650	681.1	23.4875
700	735.8	23.7607
750	791.1	24.0178
800	847.0	24.2609
900	960.5	24.7113
1000	1076.2	25.1223
1100	1194.0	25.5004
1200	1313.7	25.8511
1300	1434.9	26.1780
1400	1557.5	26.4842
1500	1681.3	26.7721
1600	1806.3	27.0438
1700	1932.1	27.3009
1800	2058.9	27.5449
1900	2186.3	27.7772
2000	2314.4	27.9986

This table is taken from Hilsenrath et al., *Tables of Thermodynamic and Transport Properties*, Elmsford, N.Y., Pergamon Press, 1960. Originally published as National Bureau of Standards Circular 564.

APPENDIX H

Standard atmosphere

Altitude Z (km)	T (K)	p (kPa)	ρ (kg/m^3)	$\mu \times 10^5$ $(N \cdot s/m^2)$
0	288.2	101.3	1.225	1.79
5	255.7	54.05	0.736	1.63
10	223.3	26.50	0.414	1.46
15	216.7	12.11	0.195	1.42
20	216.7	5.53	0.0889	1.42
25	221.6	2.55	0.0401	1.45
30	226.5	1.20	0.0184	1.48
35	236.5	0.575	0.00846	1.53
40	250.4	0.287	0.00400	1.60
45	264.2	0.149	0.00197	1.67
50	270.7	0.0798	0.00103	1.70
55	260.8	0.0425	0.000568	1.65
60	247.0	0.0220	0.000310	1.58

Reference: U.S. Standard Atmosphere 1976; NOAA, NASA, USAF, Washington, D.C., October 1976.

APPENDIX I

Conversion factors

Length	1 ft = 0.3048 m 1 m = 3.281 ft
Volume	1 ft^3 = 0.02832 m^3 1 m^3 = 35.315 ft^3 = 1000 liters 1 liter = 1.057 qt
Mass	1 lbm = 0.45359 kg 1 kg = 2.2046 lbm
Density	1 lbm/ft^3 = 16.02 kg/m^3 1 kg/m^3 = 0.06243 lbm/ft^3
Force	1 lbf = 4.448 N 1 N = 0.2248 lbf
Pressure	1 lbf/in.^2 = 6.8948 kN/m^2 1 N/m^2 = 1 Pa = 0.02089 lbf/ft^2 1 atm = 101.3 kPa
Energy	1 Btu = 1.055 kJ 1 kJ = 0.9478 Btu
Power	1 horsepower = 0.7457 kW 1 Btu/s = 1.055 kW 1 W = 3.412 Btu/h
Specific energy	1 Btu/lbm = 2.326 kJ/kg 1 kJ/kg = 0.4299 Btu/lbm
Velocity	1 ft/s = 0.3048 m/s 1 m/s = 3.281 ft/s
Temperature	°F = 1.8(°C) + 32 °C = (°F − 32)/1.8 K = °R/1.8 °R = 1.8 K °R = °F + 459.67 K = °C + 273.15

INDEX